2:00 Pm
Rm 312 TH

Test
3,4,5,6,7
5.1,2,3,4,5,6,7,8 12 H3

2nd Item Exam
11/16/73

INTERMEDIATE ALGEBRA

Library of Congress Catalog Card Number: 68–9245
SBN 471 22343 3
Printed in the United States of America

ROY DUBISCH

Professor of Mathematics
University of Washington

VERNON E. HOWES

American College in Paris

INTERMEDIATE

ALGEBRA

Second Edition

JOHN WILEY & SONS, INC.

New York • London • Sydney • Toronto

Preface

In 1960, in the first edition of this text, we said:

"The primary function of any textbook in intermediate algebra certainly should be that of developing the student's ability to solve algebraic equations and to transform algebraic expressions in various ways. To this aim we adhere, and we have tried to explain clearly, by text and by examples, this 'how to do it' aspect of the subject. Our lists of exercises are long so that the student may be assi 'ned a sufficient number of problems for drill.

"It is also our view, however, that intermediate algebra should be taught in such a way as to tie it in as closely as possible with the rest of mathematics. This means, in the first place, that the notations and methods used should be, whenever possible, those which are used in advanced work. Second, the stress should be on those aspects of intermediate algebra most useful in other courses, and the student should be told the ways in which the subject matter of intermediate algebra is used. Finally, although there cannot be, at this level, a completely rigorous approach to the subject, every effort should be made to be completely honest with the student, and it should at least be made clear to him that proofs are just as much a feature of algebra as they are of geometry."

The improvements in the mathematics curriculum since publication of the first edition—emphasizing basic principles rather than rules— have made our goals much more easily attainable. In keeping with these welcome developments we have, in this revision, paid greater attention to proof and to precision of language.

At the same time, however, we have tried to maintain avoidance of pedantic detail. We have not, for example, hesitated to assume properties of the real numbers that could be proved when, in our opinion, such proofs are, for this level, unduly sophisticated. Similarly, we have not hesitated to use, with proper explanation, such "shorthand" phrases as "the point $(1, 2)$" instead of "the point corresponding to $(1, 2)$."

In addition to many small but significant changes, as suggested above, we have made the following major modifications.

1. Made substantial — but not obsessive — use of the language of sets;

2. Provided an earlier development of the arithmetic of rational numbers;

3. Introduced the Gauss elimination technique, including matrix representation, for solving systems of linear equations; and

4. Included a discussion of the synthetic division process and the factor theorem.

As in the first edition we have provided some more difficult problems, marked with an asterisk, as challenges for the better student.

Roy Dubisch
Vernon Howes

London, England
December, 1967

Contents

INTERMEDIATE ALGEBRA

FUNDAMENTAL CONCEPTS

1

1.1 INTRODUCTION

Every student of intermediate algebra has been exposed to at least a year of high school algebra. Often, however, three years or more have elapsed since this contact, and even the better students may have forgotten much of the subject. The plight of the poorer student is much worse. Hence it is necessary to review much of the material usually presented in first-year algebra. However, simply to repeat what has already been studied—even if almost forgotten—can be a dull business and can easily kill whatever enthusiasm a student may bring to the subject of intermediate algebra. Furthermore, a student in an intermediate algebra course is entitled to a more mature treatment of the subject than are beginning students. Thus the treatment of the fundamentals of algebra given here may often differ considerably from that of a first-year algebra course, and it will pay even the better student to approach the review topics with an eye to improving his understanding of the basic concepts involved.

1.2 SETS

A term frequently used by laymen to distinguish "modern" from "old-fashioned" mathematics is the word *set*. The concept of a set, however, is far from a new one, being perhaps as old as the concept of number itself since the very act of counting the objects that make up a collection of things presupposes the existence of that collection.

A set could be defined as a collection of objects or a group of things, but such a use of synonyms would not really produce any clarification. Let us, therefore, accept the concept of a set as a basic, undefined, but intuitively understandable concept. However, although the general concept of a set is left undefined, each individual set in any discussion must be *well defined*. By this we mean that, given a set A, it must be possible to determine for any given object whether that object belongs to the set A or not. The individual objects that belong to a set are called **elements** or **members** of the set.

One way to define a particular set is to list all of its elements. We commonly do this by listing the elements between braces and separating them by commas as, for example,

$$S = \{a, b, c, d\},$$
$$T = \{-1, 0, 1\}, \text{ and}$$
$$V = \{a, e, i, o, u\}$$

To indicate that an element belongs to a particular set, we use the symbol $\in$. Thus

$$a \in S, \quad -1 \in T, \quad i \in V,$$

but

$$e \notin S, \quad \tfrac{1}{2} \notin V, \quad \text{etc.}$$

(How would you read $\notin$?)

Some sets have too many elements to list them all conveniently. However, if the set is well defined and there is a "natural order" in which the elements could be listed, it is generally permissible to list only the first few and use three dots to indicate the missing elements. Thus, the letters of the alphabet could be "listed" as the set,

$$A = \{a, b, c, \ldots, z\}$$

Even sets with an infinite number of elements are sometimes "listed" in this way. For example,

$$N = \{1, 2, 3, 4, \ldots\}$$

represents the set of counting numbers.

If every element of a set X is an element of a set Y, we say that X is a

subset of Y and we write $X \subseteq Y$ or $Y \supseteq X$. Thus $V = \{a, e, i, o, u\}$ is a subset of $A = \{a, b, c, \ldots, z\}$; $V \subseteq A$. If $X \subseteq Y$ and also $Y \subseteq X$, then X and Y are equal, and we write $X = Y$.

A convenient method of defining a set is by describing its elements. Let U be a given set. Then the subset of U consisting of those elements of U that have a certain property P can be given in "set-builder" notation as follows:

$$X = \{x | x \in U \text{ and } x \text{ has property } P\}$$

The vertical bar, $|$, is read "such that" and the whole statement is read "X is the set of elements x such that x is an element of U and x has property P." For example,

$$V = \{x | x \in A \text{ and } x \text{ is a vowel}\}$$

(where A is the set of letters of the alphabet) defines the set of vowels. Or let $E = \{2, 4, 6, \ldots\}$, the set of even counting numbers, and consider

$$X = \{x | x \in E \text{ and } x < 10\}$$

X is well defined, for to find the elements of X we have merely to choose those elements of E that are less than 10. Thus

$$X = \{2, 4, 6, 8\}$$

Now let E be the set of even numbers, as in the above example, and consider the set

$$Y = \{x | x \in E \text{ and } x \text{ is an odd number}\}$$

Can you find an element of Y? Certainly not! Nevertheless, the set Y is well defined, since it is possible to determine whether any given object belongs to Y. It happens that no object belongs to Y. The set that has no elements is called the **empty set** or the **null set** and is designated by the symbol $\emptyset$. We see that Y is the empty set: $Y = \emptyset$. There is only one empty set, although, as we shall see, it can be described in many ways.

When no confusion can result we frequently omit mention of the basic set U in the set-builder notation

$$X = \{x | x \in U \text{ and } x \text{ has property } P\}$$

Thus, for example, we might write simply

$$V = \{x | x \text{ is a vowel}\}$$

rather than

$$V = \{x|x \in A \text{ and } x \text{ is a vowel}\}$$

since we know that a vowel is a letter of the alphabet. Similarly, we might write just

$$X = \{x|x \text{ is an even number}\}$$

rather than

$$X = \{x|x \in N \text{ and } x \text{ is an even number}\}$$

since we know that an even number is certainly a counting number.

Exercise 1.2

X **1.** Let $U = \{0, 1, 2, 3, 4, 5, 6, 7, 8, 9\}$. Describe the following sets by listing their elements.

(a) $A = \{x|x \in U \text{ and } x + 1 = 5\}$
 (b) $B = \{x|x \in U \text{ and } x + 1 = x\}$
(c) $C = \{x|x \in U \text{ and } x = x\}$
 (d) $D = \{x|x \in U \text{ and } 2x = 14\}$
(e) $E = \{x|x \in U \text{ and } 2x = 15\}$
 (f) $F = \{x|x \in U \text{ and } x + 1 = 5 \text{ or } 2x = 6\}$
 (g) $G = \{x|x \in U \text{ and } x + 1 = 5 \text{ and } 2x = 6\}$

X **2.** Which of the following statements are true and which are false?

(a) $5 \in \{3, 4, 5\}$
(b) $1 \in \{1\}$
(c) $5 = \{5\}$
(d) $0 \in \varnothing$
(e) $0 = \varnothing$
(f) $\{1, 3\} \subseteq \{1, 2, 3, 4\}$
(g) $\{1, 2\} \subseteq \{1, 3\}$
(h) $\{a, b, c\} \subseteq \{b, c, a\}$
(i) $\varnothing \subseteq \{a, b, c\}$

3. List all the subsets of $\{a, b, c\}$ that have at least two elements.

4. List all the subsets of $\{a, b, c\}$ that have less than two elements.

5. How many subsets are there of the set $\{a, b, c\}$?

6. If A is a subset of B and $B = \{a, b, c, d\}$, list all the possibilities for A.

7. If $a \in A$, $b \in B$, $A \subseteq C$, and $B \subseteq C$, then

(a) Is $a \in C$?

(b) Could we have $b \in A$?

(c) Could there be an element x such that $x \notin A$, $x \notin B$, and $x \in C$?

(d) Could there be an element y such that $y \in A$, $y \in C$, and $y \notin B$?

8. Construct an example of sets A, B, and C that will illustrate Problem 7.

$A \{ aby \}$ $B = \{ b \}$

$C \{ a b x y \}$

1.3   THE POSITIVE INTEGERS

The line dividing algebra from arithmetic is a thin one. The theorems of algebra gain in clarity when they are interpreted by arithmetical examples and, vice versa, the facts of arithmetic are illuminated by a consideration of their more general algebraic counterparts. It is true that much of modern abstract algebra does not concern itself with ordinary numbers. But this is not the type of algebra that is so directly useful in such subjects as elementary engineering and physics. From our point of view, then, the beginnings of algebra can be considered the same as those of arithmetic. And the beginnings of arithmetic go back very far indeed. Many primitive tribes are found with an adequate counting vocabulary; and essentially the type of arithmetic taught in the grade schools today is found in the clay tablets of the Babylonians as long ago as 4000 B.C.

The building blocks of our systems of numbers are the elements of the set of **natural numbers** or **positive integers,**

$$N = \{1, 2, 3, 4, \ldots\}.$$

The arithmetic of the first four or five grades in grammar school is largely devoted to the addition, subtraction, multiplication, and division of the elements of N. For example, we learn that if two elements of N are added together or multiplied together, we obtain another element of N. Thus we say that N is **closed** under the operations of addition and multiplication. Symbolically, we state the **closure properties:** *If $a \in N$ and $b \in N$, then*

$$a + b \in N \text{ and } a \times b \in N.$$

For example, the result of adding 2 and 3 is $2 + 3 = 5$ which is an element of N. Likewise, $2 \times 3 = 6$ is an element of N. As self-evident as these laws might be for the positive integers, they are actually assumptions because no amount of verification of special cases (such as with 2 and 3 above) constitutes a proof in the mathematical sense. In more advanced courses in algebra these assumptions (and others to be discussed later) are carefully analyzed. Here we simply point out that N is not closed under all the operations of ordinary arithmetic. It is important to note that $a - b$ is not an element of N unless a is greater than b nor is $a \div b$ always an element of N. Furthermore, other sets of numbers can be constructed which do not have these properties. For example, the set of odd numbers is not closed under the operation of addition. Another important and useful example is the set of "vectors" where the so-called "dot-product," $a \cdot b$, of two vectors is not a vector but a number.

Continuing our review of grammar school arithmetic, we recall that if we add 2 to 3 we obtain the same result as if we add 3 to 2. Similarly, 2×3 is the same number as 3×2. In general, *If $a \in N$ and $b \in N$, then*

$$a + b = b + a \text{ and } a \times b = b \times a$$

These statements are called, respectively, the **commutative property of addition** and the **commutative property of multiplication.**

Another fact recalled from arithmetic is that when adding three or more numbers, the numbers may be grouped in different ways without changing the answer. For example,

$$2 + (3 + 4) = 2 + 7 = 9 \quad \text{C P OF A}$$

and, also,

$$(2 + 3) + 4 = 5 + 4 = 9 \quad \text{C P OF A}$$

Similarly, in multiplication, we have

$$2 \times (3 \times 4) = 2 \times 12 = 24 \quad \text{C P OF M}$$

and, also,

$$(2 \times 3) \times 4 = 6 \times 4 = 24 \quad \text{C P OF M}$$

Generalizing and formalizing these properties we have, respectively, the **associative property of addition** and the **associative property of multiplication:** *If a, b, and c are any elements of N, then*

$$a + (b + c) = (a + b) + c \quad \text{A P OF A}$$

1, 2, 3   3

and

$$a \times (b \times c) = (a \times b) \times c \quad \text{A PoF M}$$

Finally, our procedures for multiplying in arithmetic frequently involve the use of the **distributive property**. Thus, for example,

$$4 \times 12 = 4 \times (10 + 2) = (4 \times 10) + (4 \times 2) = 40 + 8 = 48 \quad \text{D P}$$

or, generalizing and formalizing: *If a, b, and c are any elements of N, then*

$$a \times (b + c) = (a \times b) + (a \times c) \quad \text{D P}$$

and, also,

$$(b + c) \times a = (b \times a) + (c \times a) \quad \text{D P}$$

In fact, by use of the distributive property, multiplication can be regarded as a process of repeated addition. For example,

$$3 \times 5 = (1 + 1 + 1) \times 5 = (1 \times 5) + (1 \times 5) + (1 \times 5) = 5 + 5 + 5$$

and we see that 3 times 5 means the sum of three fives (or, similarly, the sum of five threes).

For convenience of reference we collect these fundamental properties together:

If a, b, and c are any elements of N, then

$a + b \in N$	(closure property of addition)
$a \times b \in N$	(closure property of multiplication)
$a + b = b + a$	(commutative property of addition)
$a \times b = b \times a$	(commutative property of multiplication)
$a + (b + c) = (a + b) + c$	(associative property of addition)
$a \times (b \times c) = (a \times b) \times c$	(associative property of multiplication)
$a \times (b + c) = (a \times b) + (a \times c)$	(left-hand distributive property)
$(b + c) \times a = (b \times a) + (c \times a)$	(right-hand distributive property)

Exercise 1.3

In problems 1–12, state which of the fundamental properties are illustrated by the equality and check the equality as in the text examples.

1. $3 + (5 + 2) = (3 + 5) + 2$
2. $(13 + 10) + 2 = 13 + (10 + 2)$
3. $2 \times (5 \times 3) = (2 \times 5) \times 3$
4. $(11 \times 11) \times 10 = 11 \times (11 \times 10)$
5. $4 \times (3 + 5) = (4 \times 3) + (4 \times 5)$
6. $2 \times (1 + 2) = (2 \times 1) + (2 \times 2)$
7. $(13 + 10) \times 2 = (13 \times 2) + (10 \times 2)$
8. $6 \times (10 + 3) = (6 \times 10) + (6 \times 3)$
9. $(2 + 3) \times 5 = (2 \times 5) + (3 \times 5)$
10. $(7 + 2) \times 10 = (7 \times 10) + (2 \times 10)$
11. $(2 + 4) \times 5 = 5 \times (2 + 4)$
12. $(2 \times 3) + 7 = 7 + (2 \times 3)$

In problems 13–22, state whether or not the given set is closed with respect to the given operation.

13. Even numbers; addition
14. Even numbers; multiplication
15. Odd numbers; multiplication
16. Odd numbers; addition
17. Prime numbers; addition
18. Prime numbers; multiplication
19. The set of fractions of the form $\frac{x}{3}$ (where x is a positive integer); addition
20. The set of fractions of the form $\frac{x}{3}$ (where x is a positive integer); multiplication
21. $\{0, 1\}$; addition
22. $\{0, 1\}$; multiplication

1.4   THE ALGEBRA OF SETS

As an example of a mathematical system other than an arithmetical one, let us consider an algebra of sets. The elements of this algebra are sets or, more precisely, subsets of a given set called the *universal* set. The operations in the algebra of sets are called intersection and union and are defined as follows:

If A and B are subsets of some universal set U, the **intersection** of A and B, written $A \cap B$, is the set of elements that belong to both A *and* B. That is, $x \in A \cap B$ if and only if $x \in A$ *and* $x \in B$. The **union** of A and B, written $A \cup B$, is the set of elements that belong to *either A or B* (or both). That is, $x \in A \cup B$ if and only if $x \in A$ *or* $x \in B$ (or both).

As an illustration, let $A = \{1, 2, 5, 6\}$ and $B = \{2, 4, 6, 8\}$. Then $A \cap B = \{2, 6\}$ and $A \cup B = \{1, 2, 4, 5, 6, 8\}$.

To visualize these operations on sets, we sometimes use a **Venn diagram** in which we think of the elements as a set of points within a closed curve. If A and B have elements in common, we let the areas representing A and B overlap. For example, intersection and union are illustrated by Venn diagrams in Figure 1.1.

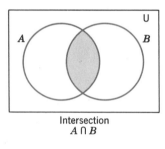

Intersection
$A \cap B$

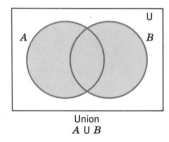

Union
$A \cup B$

Figure 1.1

We can use Venn diagrams to make plausible the following fundamental properties of this algebra of sets.

1. **Closure properties:**
$$A \cap B \subseteq U$$
$$A \cup B \subseteq U$$

2. **Commutative properties:**
$$A \cap B = B \cap A$$
$$A \cup B = B \cup A$$

3. **Associative properties:**
$$(A \cap B) \cap C = A \cap (B \cap C)$$
$$(A \cup B) \cup C = A \cup (B \cup C)$$

4. **Distributive properties:**
$$A \cap (B \cup C) = (A \cap B) \cup (A \cap C)$$
$$A \cup (B \cap C) = (A \cup B) \cap (A \cup C)$$

Figure 1.2 illustrates the first distributive property: $A \cap (B \cup C)$ is illustrated in the left-hand diagram and $(A \cap B) \cup (A \cap C)$ is shown in the right-hand diagram.

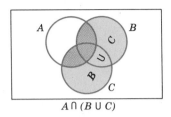

$A \cap (B \cup C)$

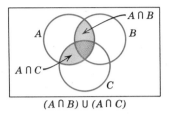

$(A \cap B) \cup (A \cap C)$

Figure 1.2

The algebra of subsets of a set is an example of a *Boolean* algebra, named in honor of its inventor, George Boole (English mathematician-logician, 1815–1864). From the definition and fundamental properties given above, and very few additional definitions, the entire structure of a Boolean algebra can now be established. The exercises will illustrate some of these results.

all problems

Exercise 1.4

In problems 1–8, let $A = \{1, 3, 4, 6, 7\}$, $B = \{2, 4, 6\}$, and $C = \{3, 5, 7\}$; and list the elements of the given set.

1. $A \cup B$
2. $A \cap B$
3. $B \cup C$
4. $A \cap C$
5. $(A \cup B) \cap C$
6. $(A \cap B) \cap C$
7. $A \cap (B \cup C)$
8. $(A \cap B) \cup (A \cap C)$

9. (a) Define $A \cap B$ in set builder notation. (b) Define $A \cup B$ in set builder notation.

10. Using Venn diagrams, illustrate $A \cap B$ and $A \cup B$

(a) if A and B are disjoint (have no elements in common)
(b) if $A \subseteq B$
(c) if $B \subseteq A$
(d) if $A = B$

11. (a) If $A \subseteq B$ and $A \cap B = \emptyset$, what can you conclude about set A?

(b) If $A \cup B = \varnothing$, what can you conclude about sets A and B?

(c) If $A \cap B = \varnothing$, does it necessarily follow that $A = \varnothing$ or $B = \varnothing$?

12. Complete each of the following statements where $A \subseteq U$.

(a) $A \cap \varnothing =$ (b) $A \cup U =$

(c) $A \cap U =$ (d) $A \cup \varnothing =$

(e) $U \cap \varnothing =$ (f) $U \cup \varnothing =$

(g) $A \cap A =$ (h) $A \cup A =$

13. If A is a subset of U, the **complement** of A is written as A' and defined to be the set of elements of U which do *not* belong to A.

(a) Define A' in set builder notation.

(b) Illustrate A' in a Venn diagram.

14. Use Venn diagrams to illustrate that

(a) $(A \cap B)' = A' \cup B'$

(b) $(A \cup B)' = A' \cap B'$

15. Use Venn diagrams to illustrate the following properties of the algebra of sets:

(a) $A \cap (B \cap C) = (A \cap B) \cap C$

(b) $A \cup (B \cup C) = (A \cup B) \cup C$

(c) $A \cup A' = U$

(d) $A \cap A' = \varnothing$

(e) $(A \cap B) \cup (A \cap B') \cup (A' \cap B) = A \cup B$

1.5   AXIOM

Just as all of the properties of a Boolean algebra can be built up from a small number of fundamental laws and definitions, so also can all of the properties of our number systems be developed from their fundamental properties and definitions. We have given the fundamental properties for addition and multiplication of positive integers in Section 1.3, and it is possible to build other number systems by defining different kinds of numbers in terms of positive integers. In fact, we shall do this in subsequent sections without, however, going too deeply into the logical structure involved.

The fundamental properties of a mathematical system that we use in developing additional properties are commonly referred to as *axioms*. We can use the axioms of Section 1.3 to develop many of the basic principles of elementary algebra. Thus, for example, to show that, for all integers a and b,

$$(a + b)(a + b) = a^2 + 2ab + b^2$$

we begin by letting $c = a + b$. Then, by the closure axiom, c is also an integer and we can write

$$(a + b)(a + b) = c \times (a + b).$$

But, by the left-hand distributive property,

$$c \times (a + b) = (c \times a) + (c \times b).$$

Replacing c by $a + b$, we have this equal to

$$(a + b) \times a + (a + b) \times b$$

Then, by the right-hand distributive property

$$[(a + b) \times a] + [(a + b) \times b] = (a^2 + b \times a) + (a \times b + b^2)$$

Since (by what property?) $a \times b = b \times a$, we have

$$(a^2 + b \times a) + (a \times b + b^2) = (a^2 + a \times b) + (a \times b + b^2)$$

Finally, by the associative property of addition,

$$a^2 + (a \times b + a \times b) + b^2 = a^2 + 2(a \times b) + b^2$$

In this development the student will note that we have used both the "+" sign and the "×" sign in order to emphasize the arithmetic nature of our work. In what follows, however, we will employ the usual algebraic custom of indicating multiplication by simple juxtaposition whenever no confusion can arise and the use of a "·" in other cases. Thus

$$a \times b = a \cdot b = ab$$

To develop algebra completely in such an axiomatic fashion would take too long and would be too difficult for a course in intermediate algebra. So we will mention these points only occasionally, especially for the benefit of those students who want to go on to advanced mathematics where such procedures form a basic part of the course work.

Exercise 1.5

Using only the axioms (properties) given in Section 1.3, show that the following equalities hold in N.

    ○1. $(a + 3)(a + 3) = a^2 + 6a + 9$

    2. $(x + 1)(x + 1) = x^2 + 2x + 1$

    3. $(a + 2)(a + 3) = a^2 + 5a + 6$

    ○4. $(y + 4)(y + 5) = y^2 + 9y + 20$

    5. $(2x + 3)(2x + 3) = 4x^2 + 12x + 9$

    6. $(3x + 2)(3x + 2) = 9x^2 + 12x + 4$

    7. $(5x + 1)(2x + 3) = 10x^2 + 17x + 3$

    8. $(4x + 5)(x + 2) = 4x^2 + 13x + 10$

    9. $(3b + 1)(5b + 7) = 15b^2 + 26b + 7$

    10. $(9a + 5)(7a + 9) = 63a^2 + 116a + 45$

1.6 INTEGERS

A great deal of algebra can be done without the use of negative numbers, and, since these form a substantial stumbling block to many students beginning algebra, it would be interesting to try a development of algebra in which negative numbers were introduced much later than they customarily are. However, you have studied negative numbers already, so it is appropriate at this time to review them here. Historically, we find that negative numbers were introduced only after considerable progress had been made in algebra. The Babylonians as early as 2000 B.C. solved problems of at least as great a difficulty as many of those which we will encounter in this course — except that they did not use negative numbers and hence never obtained the negative solutions of equations.

In Section 1.3, the operations of addition and multiplication in N, the set of positive integers, were discussed, and the fundamental properties of N were stated in terms of these operations. Subtraction can now be defined in terms of addition.

Definition 1.1. Let a and b be elements of N. Then $b - a$ is the unique solution of the equation $x + a = b$.

Thus, for example, $5 - 2 = 3$ because 3 is the solution of the equation $x + 2 = 5$.

The first question that comes to mind is "Is N closed with respect to subtraction?" The answer, of course, is "No." In fact, $b - a \in N$ if and only if b is greater than a. Let us define a new number.

Definition 1.2. If $a \in N$, the unique solution of the equation $x + a = a$ is designated by the symbol 0 (zero).

Now we can define a whole set of new numbers.

Definition 1.3. If $a \in N$, the unique solution of the equation $x + a = 0$ is designated by the symbol $-a$, which we read *"the negative of a"* or *"the additive inverse of a."*

Thus, with every positive integer, a, is associated another number, its negative, $-a$. The set of positive integers, their negatives, and zero is the set of integers, which we shall designate by the letter I. Note that the basic property of $-a$ is that $(-a) + a = 0$. The set $\{x \mid x = -a$ where $a \in N\}$ is called the set of *negative integers*.

It is convenient to consider the integers geometrically as equally spaced points on a horizontal straight line (frequently called a **number line**). We select a point as the origin, 0, and let it correspond to the integer zero. The first point to the right of 0 corresponds to 1, the second point corresponds to 2, etc. Likewise, the first point to the left of 0 corresponds to -1, the second point to -2, etc. (See Figure 1.3.)

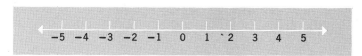

Figure 1.3

Thus every point to the right of 0 represents a positive integer (an element of N), and every point to the left of 0 represents a negative integer. Geometrically, if a is a positive integer, $-a$ is the reflection of a through 0; -3, for example, is represented by a point the same distance to the left of 0 as the point representing 3 is to the right of 0. If, now, we ask what is meant by $-(-3)$, it is geometrically intuitive to say it is 3, since the point representing 3 is the reflection through 0 of the point representing -3.

Let us justify this reasoning algebraically. First we extend Definition 1.3:

Definition 1.4. For all a in I, the unique solution of $x + a = 0$ is designated by the symbol $-a$.

Now, we know that the solution of $x + 3 = 0$ is -3. That is, $(-3) + 3 = 0$. Now if we assume that addition in I is commutative, we have $(-3) + 3 = 3 + (-3)$ and hence that $3 + (-3) = 0$. Thus we can conclude that the solution of $x + (-3) = 0$ is $x = 3$. But, on the other hand, by applying Definition 1.4, we conclude that the solution of $x + (-3) = 0$ is designated by $-(-3)$. Therefore, $-(-3) = 3$.

In general, we have $-(-a) = a$ for a any integer. Note that the number 0 is not considered as either positive or negative. However, $-0 = 0$ since $0 + 0 = 0$.

A common error is to think of "$-a$" as always representing a negative number because of the "negative sign," "$-$." This is not so. If, for example, $a = 3$, then $-a = -3$ which is a negative number. But if $a = -3$, then $-a = -(-3) = 3$. We now define the **absolute value** of a number.

Definition 1.5. The *absolute value* of a number a is designated by the symbol $|a|$ and defined as follows:

$$|a| = a \text{ if } a \text{ is a positive number,}$$
$$|a| = -a \text{ if } a \text{ is a negative number,}$$
$$|a| = 0 \text{ if } a \text{ is zero.}$$

Examples. If $a = 5$, $|a| = |5| = 5$; if $a = -5$, $|a| = |-5| = -(-5) = 5$; if $a = 0$, $|a| = |0| = 0$.

In the discussion above we are, of course, assuming the existence of unique solutions to the equations $x + a = b$ on the basis of your previous experience with numbers. In more advanced treatments of this topic, such equations can form a basis for the construction of the system of integers by a consideration of pairs, (a, b), of positive integers a and b.

$I = $ all $+d - $ neg

Exercise 1.6

all
0

1. Find the solutions in I of the following equations:

(a) $x + 3 = 5$ (b) $x + 25 = 26$

(c) $x + 5 = 3$ (d) $x + 2 = 100$

(e) $x + 26 = 25$ (f) $x + 100 = 2$

(g) $x + 10 = 10$ (h) $x + 2 = 0$

2. Which equations in problem 1 have solutions in N? ABCDG

3. If Definition 1.1 is extended to I (i.e., if a and b are any elements of I), give the solutions in the form $b - a$ of the following equations:

(a) $x + 3 = 5$ (b) $x + (-3) = 5$

(c) $x + 3 = -5$ (d) $x + (-3) = -5$

(e) $x + 0 = 0$ (f) $x + (12 - 2) = 0$

4. Find the negative (additive inverse) of each of the following numbers.

(a) 5   (b) -5

(c) 0   (d) 1

(e) -0 (f) -25

(g) 50   (h) -101

5. Find the absolute values.

(a) $|5|$ (b) $|-5|$

(c) $|0|$ (d) $|1|$

(e) $|-0|$ (f) $|-25|$

(g) $|50|$ (h) $|-101|$

1.7   ADDITION AND SUBTRACTION IN I

Let us recall once more that the set of integers, I, contains the set, N, of positive integers as a subset. Now we already know how to add two positive integers. We also know that $a + 0 = 0 + a = a$ (Definition 1.2 and the commutative property) and that $a + (-a) = (-a) + a = 0$ (Definition 1.3 and the commutative property). The problem of addition in I, then, is to find $a + b$ when either a or b is negative or when both a and b are negative.

Geometrically, the point corresponding to $a + b$ is determined by starting at the point corresponding to a on the number line (Figure 1.3) and, if b is positive, going to b units to the right of a, while, if b is negative, going to b units to the left of a. For example, the additions $1 + 2 = 3$ and $1 + (-2) = -1$ are illustrated in Figure 1.4. In Figure 1.5,

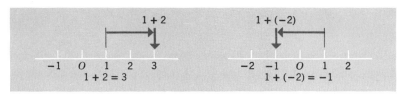

Figure 1.4

the additions $(-2) + (-3) = -5$, $2 + (-3) = -1$, and $(-2) + 3 = 1$ are illustrated.

The geometric interpretation given above can be justified algebraically by using the definitions in Section 1.6 and the assumption that the fundamental properties given in Section 1.3 for N also hold in I. For example, to verify that $(-2) + (-3) = -5$, we note first that

$$[2 + (-2)] + [3 + (-3)] = 0$$

By the associative and commutative properties, this equation can be written as

$$[(-2) + (-3)] + (2 + 3) = 0$$

Thus $(-2) + (-3)$ is the solution of the equation

$$x + (2 + 3) = 0$$

or

$$x + 5 = 0$$

But, by Definition 1.4, the unique solution of the equation $x + 5 = 0$ is -5. Therefore,

$$(-2) + (-3) = -5.$$

Similarly, to verify that $(-2) + 3 = 1$, we note that

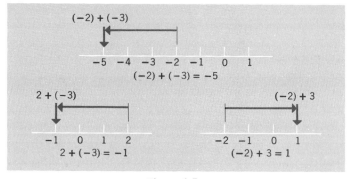

Figure 1.5

$$[2 + (-2)] + 3 = 3$$

or

$$[(-2) + 3] + 2 = 3$$

Thus $(-2) + 3$ is the solution of the equation

$$x + 2 = 3$$

But, by Definition 1.1, the unique solution of the equation $x + 2 = 3$ is $3 - 2 = 1$. Therefore,

$$(-2) + 3 = 1$$

Subtraction was defined in N by Definition 1.1. Let us define subtraction in I in the same way.

Definition 1.6. If a and b are elements of I, $b - a$ is the unique solution of the equation $x + a = b$.

Now we assumed that the fundamental properties hold in I. Applying the associative property, we have

$$[b + (-a)] + a = b + [(-a) + a]$$
$$= b + 0$$
$$= b$$

Thus we see that $b + (-a)$ satisfies the equation $x + a = b$. But Definition 1.6 stated that $b - a$ is the solution of this equation; thus we have proved:

Theorem 1.1. For all a and b in I, $b - a = b + (-a)$

That is, to subtract a from b, we add the additive inverse of a to b.

Example 1. $2 - 3 = 2 + (-3) = -1$.

Example 2. $-2 - 3 = (-2) + (-3) = -5$.

Example 3. $-2 - (-3) = (-2) + 3 = 1$.

Example 4. $2 - (-3) = 2 + 3 = 5$.

Note: The words "plus" and "positive" are often used interchangeably as are "minus" and "negative," even by professional mathematicians. It is better, however, especially for the beginning student, if these terms are not so used. Rather, "plus" should be reserved for the indication of addition and "minus" for subtraction. Thus, for example,

$$(-2) + 4 = 2$$

should be read "negative two plus four is equal to two" rather than "minus two plus four is equal to two" and

$$(-2) - (-4) = 2$$

as "negative two minus negative four is equal to two" rather than "minus two minus minus four is equal to two."

Exercise 1.7

In problems 1–20, perform the indicated operations.

1. $6 + (-4)$
2. $(-5) + (-6)$
3. $(-6) + 7$
4. $7 - 4$
5. $7 - (-4)$
6. $8 = (-2)$
7. $12 - 15$
8. $12 + (-15)$
9. $12 - (-15)$
10. $-4 - (-5)$
11. $(-7) + 8 + (-12)$
12. $(-5) + 10 + (-13)$
13. $6 - (-7) - 12$
14. $3 - (-2) - (-5)$
15. $(-1) + (-2) + (-19)$
16. $45 - 27 + (-65) - (-3)$
17. $15 - (-20) - (-26) - 15$
18. $(-7) = (-2) = (-19) + (-3)$
19. $(-16) - 4 + (-16) + (-4)$
20. $7 + (-6) + 19 - (-5) + (-17)$

In problems 21–25, verify that $a + b = b + a$ and $(a + b) + c = a + (b + c)$ if

21. $a = 25, b = -10, c = 2$
22. $a = 6, b = -5, c = -7$
23. $a = -6, b = -4, c = 8$
24. $a = -5, b = 6, c = -12$
25. $a = -1, b = 7, c = -1$

In problems 26–30, show that we have neither a commutative law for subtraction nor an associative law for subtraction by showing that $a - b$ is not equal to $b - a$ and $a - (b - c)$ is not equal to $(a - b) - c$ for the following values of a, b, and c:

26. $a = 6, b = 5, c = 3$
27. $a = 4, b = 1, c = 1$

28. $a = -3, b = -5, c = 2$ **29.** $a = 1, b = -2, c = -5$
30. $c = -7, b = -1, c = -5$

1.8   MULTIPLICATION AND DIVISION IN I

If we assume the meaning of the product of two positive numbers, it is natural to agree that the product of a positive number and a negative number will be equal in magnitude to the product of the absolute values of the numbers but that it will be negative. For example,

$$2 \cdot (-3) = -(2 \cdot 3) = -6; \qquad (-3) \cdot 2 = -(3 \cdot 2) = -6$$

That is, since -3 represents 3 units to the left of 0 on a number line, $2 \cdot (-3)$ ought to represent twice that distance to the left of 0.

It is not so obvious what, for example, we should take $(-2) \cdot (-3)$ to be. If, again, we assume or demand that the basic properties of positive integers carry over to integers, it is easy to show, however, that $(-2)(-3) \doteq 6$ and, in general, that $(-a)(-b) = ab$. Thus, for example, if we assume that

1. the distributive properties hold in I,
2. the associative property of addition holds in I,
3. $x \cdot 0 = 0 \cdot x = 0$ for all $x \in I$,
4. $x + 0 = 0 + x = x$ for all $x \in I$, and
5. $-x + x = 0$ for all $x \in I$,

we can prove the following theorem.

Theorem 1.2. If a and b are integers (positive, negative, or zero), then $(-a)(-b) = ab$.

PROOF:
$$
\begin{aligned}
(-a)(-b) &= (-a)(-b) + 0 && \text{(assumption 4)} \\
&= (-a)(-b) + a \cdot 0 && \text{(assumption 3)} \\
&= (-a)(-b) + a(-b + b) && \text{(assumption 5)} \\
&= (-a)(-b) + [a(-b) + ab] && \text{(assumption 1)} \\
&= [(-a)(-b) + a(-b)] + ab && \text{(assumption 2)} \\
&= [(-a + a)(-b)] + ab && \text{(assumption 1)} \\
&= 0 \cdot (-b) + ab && \text{(assumption 5)} \\
&= 0 + ab && \text{(assumption 3)} \\
&= ab && \text{(assumption 4)}
\end{aligned}
$$

Division of one integer by another integer other than zero is defined as follows:

Definition 1.7. Let a and $b \neq 0$ be elements of I. Then $a \div b$ is the unique solution of the equation $bx = a$.

Example 1. Find $6 \div 2$. We look for the solution of $2x = 6$. The solution is $x = 3$. Therefore, $6 \div 2 = 3$.

Example 2. Find $35 \div (-7)$. The solution of $(-7)x = 35$ is $x = -5$. Therefore, $35 \div (-7) = -5$.

Example 3. $(-36) \div (-4)$. The solution of $(-4)x = -36$ is $x = 9$. Therefore, $(-36) \div (-4) = 9$.

Note that, according to Definition 1.7, $4 \div 3$ is the solution of $3x = 4$. But the solution of $3x = 4$ is $\frac{4}{3}$ which is not an integer. Thus we see that the set I is not closed under the operation of division.

In the definition of division (Definition 1.7), it was stated that the divisor, b, should not be zero. To see why we have this requirement, let us suppose b is equal to zero. Then, if we want Definition 1.7 to continue to apply, we would have to have $0 \cdot x = a$ when $b = 0$. But $0 \cdot x = 0$ for all numbers x and so we would have to have $a = 0$ when $b = 0$. If, however, $a = 0$, then $0 \cdot x = 0$ does not have a *unique* solution since, for *all* numbers x, $0 \cdot x = 0$. For this reason we agree not to consider division by 0.

Exercise 1.8

In problems 1–24, find the indicated products and quotients.

1. $5(-3)$	**2.** $6(-2)$	**3.** $4(-5)$
4. $(-6)(-2)$	**5.** $(-9)(-3)$	**6.** $(-7)(-2)$
7. $(-3)(-5)(7)$	**8.** $4(-3)(2)$	**9.** $(-5)(-3)(-2)$
10. $(-25)(-2)(-7)$	**11.** $4(-2)(-5)(7)$	**12.** $3(-3)(-5)(-2)$
13. $(-18) \div 3$	**14.** $(-14) \div 7$	**15.** $(-15) \div 5$
16. $(-21) \div (-3)$	**17.** $(-20) \div (-4)$	**18.** $(-25) \div (-5)$
19. $[(-3)(-12)] \div (-4)$	**20.** $[(-12)(-16)] \div (-8)$	
21. $[(-6)(-9)] \div (-27)$	**22.** $[(-12)(-16)] \div [(-8)(-6)]$	
23. $[(-15)(-8)] \div [(-12)(2)]$	**24.** $[(-15)(18)] \div [(-5)(-6)]$	

In problems 25–36, verify that $a(b + c) = ab + ac$ for the following values of a, b, and c.

25. $a = 2$, $b = -5$, $c = 3$ 26. $a = 4$, $b = -2$, $c = 4$

27. $a = 6$, $b = -3$, $c = 2$ 28. $a = -2$, $b = -5$, $c = 6$

29. $a = -3$, $b = -2$, $c = 5$ 30. $a = -5$, $b = 6$, $c = 2$

31. $a = 4$, $b = -6$, $c = -3$ 32. $a = 2$, $b = -3$, $c = -2$

33. $a = 6$, $b = -4$, $c = -3$ 34. $a = -4$, $b = -6$, $c = -4$

35. $a = -3$, $b = -5$, $c = -2$ 36. $a = -4$, $b = -6$, $c = 3$

37. Prove that $0 \div b = 0$ for all integers $b \neq 0$.

38. Prove that $b \div 1 = b$ for all integers b.

39. Prove that $b \div b = 1$ for all integers $b \neq 0$.

1.9 RATIONAL NUMBERS

A rational number is any number that can be written in the form $\dfrac{a}{b}$ where a is an integer and b is a nonzero integer. Thus, for example, $\frac{2}{3}$, $\frac{-25}{2}$, and $\frac{5}{1}$ are all rational numbers. Furthermore, since $3 = \frac{3}{1}$, $0 = \frac{0}{1}$, $-3 = \dfrac{-3}{1}$ and, in general, $a = \dfrac{a}{1}$ for any integer a, all integers are rational numbers. Since $0.4 = \frac{4}{10}$, $0.57 = \frac{57}{100}$, etc., we see that these decimals also name rational numbers. What about, for example, $\frac{5}{0}$? Since we have $\dfrac{a}{b} = a \div b$ we would have, if $\dfrac{5}{0} = x$, $5 \div 0 = x$. But we have previously (Section 1.8) observed that $5 \div 0$ does not exist.

We shall designate the set of all rational numbers by the letter R. Here is a way to indicate systematically all of the rational numbers — by an "infinite rectangular array":

$$\cdots\cdots\ \frac{-4}{1},\ \frac{-3}{1},\ \frac{-2}{1},\ \frac{-1}{1},\ \frac{0}{1},\ \frac{1}{1},\ \frac{2}{1},\ \frac{3}{1},\ \frac{4}{1},\ \cdots\cdots$$

$$\cdots\cdots\ \frac{-4}{2},\ \frac{-3}{2},\ \frac{-2}{2},\ \frac{-1}{2},\ \frac{0}{2},\ \frac{1}{2},\ \frac{2}{2},\ \frac{3}{2},\ \frac{4}{2},\ \cdots\cdots$$

$$\cdots\cdots\ \frac{-4}{3},\ \frac{-3}{3},\ \frac{-2}{3},\ \frac{-1}{3},\ \frac{0}{3},\ \frac{1}{3},\ \frac{2}{3},\ \frac{3}{3},\ \frac{4}{3},\ \cdots\cdots$$

$$\cdot\ \cdot\ \cdot\ \cdot\ \cdot\ \cdot\ \cdot\ \cdot\ \cdot\ \cdot\ \cdot\ \cdot\ \cdot\ \cdot\ \cdot\ \cdot\ \cdot\ \cdot$$

$$\cdot\ \cdot\ \cdot\ \cdot\ \cdot\ \cdot\ \cdot\ \cdot\ \cdot\ \cdot\ \cdot\ \cdot\ \cdot\ \cdot\ \cdot\ \cdot\ \cdot\ \cdot$$

The dots, of course, indicate that the array extends indefinitely.

If a number is equal to one of the numbers in the above array, then it is a rational number.

If a number is not equal to one of the numbers in the above array, then it is not a rational number.

Note that none of the numbers in the above array have 0 as a denominator. For example, $\dfrac{5}{0}$ does not appear. In general, the symbol $\dfrac{a}{0}$ is undefined, which means that symbols such as $\frac{5}{0}, \frac{-3}{0}, \frac{0}{0}$ are not rational numbers and, in fact, are not numbers at all.

We shall now define the operations of addition, subtraction, multiplication, and division in R in such a way that all of the fundamental properties of I become properties of R. First we define equality in R.

Definition 1.8. If $\dfrac{a}{b}$ and $\dfrac{c}{d}$ are rational numbers, then $\dfrac{a}{b} = \dfrac{c}{d}$ if and only if $ad = bc$.

Examples. $\frac{2}{3} = \frac{4}{6}$ because $2 \cdot 6 = 3 \cdot 4$;
$\frac{3}{5} = \frac{6}{10}$ because $3 \cdot 10 = 5 \cdot 6$.

Multiplication is defined as follows:

Definition 1.9. If $\dfrac{a}{b}$ and $\dfrac{c}{d}$ are rational numbers, then $\dfrac{a}{b} \cdot \dfrac{c}{d}$ is defined to be equal to $\dfrac{ac}{bd}$.

Example. $\dfrac{-2}{3} \cdot \dfrac{4}{5} = \dfrac{(-2)(4)}{3 \cdot 5} = \dfrac{-8}{15}$.

Theorem 1.3. $\dfrac{c}{c} = 1$ for $c \neq 0$.

PROOF: $\dfrac{c}{c}$ is the solution of $cx = c$. But $c \cdot 1 = c$; therefore $x = 1$.

Theorem 1.4. If a and b are integers and $b \neq 0$, then $\dfrac{-a}{-b} = \dfrac{a}{b}$.

PROOF: Apply Definition 1.8.

Theorem 1.5. If $\dfrac{a}{b}$ is a rational number and c any integer $\neq 0$, then $\dfrac{ac}{bc} = \dfrac{a}{b}$. Also $\dfrac{ca}{cb} = \dfrac{a}{b}$.

PROOF: Apply Definition 1.8 and use the associative and commutative properties for integers.

Theorem 1.5 is commonly known as the **cancellation theorem** and the operation of eliminating c from $\dfrac{ac}{bc}$ is known as **cancellation.** Thus we often see such examples as $\dfrac{25}{20} = \dfrac{5 \cdot \cancel{5}}{4 \cdot \cancel{5}} = \dfrac{5}{4}$ where it is said that the "5's are canceled." The term "cancel" is a very handy one and is widely used. Unfortunately, however, it is also widely *misused* and most beginning students are well advised to avoid it and, instead, to employ the **Fundamental Principle:**

If a is any number, $a \cdot 1 = a$.

That is, the number 1 is a **multiplicative identity.**

Thus we recommend doing the above example as follows:

$$\frac{25}{20} = \frac{5 \cdot 5}{4 \cdot 5} = \frac{5}{4} \cdot \frac{5}{5} = \frac{5}{4} \cdot 1 = \frac{5}{4}$$

In this way, mistakes in "cancellation" such as

$$\frac{\cancel{2} + 5}{\cancel{2}} \quad \text{and} \quad \frac{\cancel{x} + 4}{\cancel{x} + 2}$$

may be easily avoided.

Definition 1.10. A rational number $\dfrac{a}{b}$ in which a and b have no common integral factor (other than 1 or -1) is said to be in **lowest terms.**

Example 1. $\frac{2}{3}$, $\frac{7}{5}$, and $\frac{-12}{19}$ are in lowest terms.

Example 2. $\frac{15}{10}$, $\frac{19}{38}$, and $\frac{8}{4}$ are not in lowest terms since

$$\frac{15}{10} = \frac{3 \cdot 5}{2 \cdot 5} = \frac{3}{2}, \qquad \frac{19}{38} = \frac{19 \cdot 1}{19 \cdot 2} = \frac{1}{2}, \qquad \text{and} \qquad \frac{8}{4} = \frac{2 \cdot 4}{1 \cdot 4} = \frac{2}{1} = 2$$

We noted in Section 1.8 that the equation $bx = a$ $(b \neq 0)$ does not always have a solution in I. But this equation always has the solution $\frac{a}{b}$ in R since

$$b \cdot \frac{a}{b} = \frac{b}{1} \cdot \frac{a}{b}$$

$$= \frac{b \cdot a}{1 \cdot b} \qquad \text{Definition 1.9}$$

$$= \frac{a \cdot b}{1 \cdot b} \qquad \begin{array}{l}\text{Commutative property}\\ \quad \text{of multiplication}\end{array}$$

$$= \frac{a}{1} \qquad \text{Theorem 1.5}$$

$$= a$$

Thus, since $\frac{a}{b}$ is a solution of $bx = a$ and $a \div b$ is also a solution (by Definition 1.7), we see that we are justified in saying that

$$\frac{a}{b} = a \div b$$

Exercise 1.9

1. Which of the following are rational numbers?

$5, -1, \frac{1}{2}, \frac{-3}{4}, \frac{-100}{0}, 72, 1.72, 3, \frac{9}{5}, 0.4, -0.23, 0.345, \frac{10}{-6}, \frac{7}{4}, 4, 2.$

2. Which of the following rational numbers are equal?

$\frac{1}{4}, -0.7, \frac{50}{48}, \frac{3}{-2}, \frac{-21}{30}, \frac{100}{96}, 0.5, \frac{700}{-1000}, -\frac{0.119}{70}, \frac{-15}{30}, \frac{125}{120}, -1.5, \frac{-7}{14}, -1.7.$

In problems 3–14, use the fundamental principle to reduce each of the following fractions to lowest terms.

3. $\frac{3}{6}$ 4. $\frac{-15}{21}$ 5. $\frac{48}{6}$

6. $\frac{-4}{4}$ 7. $\frac{10}{14}$ 8. $\frac{10}{15}$

9. $\frac{18}{36}$ 10. $\frac{12}{24}$ 11. $\frac{25}{10}$

12. $\frac{35}{7}$ 13. $\frac{72}{12}$ 14. $\frac{42}{24}$

In problems 15–28, multiply as indicated and express the answer in lowest terms.

15. $\frac{1}{2} \cdot \frac{1}{3}$ 16. $\frac{2}{3} \cdot \frac{-2}{5}$ 17. $\frac{1}{2} \cdot \frac{4}{3}$

18. $\frac{-3}{4} \cdot \frac{2}{3}$ 19. $\frac{10}{3} \cdot \frac{6}{-15}$ 20. $\frac{21}{12} \cdot \frac{16}{15}$

21. $\frac{-48}{35} \cdot \frac{49}{-12}$ 22. $\frac{2}{15} \cdot \frac{-2}{3} \cdot \frac{9}{4}$ 23. $\frac{8}{25} \cdot \frac{5}{12} \cdot \frac{15}{2}$

24. $5 \cdot \frac{3}{25}$ **25.** $(-1) \cdot \frac{3}{5} \cdot \frac{-2}{7}$ **26.** $\frac{14}{9} \cdot (-3) \cdot \frac{1}{7}$

27. $\frac{-3}{5} \cdot \frac{25}{-6} \cdot \frac{2}{-5}$ **28.** $\frac{-3}{4} \cdot (-2) \cdot \frac{28}{27}$

1.10 ARITHMETIC OF RATIONAL NUMBERS

We have already (Definition 1.9) defined the product of two rational numbers. Now we define addition for fractions with the same denominators.

Definition 1.11. If $\dfrac{a}{c}$ and $\dfrac{b}{c}$ are rational numbers, then $\dfrac{a}{c} + \dfrac{b}{c} = \dfrac{a+b}{c}$.

Example. $\dfrac{-7}{3} + \dfrac{5}{3} = \dfrac{-7+5}{3} = \dfrac{-2}{3}$.

The following theorem shows how to add fractions with different denominators.

Theorem 1.6. If $\dfrac{a}{b}$ and $\dfrac{c}{d}$ are rational numbers, then $\dfrac{a}{b} + \dfrac{c}{d} = \dfrac{ad+bc}{bd}$.

PROOF: $\dfrac{a}{b} + \dfrac{c}{d} = \dfrac{ad}{bd} + \dfrac{cb}{db}$ by Theorem 1.4.

$$= \dfrac{ad+bc}{bd} \text{ by Definition 1.11 (and the commutative}$$

property of multiplication for integers).

Example 1. $\dfrac{5}{7} + \dfrac{3}{11} = \dfrac{5 \cdot 11 + 3 \cdot 7}{7 \cdot 11} = \dfrac{76}{77}$.

Example 2. $\dfrac{-3}{2} + \dfrac{12}{7} = \dfrac{(-3)(7) + (2)(12)}{14} = \dfrac{3}{14}$.

Example 3.

$$\frac{3}{10} + \frac{2}{15} = \frac{3 \cdot 15 + 2 \cdot 10}{10 \cdot 15} = \frac{45 + 20}{150} = \frac{65}{150} = \frac{13 \cdot 5}{30 \cdot 5} = \frac{13}{30}.$$

Note, however, in these three examples, that we are simply illustrating Theorem 1.6 and are not suggesting its use in practice.

Rather, we suggest that the student do these three problems by the use of the fundamental principle and Definitions 1.9 and 1.11 as follows:

$$\tfrac{5}{7} + \tfrac{3}{11} = \tfrac{5}{7} \cdot \tfrac{11}{11} + \tfrac{3}{11} \cdot \tfrac{7}{7} = \tfrac{55}{77} + \tfrac{21}{77} = \tfrac{76}{77}$$

$$\tfrac{-3}{2} + \tfrac{12}{7} = \tfrac{-3}{2} \cdot \tfrac{7}{7} + \tfrac{12}{7} \cdot \tfrac{2}{2} = \tfrac{-21}{14} + \tfrac{24}{14} = \tfrac{3}{14}$$

$$\tfrac{3}{10} + \tfrac{2}{15} = \tfrac{3}{10} \cdot \tfrac{3}{3} + \tfrac{2}{15} \cdot \tfrac{2}{2} = \tfrac{9}{30} + \tfrac{4}{30} = \tfrac{13}{30}$$

As in the case of integers, the negative of a rational number is defined in terms of the solution of an equation.

Definition 1.12. If $\dfrac{a}{b}$ is a rational number, the unique solution of the equation $x + \dfrac{a}{b} = 0$ is $-\dfrac{a}{b}$.

Theorem 1.7. If a and b are integers and $b \neq 0$, then $-\dfrac{a}{b} = \dfrac{-a}{b} = \dfrac{a}{-b}$.

PROOF: From Definition 1.12, $-\dfrac{a}{b}$ is the solution of $x + \dfrac{a}{b} = 0$. But $\dfrac{-a}{b}$ also satisfies this equation since $\dfrac{-a}{b} + \dfrac{a}{b} = \dfrac{(-a) + a}{b} = \dfrac{0}{b} = 0$. Hence, by uniqueness, $\dfrac{-a}{b} = -\dfrac{a}{b}$. Now we use Definition 1.8 and Theorem 1.2 to conclude that $\dfrac{-a}{b} = \dfrac{a}{-b}$.

Again, as in the case of integers, subtraction is defined in terms of the solution of an equation.

Definition 1.13. If $\dfrac{a}{b}$ and $\dfrac{c}{d}$ are rational numbers, then $\dfrac{a}{b} - \dfrac{c}{d}$ is the unique solution of the equation $x + \dfrac{c}{d} = \dfrac{a}{b}$.

Theorem 1.8. If $\dfrac{a}{b}$ and $\dfrac{c}{d}$ are rational numbers, then $\dfrac{a}{b} - \dfrac{c}{d} = \dfrac{ad - bc}{bd}$.

PROOF: $\dfrac{a}{b} - \dfrac{c}{d}$ is the solution of $x + \dfrac{c}{d} = \dfrac{a}{b}$.

But $\dfrac{ad-bc}{bd}+\dfrac{c}{d}=\dfrac{ad-bc}{bd}+\dfrac{bc}{bd}$ by Theorem 1.5

$$=\dfrac{(ad-bc)+bc}{bd}$$ by Definition 1.11

$$=\dfrac{ad+(bc-bc)}{bd}$$ by the properties of integers

$$+\dfrac{ad}{bd}=\dfrac{a}{b}$$ by Theorem 1.5

Thus $\dfrac{ad-bc}{bd}$ also satisfies the equation $x+\dfrac{c}{d}=\dfrac{a}{b}$ and hence, by uniqueness, $\dfrac{ad-bc}{bd}=\dfrac{a}{b}-\dfrac{c}{d}.$

Example. $\dfrac{5}{7}-\dfrac{3}{4}=\dfrac{(20-21)}{28}=\dfrac{-1}{28}=-\dfrac{1}{28}.$

Again, the student is urged not to apply blindly the rule for subtraction of fractions but, rather, to appeal to basic principles. Thus

$$\frac{5}{7}-\frac{3}{4}=\frac{5\cdot4}{7\cdot4}-\frac{3\cdot7}{4\cdot7}=\frac{20}{28}-\frac{21}{28}=\frac{20-21}{28}=-\frac{1}{28}$$

Theorem 1.9. If $\dfrac{a}{b}$ and $\dfrac{c}{d}$ are rational numbers and $c\neq0,\dfrac{a}{b}\div\dfrac{c}{d}=\dfrac{a}{b}\cdot\dfrac{d}{c}.$

PROOF: $\dfrac{a}{b}\div\dfrac{c}{d}=\dfrac{\dfrac{a}{b}\cdot\dfrac{d}{c}}{\dfrac{c}{d}\cdot\dfrac{d}{c}}=\dfrac{\dfrac{a}{b}\cdot\dfrac{d}{c}}{\dfrac{cd}{dc}}=\dfrac{a}{b}\cdot\dfrac{d}{c}.$

where, once again, we apply the Fundamental Principle that a number is unchanged by multiplication by 1.

Example 1.

$$\frac{\left(\dfrac{5}{-7}\right)}{\left(\dfrac{-15}{28}\right)}=\left(\frac{5}{-7}\right)\left(\frac{28}{-15}\right)=\frac{5\cdot28}{(-7)(-15)}=\frac{5\cdot(7\cdot4)}{7\cdot(5\cdot3)}=\frac{4}{3}$$

Example 2.

$$\frac{-12}{21} \div \left(-\frac{-3}{14}\right) = \frac{-12}{21} \div \frac{3}{14} = \left(\frac{-12}{21}\right)\left(\frac{14}{3}\right) = \frac{(-12)(14)}{(21)(3)}$$

$$= \frac{(-4) \cdot 3 \cdot 2 \cdot 7}{3 \cdot 7 \cdot 3} = \frac{-8}{3} = -\left(\frac{8}{3}\right)$$

Although there is more to be said concerning the system of rational numbers, and our treatment is incomplete, we shall conclude this section with the observation that this system also possesses the associative, commutative, and distributive properties. That is, if $\frac{a}{b}, \frac{c}{d}$, and $\frac{e}{f}$ are rational numbers, then

$$\frac{a}{b} + \frac{c}{d} = \frac{c}{d} + \frac{a}{b}, \qquad \frac{a}{b} \cdot \frac{c}{d} = \frac{c}{d} \cdot \frac{a}{b}$$

$$\left(\frac{a}{b} + \frac{c}{d}\right) + \frac{e}{f} = \frac{a}{b} + \left(\frac{c}{d} + \frac{e}{f}\right), \qquad \left(\frac{a}{b} \cdot \frac{c}{d}\right) \cdot \frac{e}{f} = \frac{a}{b} \cdot \left(\frac{c}{d} \cdot \frac{e}{f}\right)$$

$$\frac{e}{f} \cdot \left(\frac{a}{b} + \frac{c}{d}\right) = \frac{e}{f} \cdot \frac{a}{b} + \frac{e}{f} \cdot \frac{c}{d}, \qquad \left(\frac{a}{b} + \frac{c}{d}\right) \cdot \frac{e}{f} = \frac{a}{b} \cdot \frac{e}{f} + \frac{c}{d} \cdot \frac{e}{f}$$

Exercise 1.10

In problems 1–14, perform the indicated addition or subtraction and reduce the answer to lowest terms by first using the Fundamental Principle to write each fraction with the same denominator.

1. $\frac{1}{2} + \frac{1}{3} + \frac{1}{4}$

2. $\frac{5}{6} - \frac{1}{9} + \frac{13}{3}$

3. $\frac{3}{10} - \frac{17}{15} - \frac{1}{6}$

4. $\frac{5}{18} + \frac{1}{12} - \frac{15}{4}$

5. $\frac{1}{20} + \frac{2}{15} - \frac{1}{10}$

6. $\frac{1}{2} - \frac{1}{3} + \frac{2}{15}$

7. $\frac{17}{15} - \frac{1}{3} + \frac{2}{9}$

8. $\frac{15}{8} - \frac{1}{12} - \frac{-1}{24}$

9. $\frac{-5}{6} - \frac{13}{-3} + \frac{5}{12}$

10. $\frac{-5}{12} - \frac{-4}{5} + \frac{2}{15}$

11. $\frac{3}{20} - \frac{-4}{15} + \frac{7}{6}$

12. $\frac{3}{4} - \frac{4}{5} + \frac{-8}{15}$

13. $\frac{13}{24} - \frac{-1}{8} + \frac{1}{-12}$

14. $-\frac{3}{5} + \frac{5}{12} - \frac{8}{3}$

In problems 15–26, use the Fundamental Principle as in the proof of Theorem 1.9 to perform the indicated divisions.

15. $\frac{1}{2} \div \frac{1}{3}$

16. $\frac{3}{6} \div \frac{9}{2}$

17. $\frac{17}{5} \div \frac{10}{3}$

18. $\frac{13}{5} \div 5$

19. $-\frac{18}{5} \div -15$

20. $\frac{-2}{15} \div \frac{-1}{10}$

21. $\frac{-5}{12} \div 5$

22. $\frac{2}{-3} \div \frac{7}{-5}$

23. $\frac{-5}{6} \div \frac{13}{3}$

24. $\frac{-15}{8} \div -\frac{1}{12}$

25. $\frac{-5}{12} \div \frac{-5}{3}$

26. $\frac{17}{-15} \div \frac{1}{6}$

In problems 27–38, perform the indicated operations and reduce the answer to lowest terms:

27. $\frac{5}{3}(\frac{7}{15} - \frac{21}{20})$

28. $\frac{7}{8}(\frac{4}{21} - \frac{9}{14})$

29. $\dfrac{\frac{3}{5} \cdot \frac{3}{5}}{\frac{3}{5}} \div 5$

30. $\frac{2}{5} \div \dfrac{\frac{5}{8}}{\frac{10}{9}}$

31. $(\frac{2}{3})^2 \cdot \frac{3}{4}$

32. $\dfrac{\frac{2}{3}}{\frac{5}{8}} \div \frac{10}{9}$

33. $\frac{1}{2} \div \dfrac{\frac{4}{3}}{\frac{2}{9}}$

34. $(4 + \frac{7}{8}) \div (\frac{2}{3} - \frac{1}{6})$

35. $(5 + \frac{2}{3}) \div (\frac{2}{9} - \frac{3}{5})$

36. $\dfrac{\frac{5}{7}}{\frac{9}{14}} \cdot \frac{6}{15}$

37. $(\frac{9}{10} \div \frac{4}{15}) \div \frac{5}{18}$

38. $\dfrac{5}{\frac{9}{4}} - \dfrac{6}{\frac{7}{3}}$

39. Check the associative property for addition of rational numbers by showing that

$$(\tfrac{1}{2} + \tfrac{3}{5}) + \tfrac{5}{6} = \tfrac{1}{2} + (\tfrac{3}{5} + \tfrac{5}{6})$$

40. Check the left-hand distributive property for rational numbers by showing that

$$\tfrac{1}{2} \cdot (\tfrac{1}{3} + \tfrac{1}{4}) = \tfrac{1}{2} \cdot \tfrac{1}{3} + \tfrac{1}{2} \cdot \tfrac{1}{4}$$

41. Check the right-hand distributive property for rational numbers by showing that

$$(\tfrac{3}{5} + \tfrac{2}{15}) \cdot \tfrac{5}{6} = \tfrac{3}{5} \cdot \tfrac{5}{6} + \tfrac{2}{15} \cdot \tfrac{5}{6}$$

*42. Prove the commutative property of multiplication for rational numbers.

*43. Prove the commutative property of addition for rational numbers.

*44. Prove the associative property of addition for rational numbers.

*45. Prove the associative property of multiplication for rational numbers.

*46. Prove the left-hand distributive property for rational numbers.

*47. Prove the right-hand distributive property for rational numbers.

*48. Prove that if x and y are rational numbers and $xy = 0$, then $x = 0$ or $y = 0$.

*49. Prove that if x, y, and z are rational numbers with $z \neq 0$ and $xz = yz$, then $x = y$.

1.11   REAL NUMBERS

A geometric representation of integers was given in Section 1.6 as equally spaced points on a straight line. There are many points on such a number line, however, that do not correspond to integers. Some of these points correspond to rational numbers. For example, Figure 1.6 indicates a procedure for locating geometrically the point corresponding to $\frac{5}{3}$.

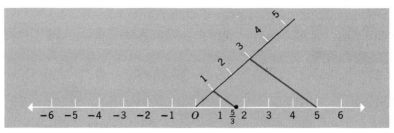

Figure 1.6

In general, the point corresponding to any rational number, $\frac{a}{b}$, can be constructed in the same way, as suggested in Figure 1.7. We draw a diagonal number line through 0, locate the point A corresponding to a on the horizontal line and the point B corresponding to b on the diagonal line. Now we draw $\overline{AB}$ and draw the line parallel to $\overline{AB}$ through C, the point corresponding to 1 on the diagonal line. Since $\triangle OPC$ is similar to $\triangle OAB, \dfrac{OP}{OC}=\dfrac{OA}{OB}$. But $OC=1, OA=a$, and $OB=b$; therefore $OP=\dfrac{a}{b}$ and P corresponds to $\dfrac{a}{b}$.

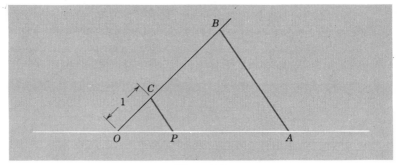

Figure 1.7

Thus to every rational number there corresponds a point on the number line. It may appear that, conversely, to every point on the number line there corresponds a rational number. Indeed, this belief was held for a long time in history. Up to about 500 B.C., it was apparently not realized that there were other numbers besides fractions, for it was thought that every length on a line could be measured exactly by a fraction. To prove that this is not true, let us construct a particular point on the number line and prove that this point cannot correspond to a fraction. We lay off, starting at 0, a length equal to the length of the hypotenuse of an isosceles right triangle with sides of unit length. (See Figure 1.8.) This hypotenuse will have, by the Pythagorean Theorem, a length c given by

$$c^2 = 1^2 + 1^2 = 1 + 1 = 2$$

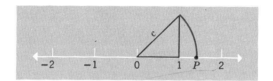

Figure 1.8

$$OC = 1$$
$$OP = \frac{A}{B}$$

Thus, if rational numbers are to suffice for the measurement of lengths, we must have a rational number whose square is equal to 2.

Now in many ways we may obtain a close rational number approximation to $\sqrt{2}$, but we can *never* find a rational number whose square is precisely 2. Thus 1.4, 1.41, 1.414, etc., are successively better rational number approximations to $\sqrt{2}$, but $(1.4)^2 \neq 2$, $(1.41)^2 \neq 2$, $(1.414)^2 \neq 2$, etc.

However, no matter how many rational numbers we may try, there are certainly others left to try, so how can we be sure that we can *never* find one that will work? The proof that none can be found is of the *reductio ad absurdum* type so frequently used in mathematics.

Assume that $\sqrt{2}$ is a rational number. Then $\sqrt{2} = \frac{a}{b}$ where a and b are positive whole numbers, not both even. The assumption that a and b are not both even is no restriction, for if x is the quotient of two even integers, it is also the quotient of two integers, one of which, at least, is not even. For example, if $x = \frac{14}{10}$, x is also equal to $\frac{7}{5}$; if $x = \frac{16}{10}$, x is also equal to $\frac{8}{5}$; etc. It will now be shown that the assumption that $x^2 = 2$ implies that both a and b are even, and hence the assumption must be false. For if $x^2 = 2$, then

$$x^2 = \left(\frac{a}{b}\right)^2 = \frac{a^2}{b^2} = 2$$

so that

$$a^2 = 2b^2$$

Thus, a^2 is certainly even. But this implies that a is even, since if a were odd its square would be odd.[1] However, the "evenness" of a implies $a = 2d$, where d is an integer ($4 = 2 \times 2$, $6 = 3 \times 2$, etc.). But then

$$a^2 = (2d)^2 = 4d^2 = 2b^2$$

so that $b^2 = 2d^2$; and hence, by the same argument as used for a, we conclude that b is also even. Thus there emerges the desired contradiction to the assumption that $\frac{a}{b} = \sqrt{2}$, since it could be assumed that at least one of a and b was odd.

There are many other points on the number line that do not correspond to rational numbers. Let us define a **real number** as a number representing a distance on the number line, which we shall henceforth call the **real line.** Then every point on the real line corresponds to a real number and conversely every real number corresponds to a point on the real line. Any real number which, like $\sqrt{2}$, is not a rational number, is called an **irrational number.** Examples of other irrational numbers are $\sqrt{5}$, $\sqrt{10}$, $\sqrt[3]{3}$, and π (the ratio of the circumference to the diameter of a circle). Notice, however, that rational numbers certainly correspond to a distance on the real line and, therefore, rational numbers are also real numbers. We shall designate the set of real numbers by the symbol R^*.

Real numbers may be added, subtracted, multiplied, and divided (except by 0), and the result is a real number. The operations of addition and multiplication on the set of real numbers possess the commutative, associative, and distributive properties.

---

[1] For any odd number, a, can be written in the form $a = 2n + 1$ where n is an integer. Then $a^2 = (2n + 1)^2 = 4n^2 + 4n + 1 = 4(n^2 + n) + 1$ and is certainly odd, since $4(n^2 + n)$ is even.

Exercise 1.11

1. Construct the following rational numbers on the real line. (Note: The units used on the diagonal line need not be the same as the units on the coordinate line.)

(a) $\frac{1}{2}$ (b) $\frac{13}{3}$ (c) $-\frac{5}{3}$ (d) $-\frac{25}{4}$ (e) $\frac{2}{11}$

2. Construct the following irrational numbers on the real line:

(a) $\sqrt{5}$ (Use a right triangle with sides of lengths 2 and 1.)
(b) $\sqrt{3}$ (Use a right triangle with sides of lengths $\sqrt{2}$ and 1.)
(c) $-\sqrt{5}$ (d) $-\sqrt{3}$
(e) $\sqrt{13}$ (f) $1 + \sqrt{2}$ (Construct $\sqrt{2}$ and add 1 geometrically.)

(g) $\sqrt{3} - 3$ (h) $3\sqrt{2}$

1.12 THE USE OF LETTERS IN ALGEBRA

In the minds of many students it is the use of letters which distinguishes algebra from arithmetic. This, however, is not a fundamental point. In fact, as previously suggested, the mathematician refuses to draw a hard and fast line between arithmetic and algebra, and, if he were forced to give a definition of elementary algebra, he would probably say it consists of a study of the *general* principles of arithmetic. Thus arithmetic is concerned with, for example, the facts that $3 \times 2 = 6$ and $2 \times 3 = 6$, whereas algebra is concerned with the fact that $a \times b = b \times a$ for *all* numbers a and b.

Now although it is convenient to write the commutative property for multiplication in this symbolic fashion, one might just as well say, "If one number is multiplied by a second the result is the same as when the second is multiplied by the first." There is no denying, however, that there is a great gain in clarity when statements concerning the properties of numbers are written in symbols—*provided that these symbols are clearly understood and that it is always realized that these symbols represent numbers.* For example, a student who evaluates (correctly) $2^2 + 2^3$ as $4 + 8 = 12$ ($\neq 2^5 = 32$) but who writes (incorrectly) $x^2 + x^3 = x^5$ is simply not thinking of x as representing a number but as a mysterious letter that moves according to mystifying rules.

The use of letters to aid in the solving of problems is very old, dating back to the ancient Egyptians. The evolution of algebraic symbolism to its present form, however, was a very slow one and was not fully accomplished until the seventeenth century.

In beginning algebra the student is frequently confused by the different ways letters are used to represent numbers. There are essentially three different ways:

1. The use of letters to *designate* numbers. If we, for example, write down

$$x + y$$

we are not stating anything about the numbers x and y, nor are we asking a question about the numbers x and y. Rather the writing of $x + y$ is a shorthand for: Let x be a number and y be a number. Then $x + y$ is the number obtained by adding x and y.

2. The use of letters to describe *universal* truths of arithmetic. Examples of this have already been given:

$$ab = ba; \qquad a(bc) = (ab)c; \qquad a(b + c) = ab + ac$$

Thus, for example, $ab = ba$ is taken to mean that for all numbers a and b we have $ab = ba$.

3. The use of letters to describe *conditional* truths of arithmetic. Thus the statement that

$$x + 2 = 5$$

does not mean that this equality holds for all numbers x but for, in this case, only one number: $x = 3$.

Although it is vitally necessary for the student to distinguish between *universal* statements and *conditional* ones, there is no commonly used notation to describe this difference. It is true that the sign "$\equiv$" is sometimes used in place of the equality sign when a universal statement is made. Thus we may write

$$5(x + 1) \equiv 5x + 5 \qquad \text{and} \qquad x + 2 = 5$$

However great the advantages of using the two separate signs might be, one must face the fact that the equality sign is commonly used for both purposes. And so in this book we, too, will use only the equality sign with the correct interpretation to be inferred from the context.

Exercise 1.12

1. Give three additional examples of designatory expressions in algebra.

2. Give three additional examples of the use of letters to express universal truths of arithmetic.

3. Give three additional examples of the use of letters to describe conditional truths of arithmetic.

In problems 4–15, show that the stated equality is a conditional one rather than a universal one by finding number replacements for the letters so that the equality does not hold. Also, if possible, determine in each problem number replacements for the letters that do give an equality.

4. $x^2 x^3 = x^6$

5. $\dfrac{a + b}{a} = 1 + b$

6. $(x + y)^2 = x^2 + y^2$

7. $3x^2 + 3x + 3 = 3(x^2 + x)$

8. $2(x + 1) = 2x$

9. $\dfrac{x + 4}{2} = x + 2$

10. $x - 1 = -x + 1$

11. $(x^2)^3 = x^5$

12. $\dfrac{x^6}{x^3} = x^2$

13. $-x + y = -(x + y)$

14. $a - (b - c) = a - b - c$

15. $x^2 - (x - 1) = x^2 - x - 1$

1.13 TERMS, FACTORS, AND SIGNS OF GROUPING

The student has noticed, in this and other books on algebra, the use of two different "kinds" of letters to represent numbers; namely, those at the beginning of the alphabet and those at the end of the alphabet. The use of one type seems to be exclusive of the use of the other type. René Descartes (1596–1650) suggested the present-day usage of the letters at the beginning of the alphabet to represent "known constants" and the last letters of the alphabet to represent "unknowns." Some mystery has probably developed in the minds of the students as to which letter should be used in a certain context. The fact is, there is no fundamental difference between these two types as all letters of the alphabet are used to represent numbers, and this means that we do not have two separate sets of rules for algebraic manipulations (as some students have come to suspect!) but only one set.

When two or more numbers are added together, the result is called a **sum.** The numbers making up the sum are called **terms.** Thus, if 3 and -5 are added together, the result is the sum $3 + (-5)$ (or -2); 3 and -5 are terms. The expression $3x + 5b - c$ can be written as a sum, $3x + 5b + (-c)$, which consists of three terms, $3x$, $5b$, and $-c$.

If two or more numbers are multiplied together, the result is called a **product.** The numbers making up the product are called **factors.** Thus, if 3 and 7 are multiplied together, the result is the product $3 \cdot 7$ (or 21); 3 and 7 are factors. In the product $-3xy$, the factors are -3, x, and y.

If any factor or group of factors of a product is specified, then the product of the other factors constitutes the **coefficient** to the specified set. For example, in the expression $3axy$, the coefficient of x is $3ay$, the coefficient of xy is $3a$, the (numerical) coefficient of axy is 3, and so on.

By the introduction or elimination of **parentheses ( )**, **brackets [ ]**, or **braces { } (signs of grouping),** we may decrease or increase the number of terms or factors in an expression. Thus $3x + 5b - c$ has three terms but it may be written, for example, as $(3x + 5b) - c$ and, in this form it has only the two terms $(3x + 5b)$ and $-c$. Similarly, $-3xy$ has three factors, but it may be written with only two factors in the form $(-3x)y$, $-3(xy)$, $(-3y)x$, etc. More will be said about products in the next chapter. For the moment, let us examine the use of grouping symbols as applied to addition and subtraction.

According to the associative property of addition, it does not matter how the terms of a sum are grouped together. Thus we can write

$$a + (b + c) = (a + b) + c = a + b + c$$

In other words, if a parenthesis is preceded by a plus sign (or no sign), it may simply be removed. If a parenthesis is preceded by a negative sign, however, we use the principle illustrated by the following theorem.

Theorem 1.10. $-(a + b - c) = -a - b + c$.

PROOF: Recall first (Definition 1.4) that $-(a + b - c)$ is the unique solution of the equation $x + (a + b - c) = 0$. Now note that $(-a - b + c) + (a + b - c) = (-a + a) + (-b + b) + c + (-c) = 0 + 0 + 0 = 0$. (Give reasons.) Thus $-a - b + c$ is a solution of $x + (a + b - c)$ so that, by uniqueness, $-(a + b - c) = -a - b + c$.

Examples.
$$a + (b - c) = a + b - c$$
$$ax - (by - c) = ax - by + c$$
$$2 - (-x + y) = 2 + x - y$$

In case several signs of grouping are involved, one within another, they may be removed by removing the innermost ones first, according to the procedures just indicated.

Example 1.
$$3s - [5t + (2s - 5)] = 3s - [5t + 2s - 5]$$
$$= 3s - 5t - 2s + 5$$
$$= s - 5t + 5$$

Example 2. $2 - \{5 + (2 - 3) + [2 - (3 - 4)]\}$

SOLUTION: Here, of course, we can do the problem by actually combining numbers as we go along. Let us see if we get the same answer by doing the problem this way as we do by following the process used in Example 1. By the method used in Example 1 we have

$$2 - \{5 + (2 - 3) + [2 - (3 - 4)]\} = 2 - \{5 + 2 - 3 + [2 - 3 + 4]\}$$
$$= 2 - \{5 + 2 - 3 + 2 - 3 + 4\}$$
$$= 2 - 5 - 2 + 3 - 2 + 3 - 4 = -5$$

whereas by direct arithmetic computation we have

$$2 - \{5 + (2 - 3) + [2 - (3 - 4)]\} = 2 - \{5 + (-1) + [2 - (-1)]\}$$
$$= 2 - \{4 + 3\} = 2 - 7 = -5$$

The distributive property is useful in rewriting a sum of two or more terms which involve the same letter or letters.

Examples.
$$3a + 5a = (3 + 5)a = 8a$$
$$5ac + 2bc = (5a + 2b)c$$

Since $ba - ca = ba + [-(ca)] = ba + [(-c)a] = [b + (-c)]a = (b - c)a$ we can write, for example,

$$3a - 5a = (3 - 5)a = (-2)a = -2a$$

and

$$5ac - 2bc = (5a - 2b)c$$

Exercise 1.13

In problems 1–12, compute the answer in two ways: (1) by removing all grouping symbols before adding or subtracting, and (2) by combining inside each grouping symbol before removing the symbol.

1. $(2 + 1) + (4 - 2)$
2. $(-2 + 5) + (3 + 2)$
3. $(-4 - 5) - (2 - 3)$
4. $(-2 + \frac{1}{4}) + (\frac{3}{2} - 2)$
5. $(\frac{1}{2} - \frac{1}{3}) - (\frac{5}{6} + 1)$
6. $[3 - (4 + 5)] + [(2 - 3) + 1]$
7. $[-7 - (3 - 2)] - [-5 + (2 - 3)]$
8. $[\frac{3}{2} + (1 - \frac{1}{2})] + [(3 + \frac{1}{4}) - (1 - \frac{3}{4})]$
9. $3 - [(\frac{5}{8} + 1) - \frac{3}{4}] - (\frac{7}{8} - 2)$
10. $4 - \{4 + [5 - (7 - 3)] + 5\}$
11. $-(2 - \frac{5}{2}) + 1 - \{\frac{3}{2} + [4 - (5 + \frac{5}{2})]\}$
12. $\{1 - \frac{9}{5} + [\frac{1}{3} - \frac{1}{5} - (\frac{4}{5} - \frac{6}{5})]\} - 1$

In problems 13–24, remove the grouping symbols and simplify the result.

13. $(2a - 1) + (5a - 6)$
14. $(3x + y) - (7x - 2y)$
15. $-(\frac{3}{2}a - 2b) + \frac{1}{2}(5a + 2b)$
16. $2(a - \frac{1}{3}) - 3(a + \frac{2}{3})$
17. $3(1 - 2a) + 5(a - 3)$
18. $3x - \frac{1}{2}[2y - 2(\frac{3}{2}x + 4)]$
19. $\frac{1}{3}x - 3[\frac{1}{6}x - \frac{1}{6}(2y + 3x)]$
20. $-2(a - 2) + [3 - (a - 2)]$
21. $3x - 2[y - (3x + 4y)]$
22. $[2x - (x - y)] + 2\{x - [y - (2x - y)]\}$

23. $\frac{3}{4}[x - \frac{1}{2}(x - y)] + \frac{3}{4}\{2x - [4y - (x - \frac{1}{2}y)]\}$

24. $-2(3x - \frac{1}{2}) - \{3x - \frac{1}{2}[4x - 2 + \frac{3}{2}(x - 4)]\}$

In problems 25–36, two algebraic expressions are given. State whether the second is a factor or a term of the first.

25. $3ax + b$; b

26. $-2x - 4$; $-2x$

27. $-4x + 5$; 5

28. $3x^2 - 2x + 1$; $-2x$

29. $4a^2 - 2a - 4$; -4

30. $-5y^2 + 6y - 3$; $-5y^2$

31. $3ab$; a

32. $9a^2b$; $9b$

33. $7x^3y^3$; x

34. $(x - 2)(x - 4)$; $x - 2$

35. $(a + 2)(a - 3)$; $a + 2$

36. $(x - y)(x + y)$; $x + y$

In problems 37–48, two algebraic expressions are given. Give the coefficient of the second when it is considered as a factor of the first.

37. $9ab$; a

38. $8x^2y$; x^2

39. $7cd$; $7c$

40. $9x^2y^3$; y^3

41. $2xy$; $2x$

42. $14z^3$; z^3

43. $7(x - 2)$; $x - 2$

44. $11(x + y)$; $x + y$

45. $4a(a - b)$; $a - b$

46. $7(x - 2)(x + 3)$; $x + 3$

47. $9(x + y)(x - y)$; $x - y$

48. $4x(x^2 - 1)$; $x^2 - 1$

POLYNOMIALS

2

2.1 DEFINITIONS

In Chapter 1 we discussed operations on the sets of natural numbers, integers, rational numbers, and real numbers without attempting precise definitions of these numbers. In subsequent chapters we shall further discuss real numbers and complex numbers. In other words, it is possible to consider operations in number systems without ever defining precisely the word "number." Similarly, we have been working with algebraic expressions without, however, defining what is meant by the term "algebraic expression." The algebraic expressions to be studied in this chapter are called polynomials.

Let us first recall that we define $x^1 = x$, $x^2 = x \cdot x$ and, in general, $x^n = x \cdot x \cdot \cdots \cdot x$ (n factors) for any positive integer n. Furthermore, for $x \neq 0$ we define $x^0 = 1$. (The reason for defining $x^0 = 1$ will be discussed in Chapter 8.)

Definition 2.1. A **polynomial** in x is an algebraic expression of the form

$$a_n x^n + a_{n-1} x^{n-1} + \cdots + a_1 x + a_0 \ (a_n \neq 0)$$

where the coefficients a_n, $a_{n-1}, \ldots, a_1$, a_0 are numbers [1] and n is a positive integer or zero.

Thus a polynomial is a sum and the terms of the sum are called

---

[1] For the time being, these numbers will be restricted to rational numbers. Later we shall allow any real or complex numbers as coefficients of a polynomial.

terms of the polynomial. Each term contains x to some power. The highest power to which x is raised is called the **degree** of the polynomial. The numbers $a_n, a_{n-1}, \ldots, a_1, a_0$ are the **coefficients** of x^n, $x^{n-1}, \ldots, x^1$, and x^0, respectively, if we agree that $x^0 = 1$. Thus $a_0 x^0 = a_0 \cdot 1 = a_0$ and a_0 is frequently called the **constant term.**

Examples. $3x^2 - 5x + 1$ is a polynomial in x of degree 2. Its terms are $3x^2$, $-5x$, and 1. We have $a_n = a_2 = 3$, $a_{n-1} = a_1 = 5$, $a_0 = 1$. $y^5 - \frac{3}{2}$ is a polynomial in y of degree 5. Its terms are y^5 and $-\frac{3}{2}$. We have $a_5 = 1$ (since $y^5 = 1 \cdot y^5$), $a_4 = 0$, $a_3 = 0$, $a_2 = 0$, $a_1 = 0$, and $a_0 = -\frac{3}{2}$.

With the agreement that $x^0 = 1$, any nonzero number, a_0, can be written as a polynomial of degree 0: $a_0 = a_0 \cdot 1 = a_0 x^0$. For example, $\frac{1}{2} = \frac{1}{2}x^0$ is a polynomial of degree 0 where $a_n = a_0 = \frac{1}{2}$. Thus *nonzero numbers are polynomials* of degree zero.[2]

Polynomials with one, two, or three terms occur frequently and are given special names. They are called **monomials, binomials,** and **trinomials** respectively. Another classification of polynomials is in terms of the degree. Polynomials of degree 0, 1, 2, 3, or 4 are called **constants, linear, quadratic, cubic,** or **quartic** polynomials respectively.

Examples. $x^2 - 5x + 6$ is a quadratic trinomial.

$\frac{13}{2}$ is a constant monomial.

$\frac{1}{2}x + 2$ is a linear binomial.

$5x^4$ is a quartic monomial.

We can also consider polynomials in more than one letter. For example,

$$3ax^2y - 5xy^2$$

is a polynomial in a, x, and y where the (numerical) coefficient of ax^2y is 3 and the (numerical) coefficient of xy^2 is -5. But this polynomial may also be considered as a polynomial in just x and y:

$$(3a)x^2y - 5xy^2$$

where the coefficient of x^2y is $3a$ and the coefficient of xy^2 is -5. The same polynomial may also be considered as a polynomial in x:

$$(3ay)x^2 - (5y^2)x$$

---

[2] The number 0 is also regarded as a polynomial, although it is customary not to assign a degree to it.

where the coefficient of x^2 is $3ay$ and the coefficient of x is $-5y^2$. Finally, let us consider this polynomial as being a polynomial in y:

$$(-5x)y^2 + (3ax^2)y$$

The coefficient of y^2 is $-5x$ and the coefficient of y is $3ax^2$.

The degree of a monomial in several letters is the sum of the powers of the letters involved. Thus $3ax^2y$ is of degree 4. (Recall that $a = a^1$.) It is also of degree 3 in x and y alone, of degree 2 in x alone, etc. The degree of a polynomial in several letters is the highest degree of the terms of the polynomial. Thus $3ax^2y - 5xy^2$ is of degree 4. It is also of degree 3 in x and y alone, of degree 2 in x alone, of degree 2 in y alone, etc.

Exercise 2.1

Give the degree of each of the polynomials in problems 1–8 and state whether it is a monomial, binomial, or trinomial.

1. x^2
2. 1
3. $x + 1$
4. $x^2 - 5x + 6$
5. $x^2 + 2xy + y^2$
6. x^2y^3
7. $x^2y^3 + x^3y^2$
8. $1 + \sqrt{2}$

In problems 9–18, put the polynomial in the standard form of Definition 2.1 and then identify $a_n, a_{n-1}, \ldots, a_1$, and a_0.

9. $x + 1$ $x_1 = 0$ $x_0 = 1$
10. $x^2 + 1 + 2x$
11. $1 - 3x + x^2 - 3x^3$
12. $x^4 - 16$
13. $\frac{1}{2}$
14. $\frac{1}{2}x^2 - x^3$
15. $1 - 4x^2 + 4x^4$
16. $\frac{1}{3}x + x^2 - \frac{2}{3}$
17. $\frac{7}{2}x^3 - \frac{1}{2}x + 1$
18. $5 - 27x^2 - x^3$

In problems 19–23, write the polynomial as a polynomial in x and identify $a_n, a_{n-1}, \ldots, a_1$, and a_0.

19. $x^2y + x - 3y$
20. x^2y
21. $ax - 2x^2 + \frac{1}{2}$
22. $x^2y + 2xy - xx^2y + (2y - 1)x$
23. $(y^2 + y)x^3 - (y + 1)x^2 + xy - y^2 + 2y - 1$

2.2 ADDITION AND SUBTRACTION OF POLYNOMIALS

Monomials which contain the same letters and are of the same degree in each of these letters are said to be **like terms.** The distributive property gives us a method for adding or subtracting like terms. Thus, for example,

$$3x + 2x = (3 + 2)x = 5x$$
$$5xy^2 + 3xy^2 = (5 + 3)xy^2 = 8xy^2$$
$$5a - 3a = (5 - 3)a = 2a$$

(Recall that $5a - 3a = 5a + (-3a) = [5 + (-3)]a = (5 - 3)a$.) We also use the distributive property to add terms that are like in certain letters. For example, ax and bx are like terms in x. Thus

$$ax + bx = (a + b)x$$

Likewise,

$$25ax^2 - 20bx^2 = (25a - 20b)x^2$$

but $3x + 4y$ cannot be so expressed as a product.

To add two or more polynomials together we simply group the like terms together and add the like terms.

Example 1. To add $3a + 4b - 2c$ and $-3a - d + b$ we write

$$(3a + 4b - 2c) + (-3a - d + b) = 3a + 4b - 2c - 3a - d + b$$
$$= (3a - 3a) + (4b + b) - 2c - d$$
$$= 5b - 2c - d$$

Example 2.

$$(4x^2y + xy^2 - 2y) + (6x^2y - 2xy^2 + 4y) = 10x^2y - xy^2 + 2y$$

The student should note that extensive use has been made of the commutative property of addition in doing these problems.

Subtraction of polynomials requires the use of Theorem 1.10: $-(a + b - c) = -a - b + c$.

Example 3.

$$(2x - 4by - 5c) - (ax + 3by - 4c) =$$
$$2x - 4by - 5c - ax - 3by + 4c = (2x - ax) +$$
$$(-4by - 3by) + (-5c + 4c) + (2 - a)x - 7by - c.$$

Exercise 2.2

In problems 1–12, add the polynomials listed.

1. $3x^2 - 2x + 1, 5x^2 + 5x - 2$
2. $-5a^2 - 3a - 2, 3a^2 - 3a + 3$
3. $-6y^2 + 5y + 3, -4y^2 - 4y + 2$
4. $3x - 2y - z + 5, x + 8y + 3z - 4$
5. $3x - 2y - 5, -5x + 7y - 8$
6. $4a - 2b + c, -6a + 7b - 3c$
7. $2a - 4b - 5c, -7a + 6b - c, -2a - 5b + c$
8. $7x - 2y + 3z, 5x - 2y + z, -6x + 2y - z$
9. $-4a - 3b + c, -3a + 7b - c, -6a + 4b - 2c$
10. $2x^2 - 3x + 4, -3x^2 - 2x - 4, -5x^2 + x - 2$
11. $a^2 - 5a - 2, 3a^2 + 6a - 2, 7a^2 - 8a + 5$
12. $-3a^2 + 7a + 5, -6a^2 - 7a + 4, 8a^2 - 9a + 1$

In problems 13–21, subtract the second polynomial from the first.

13. $3x^2 + 2x - 1, -5x^2 + 5x + 2$
14. $-5a^2 + 3a + 2, 3a^2 + 3a - 3$
15. $-6y^2 - 5y + 3, 4y^2 - 4y - 2$
16. $3x + 2y - z + 1, x - 8y + 3z + 4$
17. $4a + b - c, -6a + b - 3c$
18. $2a - 3b + 5c, -7a - 6b + c$
19. $7x + 4y - 3z, -6x - 2y + z$
20. $-3a^2 + 7a - 5, -8a^2 + 9a - 1$
21. $a^2 - 5a + 2, -7a^2 - 8a - 5$

22. Subtract the sum of $2a - 5b + 2c$ and $6a - 3b + 5c$ from the sum of $8a - 2b + 2c$ and $3a - 9b + 4c$.

23. Subtract $-5x^3 + 3x^2 - x + 4$ from the sum of $2x^3 - 3x + 5$ and $7x^2 + 4x - 1$.

24. Subtract the sum of $5x^2 + 7x + 2$ and $3x^2 - 6x - 4$ from the sum of $3x^2 - 8x + 10$ and $x^2 - 3x + 4$.

25. Add $2x - 5$, $3x - 2$, and $3 - x$, then subtract this sum from the sum of $2x - 3$, $2x + 7$, and $-3x - 5$.

2.3 MULTIPLICATION OF MONOMIALS

As mentioned in Section 2.1, if a given number is multiplied by itself a certain number of times, we have a shorthand method of indicating this fact. For example, instead of writing $a \cdot a \cdot a \cdot a \cdot a$, we simply write a^5.

Note here that, in contrast to the use of 5 in $5a$ (where it indicates the *addition* of five a's), the 5 in a^5 indicates the *multiplication* of five a's. We shall restrict ourselves in this chapter to exponents n, m, etc., to be positive integers: a^n means that n a's are to be multiplied together, and we say that a is raised to the nth power. Hence, we have

$$a^n a^m = \underbrace{(a \cdot a \cdot a \cdot \cdots \cdot a)}_{n \text{ factors}}\underbrace{(a \cdot a \cdot a \cdot \cdots \cdot a)}_{m \text{ factors}}$$

$$= \underbrace{(a \cdot a \cdot a \cdot \cdots \cdot a)}_{(n+m) \text{ factors}} = a^{n+m} \tag{1}$$

It follows easily that when a product is raised to some power, each factor of the product is raised to that power. For example, $(ab)^n$ means n factors ab resulting in n factors a times n factors b or, in symbols,

$$(ab)^n = \underbrace{ab \cdot ab \cdot ab \cdot \cdots \cdot ab}_{n \text{ factors}}$$

$$= \underbrace{(a \cdot a \cdot a \cdot \cdots \cdot a)}_{n \text{ factors}}\underbrace{(b \cdot b \cdot b \cdot \cdots \cdot b)}_{n \text{ factors}} \tag{2}$$

$$= a^n b^n$$

(By what axioms can we pass from the first equality to the second?) Similarly, $(abc)^n = a^n b^n c^n$ etc.

A special case of this product rule is given by

$$(a^m)^n = a^{mn} \tag{3}$$

since

$$(a^m)^n = \underbrace{(a \cdot a \cdot a \cdot \cdots \cdot a)}_{m \text{ factors}}{}^n = \underbrace{a^n a^n a^n \cdot \cdots \cdot a^n}_{m \text{ factors}}$$

where the last equality is an application of (3). But each a^n of the last expression consists of n factors of a. Thus

$$(a^m)^n = \underbrace{(a \cdot a \cdot \cdots \cdot a)}_{n \text{ factors}}\underbrace{(a \cdot a \cdot \cdots \cdot a)}_{n \text{ factors}} \cdots \underbrace{(a \cdot a \cdot \cdots \cdot a)}_{n \text{ factors}}$$

$$\underbrace{}_{m \text{ groups of } n \text{ factors}}$$

$$= a^{mn}$$

since this gives a total of mn factors.

We may now state a rule for multiplying one monomial by another: Simply combine the like letters in accordance with the preceding rules for exponents and perform the arithmetic operations on the numerical factors.

Example 1. $(3a^2xy^5)(-5bx^2yz^2) = 3(-5)a^2bxx^2y^5yz^2 = -15a^2bx^3y^6z^2$

Example 2. $(3a^2b)(-2ab^2)(-ab) = 3(-2)(-1)a^2aabb^2b = 6a^4b^4$

The justification for these rules lies not only in the rules for exponents previously obtained but also in a liberal use of the associative and commutative properties for multiplication. The commutative property is rather obviously used as, without it, we could not take the first step of moving, for example, the x and the x^2 in Example 1 next to each other. The use of the associative property is somewhat less obvious but, actually, it is implied in the very writing of such expressions as $3a^2xy^5$ because, without the associative property, we would have to have a liberal use of parentheses in order to know whether we meant $3[a^2(xy^5)]$, $(3a^2)(xy^5)$, or still other possibilities. Because of the associative property, however, we know that it makes no difference which grouping we mean, since they all yield equal results.

Exercise 2.3

Simplify the following expressions by use of the laws of exponents.

1. $(-3)^3$
2. 5^2
3. $(2a)^3$
4. $(-3x^2)^2$
5. $2^2 \cdot 3^4$
6. x^2x^5
7. $x^2(xy)^4$
8. $(ab^2)^3b^5$
9. $(-3x)(xy)y^2$
10. $(3x^2)(2x)^3$
11. $(6a^2b)(-2ab^2)$
12. $(6a^4b^5c^2)^2(2a^2b^2c)^2$
13. $(-3v^2s^5)^2(-vs^3)^3$
14. $(-2a)(-3b)(-5ab)$
15. $(-x^2y)(-xy^2)(-xy)$
16. $(2a^2b)(-7a^2b)(5ab)$

17. $(5xy)(-6x^2yz^2)$ **18.** $(3a^3b^2c^5)(5a^pb^qc^m)$

19. $(3xy^3)(-4x^{a+3}y^{b+1})$ **20.** $(-2x^my^{2m})(-6x^2y^3)$

2.4 MULTIPLICATION OF POLYNOMIALS

A polynomial may be multiplied by a monomial by using the distributive property followed by the use of the rules developed in the preceding section.

Example 1.

$$3ax(ax^2 - 5bx + c) = (3ax)(ax^2) + (3ax)(-5bx) + (3ax)c$$
$$= 3a^2x^3 - 15abx^2 + 3acx$$

A polynomial may be multiplied by a polynomial by repeated use of the distributive property.

Example 2. $(5ax + by)(2ax - by) = 5ax(2ax - 3y) + by(2ax - 3y)$
$$= [(5ax)(2ax) - (5ax)(3y)] +$$
$$[(by)(2ax) - (by)(3y)]$$
$$= 10a^2x^2 - 15axy + 2abxy - 3by^2$$

It is often convenient to arrange the two factors vertically as we do in ordinary arithmetic. Thus, for example, to find the product of $3x^3 - 2x^2 + x - 1$ by $x^2 - x + 5$ we can write

$$
\begin{array}{r}
3x^3 - 2x^2 + x - 1 \\
x^2 - x + 5 \\
\hline
3x^5 - 2x^4 + \quad x^3 - \quad x^2 \\
-3x^4 + \quad 2x^3 - \quad x^2 + \quad x \\
+ 15x^3 - 10x^2 + 5x - 5 \\
\hline
3x^5 - 5x^4 + 18x^3 - 12x^2 + 6x - 5
\end{array}
$$

In fact the consideration of multiplication in this fashion sheds considerable light on the mysterious rule for multiplication of numbers of more than one digit that you learned in grammar school. Suppose, for example, that the product 254×32 is desired. We first note that 254 is an abbreviated way of writing $2 \cdot 100 + 5 \cdot 10 + 4 = 2(10)^2 + 5(10) + 4$. Similarly, $32 = 3(10) + 2$. Thus we have, with "10's" in place of "x's," a problem similar to the algebraic one just presented:

$$\begin{array}{r} 2(10)^2 + \ 5(10) + 4 \\ 3(10) + 2 \ \hline 6(10)^3 + 15(10)^2 + 12(10) \\ + \ \ 4(10)^2 + 10(10) + 8 \ \hline 6(10)^3 + 19(10)^2 + 22(10) + 8 \end{array}$$

There is this difference, however: if the answer to an algebraic problem were $6x^3 + 19x^2 + 22x + 8$, we would be completely satisfied, but, in our arithmetic notation, each coefficient of a power of 10 must be less than 10. Hence we write

$$6(10)^3 + 19(10)^2 + 22(10) + 8$$
$$= 6(10)^3 + (10 + 9)(10)^2 + (2 \cdot 10 + 2)(10) + 8$$

since $19(10)^2 = (10 + 9)(10)^2$ and $22(10) = (2 \cdot 10 + 2)(10)$. Then, using the distributive property, $(10 + 9)(10)^2 = (10)^3 + 9(10)^2$ and $(2 \cdot 10 + 2)(10) = 2(10)^2 + 2(10)$; and we have

$$6(10)^3 + [(10)^3 + 9(10)^2] + [2(10)^2 + 2(10)] + 8$$

Finally, adding together the like powers, we have

$$7(10)^3 + 11(10)^2 + 2(10) + 8$$

and, since $11(10)^2 = (10 + 1)(10)^2 = (10)^3 + (10)^2$, we have

$$7(10)^3 + 10^3 + 10^2 + 2(10) + 8$$

or

$$8(10)^3 + 1 \cdot (10)^2 + 2(10) + 8 = 8128$$

The student should check this answer by the ordinary method of multiplication and see how this explains "carrying."

Products of two binomials occur so frequently that the student should be able to do these by forming the four products involved and doing the addition mentally. This is particularly true of such products as $(2x + 3)(x - 4) = 2x^2 - 5x - 12$ and $(x - 3y)(2x - 5y) = 2x^2 - 11xy + 15y^2$.

Exercise 2.4

In problems 1–40, perform the indicated multiplication.

1. $3(4x - a)$ 2. $4(3a - 5b)$
3. $x(x^2 - x^3)$ 4. $-3(-5x - 4y)$

5. $-3y(4x - 7y)$

6. $2x(x - 3y + 2z)$

7. $abc(3a + 2b - c)$

8. $2a^2b(2ab - 4a^2)$

9. $mn^2(5m + 4n - mn)$

10. $-x^2y^2(-x^3 - 2x^2y - xy^2 - 4y^2)$

11. $(2a - b)(3a - 4b)$

12. $(x - 1)(2x - 1)$

13. $(2x - 5)(6x - 1)$

14. $(x - 3y)(4x - 2y)$

15. $(a + b)(a - b)$

16. $(2c^2 - 3)(2c^2 + 3)$

17. $(5a^2 + 4)(5a^2 - 4)$

18. $(y + 7)(y - 7)$

19. $(a^2 - 2)(2a^2 - 5)$

20. $(x^3 - 2)(x^3 + 5)$

21. $(a - 2)(a^2 + 2a + 4)$

22. $(2 - x)(3 - 4x - x^2)$

23. $(x - 4)(2x^2 + 1 - x)$

24. $(2x^2 + 5x - 1)(x - 4)$

25. $(x^2 + 2x + 3)(x + 2)$

26. $(3a^2 - 7a - 9)(2a - 3)$

27. $(a^2 + ab + b^2)(a + b)$

28. $(5x^2 - x + 3)(3x^3 - x^2 + 2x - 4)$

29. $(x^2 + 3x - 2)(x^2 - 2x + 4)$

30. $(1 - 4x)(x - 2x^2 + 7)$

31. $(5a - 2b)(7a - 3b)$

32. $(7a - b)(7a + b)$

33. $(a + b)(a^2 - ab + b^2)$

34. $(a - b)(a^2 + ab + b^2)$

35. $(x + y)(x^2 + 3xy + y^2)$

36. $(2a - b)(4a^2 - 4ab + b^2)$

37. $(4x^3 - 7xy^2 - 5y^3)(2x^2 + 3xy - y^2)$

38. $(2x^2y - 3z^2)^2$

39. $(2x - y + 4z)^2$

40. $(4x - 3y)^3$

In problems 41–54, multiply out and collect like terms.

41. $2a[a - 2a(2a - 3) - 4a]$

42. $x(y + z) + 2z(x + y) + 3y(z + x)$

43. $x^2(y + z) + y^2(x + z)$

44. $3a^2 + 2[a + 2a(a + 4)]$

45. $x(x + 2) + 2(x^2 + x + 1)$

46. $x(x^2 - 2) - 2x^2(x - 1)$

47. $-a(2a^2 - 3) + 3a^2(a + 2)$

48. $-2x^3[3 - x^2(3 - 2x) - x^2 + 2x]$

49. $3x - 2\{2x - 3[2(2x - 3) - (4x - 3) + x - 2] - 10\}$

50. $x^3 - \{x^2y + y[yz - x(2y - z) - z(2x + y) - xz] + 2xy^2\}$

51. $(x + 1)(x + 2) + (x + 2)(x + 3)$

52. $(2x - 3)(2x + 3) - (4x + 3)(x - 3)$

53. $(x + y)^2 - (x - y)^2$

54. $(x - 1)^3 - x(x + 1)^2$

In problems 55–60, perform the arithmetic computation by the method shown in the text.

55. 5×36

56. 6×19

57. 7×42

58. 12×18

59. 21×32

60. 14×26

2.5   THE CONCEPT OF FACTORIZATION

To **factor** a number means to write it as a product of two or more numbers. For example, since $6 = 3 \cdot 2$ we speak of 2 and 3 as factors of 6. If we confine ourselves to positive integers, neither 2 nor 3 have any factors except themselves and 1; such integers are called **prime numbers.**

If we allow fractions, however, both 2 and 3 can be factored, since $2 = \frac{1}{2} \cdot 4$ and $3 = \frac{2}{3} \cdot \frac{9}{2}$. Thus, whether or not a number can be factored depends on what type of numbers are allowed as factors. Similarly, in the factoring of algebraic expressions, $x^2 - 2y^2$ is unfactorable if we allow only rational numbers, but is factorable into $(x + \sqrt{2}y)(x - \sqrt{2}y)$ if we allow irrational numbers.

We shall not attempt to give a precise statement of the problem of algebraic factoring, since the factoring actually needed for the great majority of cases in both mathematics and the sciences is of a very elementary type.

The expressions whose factorization we will study will be polynomials — usually in one or two letters.

2.6   FACTORING OUT A COMMON FACTOR

In seeking to factor a polynomial expression, the first thing to look for is a factor **common** to all the terms. For example, in

$$5ax^2 - 15ax + 20a$$

the factor $5a$ is common to all the terms. Hence we may write, by the distributive property

$$5ax^2 - 15ax + 20a = 5a(x^2 - 3x + 4)$$

Similarly,

$$4b^2x^3 + 12b^3x^2 + 8b^4x = 4b^2x(x^2 + 3bx + 2b^2)$$

The common factor need not be a monomial. Thus, for example,

$$6a(a + b) - 7b(a + b) = (a + b)(6a - 7b)$$

and

$$x^2(x - 1) + (x - 1) = (x - 1)(x^2 + 1)$$

Exercise 2.6

In problems 1–9, factor the given number into prime factors.

1. 15	2. 14	3. 24
4. 36	5. 100	6. 110
7. 182	8. 365	9. 142

In problems 10–19, factor out any common factors before evaluating.

10. $(20 - 15)$	11. $(36 - 84)$
12. $(180 - 72 - 108)$	13. $(105 - 42 + 63)$
14. $5a - 3a$	15. $15x - 10x$
16. $27x - 18x - 6x$	17. $4x - 3x + 7x - 6x - 2x$
18. $6b - 3b + 9b$	19. $8y + 10y - 14y$

In problems 20–55, factor out any common factors.

20. $6a - 9$	21. $24x - 16$
22. $15x^2 - 5x - 10$	23. $3x^2 - 3x + 3$
24. $3x^2 + 6x$	25. $8a^2 - 6a$
26. $ax + ay$	27. $5a - 5b$
28. $xyz - xz$	29. $21xy - 3x$
30. $ax + 2ay + 3az$	31. $x^4 - 2x^3 - 5x^2$
32. $9xy^2z^2 + 27x^3y^3 - 9xy^3z^3$	33. $3a^2b^3c^4 - 5a^2b - 7a^2c^2$
34. $12x^3y^3 - 8x^2y + 16x^3yz^5$	35. $4b^2x^3 - 6bx^2 + 8bcx^2$
36. $5x^2 + 5x$	37. $27a^2b + 9ab$
38. $14x^2y^2 - 21x^2y + 7xy$	39. $24x^2y - 12xy^2 - 6xy$
40. $a(x + y) + b(x + y)$	41. $x(x + y) - 3(x + y)$
42. $b(2x - y) - 2a(2x - y)$	43. $2h(x - 2) - 3k(x - 2)$
44. $3x(x - y) + 2y(x - y)$	45. $7x(y - 2) + 10z(y - 2)$
46. $3x(x - y) - (x - y)$	47. $7(x - 2y) - 4a(x - 2y)$

48. $a^2(a + 1) + (a + 1)$
49. $a^2(x - y) + b^2(x - y) - 3(x - y)$
50. $2x(a - 2b) + 2y(a - 2b) - 5(a - 2b)$
51. $a(x^2 - x + 1) + b(x^2 - x + 1)$
52. $2x^2(a^2 - 3b + 5) - 7(a^2 - 3b + 5)$
53. $3x^2(x + 2y) - 9x(x + 2y) + 17(x + 2y)$
54. $p^2(p - 2) - (3p + 1)(p - 2)$
55. $x^2(x^2 - 3x + 5) - 7(x^2 - 3x + 5)$

$$\left(2x - y\right)\left[b - 2a\right]$$

2.7 THE DIFFERENCE OF SQUARES AND SUM OR DIFFERENCE OF CUBES

After removing common factors (if any) from the terms of a polynomial
it may well be that no further factorization is possible. On the other
hand, although further factorization may be possible the actual carrying out of the factorization may be extremely difficult. Thus, for example, if one were to multiply together

$$(x^2 + x + 1)(2x^2 + x - 3)$$

the resulting fourth degree polynomial would certainly be factorable
but a discussion of the finding of the factors of such polynomials is beyond the scope of a course in intermediate algebra.

For most applications up through the calculus, a two-term expression having no common factors may be considered as factorable if
and only if it is "the difference of two squares" (that is, of the form
$a^2 - b^2$) or "the sum or difference of two cubes" (that is, of the form
$a^3 + b^3$ or $a^3 - b^3$). This does not mean, however, that other two-term
expressions cannot be factored. Thus, for example,

$$a^5 - b^5 = (a - b)(a^4 + a^3b + a^2b^2 + ab^3 + b^4)$$

Such factorizations, however, are rarely needed in the first two years
of college mathematics and are also considered in a course in college
algebra.

To begin with, expressions of the form $a^2 - b^2$ may be easily handled by an application of the identity

$$a^2 - b^2 = (a + b)(a - b)$$

which is checked by multiplying $a + b$ by $a - b$.

In order, however, to utilize this formula to its fullest extent we
need to think of it something like this: (anything squared) − (anything else squared) is equal to (the sum of the two things) × (the difference of the two things in the same order).

Example 1. Factor $4x^2 - 9y^2$.

Solution: $4x^2 - 9y^2 = (2x)^2 - (3y)^2 = (2x + 3y)(2x - 3y)$.

In this case, what we have is ($2x$, the quantity squared) − ($3y$, the
quantity squared).

Example 2. $(z + 1)^2 - b^2 = [(z + 1) + b][(z + 1) - b]$.

In this case, what we have is $(z + 1$, the quantity squared$) - (b$ squared$)$.

In a similar fashion we may handle the sum and difference of two cubes by the formulas

$$a^3 - b^3 = (a - b)(a^2 + ab + b^2)$$

and

$$a^3 + b^3 = (a + b)(a^2 - ab + b^2)$$

Example 3.

$$27x^3 - 8y^3 = (3x)^3 - (2y)^3 = (3x - 2y)(9x^2 + 6xy + 4y^2)$$

In this case, what we have is $(3x$, the quantity cubed$) - (2y$, the quantity cubed$)$.

Example 4. $(x + y)^3 + z^3 = [(x + y) + z][(x + y)^2 - (x + y)z + z^2]$

In this case, what we have is $(x + y$, the quantity cubed$) + (z$ cubed$)$.

Example 5. $2x^2y - 8y^3 = 2y(x^2 - 4y^2) = 2y(x + 2y)(x - 2y)$

This example illustrates the importance of looking first for common factors.

Example 6. $a^4 - b^4 = (a^2)^2 - (b^2)^2 = (a^2 + b^2)(a^2 - b^2)$
$$= (a^2 + b^2)(a + b)(a - b)$$

Note here that our first factorization is not a complete one.

Exercise 2.7   Test p -10

Factor the following expressions. If any common factors occur, they should be removed first.

1. $r^2 - s^2$	2. $4p^2 - q^2$	3. $18x^2 - 8y^2$
4. $49x^2y^2 - 36a^2$	5. $4c^2 - 9d^2$	6. $81a^2 - 1$
7. $1 - 49x^2y^2$	8. $ax^2 - ay^2$	9. $8a^2 - 2b^2$
10. $2x^3 - 2xy^2$	11. $100a^2 - x^4$	12. $36x^4 - z^2$

13. $27a^2x^2 - 3$ 14. $a^6b^4 - 25$ 15. $m^4n^2 - r^6s^8$

16. $16a^4 - a^2$ 17. $(a + 2b)^2 - c^2$ 18. $(x - 1)^2 - 16y^2$

19. $(x + 3y)^2 - 25z^2$ 20. $(2x + y)^2 - (z - 3y)^2$

21. $(a + b)^2 - (c + d)^2$ 22. $4(a + b)^2 - 1$

23. $a^2(a + b)^2 - a^2$ 24. $27 - 3(a - b)^2$

25. $4x^2y^2 - (z^2 + 1)^2$ 26. $4x^2(p + q) - 9y^2(p + q)$

27. $36a^2x^2(x^2 - y^2) + 25b^2y^2(x^2 - y^2)$

$3(9 - p^2)$

28. $(a^2 - 3a + 5)^2 - (x^2 + 5x - 4)^2$ $3(3^2 - p^2)$

29. $(6x - 7y - 1)^2 - x^2(6x - 7y - 1)^2$ $3 \cdot (3 - p)(3 + p)$

30. $p^2q^2(z^2 - 1) - r^2s^2(z^2 - 1)$

31. $r^3 - s^3$ 32. $p^3 + q^6$

33. $27x^3 - 8y^3$ 34. $y^3 - 125$

35. $m^3 + 64$ 36. $125x^3 - 64y^3$

37. $1 - 27y^6$ 38. $64m^3n^6 + 27$

39. $24x^3 - 3$ 40. $8x^3 - x^3y^3$

41. $27x^3y^3 + (z^2 + 1)^3$ 42. $(c - d)^3 + a^3$

43. $a^3 - (b - c)^3$ 44. $(h - x)^3 - (y - x)^3$

45. $a^3(x - 3)^2 + b^3(x - 3)^2$ 46. $27r^3 + (2s - 3r)^3$

47. $x^3y^6 - (xy - z^2)^3$ 48. $(x^2 - 3z)^3 + 1$

49. $a^3(a - b)^3 + a^3$ 50. $(x^2y^2 - 4) - a^3(x^2y^2 - 4)$

2.8 FACTORING TRINOMIALS

Three-term, or trinomial, expressions will now be considered. Again, there are types of trinomials whose factorization is beyond the scope of this course, but, in practice, most trinomials (without a common factor) will either be unfactorable or, if factorable, will factor into the product of two binomials. It will help to see the reasoning behind such factorization if we begin by reviewing the multiplication of binomials discussed in Section 2.4.

For example, we have $(3x - 4)(2x + 3) = 6x^2 + x - 12$, where the first term of the answer ($6x^2$) is the product of the first terms of the two binomials, the last term (-12) is the product of the second terms of the binomials, and the middle term (x) is the sum of the two products [$(2x)(-4)$ and $(3x)(3)$] of the first and last terms.

Our problem here, of course, is the reverse: given $6x^2 + x - 12$ to express it as a product of two binomials with integral coefficients. Since, as we have seen, the product of the two first terms must be $6x^2$, the first terms must be either $6x$ and x or $2x$ and $3x$. Likewise, since the product of the two second terms must be -12 the second terms must be either, numerically, 1 and 12, 2 and 6, or 3 and 4 (and one of the two second terms must be negative to get -12). Our problem is to pick one of these combinations to yield x as a middle term.

The process is one of trial. For example, we might suppose (erroneously) that $6x$ and x will do the job. Then we would have $(6x + ?)(x - ?)$ or $(6x - ?)(x + ?)$, and we try all the ways of putting the possibilities 1 and 12, 2 and 6, and 3 and 4 in place of the question marks, as for example, $(6x + 1)(x - 2)$, $(6x - 1)(x + 6)$, etc.

Since none of these work, we try $(2x + ?)(3x - ?)$ or $(2x - ?)(3x + ?)$ and eventually arrive at the answer.

We have deliberately chosen about as complicated an example as is normally found in order to emphasize the generality of the method. Often, as in the factorization of $x^2 - 5x - 6$, for example, few trials are needed.

Once again, too, it should be emphasized that we are, here, considering only a restricted type of factorization of trinomials. Thus, from what has been said, the student would (and should!) conclude that the trinomial $x^2 - x - 3$ is unfactorable, because all possible combinations, $(x - 1)(x + 3)$, $(x - 3)(x + 1)$, etc., do not yield a factorization. This, however, is only in terms of a restriction to rational numbers, and, as a by-product of our later work in the solution of quadratic equations, the student will learn that, actually,

$$x^2 - x - 3 = \left(x - \frac{1 + \sqrt{13}}{2}\right)\left(x - \frac{1 - \sqrt{13}}{2}\right)$$

Two important special types of factorable trinomials are the **perfect square trinomial** types characterized by

$$a^2 + 2ab + b^2 = (a + b)^2$$

and

$$a^2 - 2ab + b^2 = (a - b)^2$$

To recognize a perfect square trinomial we must first notice that two of the terms must be perfect squares and that the other term must be (except possibly for sign) twice the product of the square roots of these two terms.

Example 1.

$$4x^2 - 12xy + 9y^2 = (2x)^2 - 2(2x)(3y) + (3y)^2 = (2x - 3y)^2.$$

Example 2. $16a^2 + 40a + 25 = (4a)^2 + 2(4a)(5) + 5^2 = (4a + 5)^2.$

Sometimes the terms of a trinomial will not be monomials.

Example 3. $(x + 2y)^2 - (x + 2y) - 6$
$$= [(x + 2y) - 3][(x + 2y) + 2]$$
$$= (x + 2y - 3)(x + 2y + 2)$$

Example 4. $(x + 2y)^2 + 8a(x + 2y) + 16a^2$
$$= (x + 2y)^2 + 2 \cdot 4a(x + 2y) + (4a)^2$$
$$= [(x + 2y) + 4a]^2 = (x + 2y + 4a)^2$$

Exercise 2.8 $10-8$ Test $p-10$

Factor the following expressions. If any common factors occur, they should be removed first.

1. $x^2 - 4x + 3$
2. $a^2 + 5a + 6$
3. $ax^2 - 4ax + 4a$
4. $a^2 - a - 6$ $4(x^2 - 2x - 1)$
5. $2x^2 - 8x - 10$
6. $4x^2 - 8x - 4$
7. $a^2 + 8a + 16$
8. $x^2 + 2x + 1$
9. $x^2 - 4x + 4$
10. $9x^2 - 24x + 16$
11. $x^2 - 6x + 9$
12. $3x^2 - 24x - 27$
13. $x^3 + 3x^2 - 4x$
14. $5x^2 - 5x - 60$
15. $a^2 + 7ab + 10b^2$
16. $r^2 + 6rs + 8s^2$
17. $9x^2 + 12xy + 4y^2$
18. $5b + 2by - 3by^2$
19. $1 - 14x + 49x^2$
20. $4x^2 + 12x + 9$
21. $6x^2 + 15x - 9$
22. $25x^3 - 10x^2 + x$
23. $2x^2 + 2xy - 12y^2$
24. $a^2 - 12a + 36$ $(a - 6$ $)(a - 6)$
25. $ax^2 - 10ax + 25a$
26. $1 - 8xy + 16x^2y^2$
27. $6x^2 - x - 7$
28. $10x^2 + 15x + 5$
29. $15 - 2y - y^2$
30. $27x^2 + 18xy + 3y^2$
31. $25x^2 + 30xy + 9y^2$
32. $49x^2 - 56xy + 16y^2$
33. $10a^2 - ab - 21b^2$
34. $8x^2 + 18xy + 10y^2$ —
35. $m^3 + 4m^2 + 4m$
36. $ax^3 - 8ax^2 + 16ax$
37. $6x^3y - 26x^2y^2 - 20xy^3$
38. $a^4 + 8a^2b^2 + 16b^4$

39. $a^6 - 7a^3b + 12b^2$ **40.** $y^8 - 9y^4z^2 - 10z^4$

41. $(x + y)^2 - (x + y) - 20$ **42.** $(a + 2b)^2 - 10(a + 2b) + 25$

43. $(z - 1)^2 - 2(z - 1) + 1$

44. $x(x - 2y)^2 + 12x(x - 2y) + 36x$

45. $2x^2(a + b) - 2x(a + b) - 60(a + b)$

46. $4a^2(x - y)^2 + 12ab(x - y)^2 + 9b^2(x - y)^2$

47. $2x^3 - 11x^2 + 12x$

48. $6x^2y^2 + 5xy^3 - 4y^4$

49. $4(x + y)^2 - 4(x + y) + 1$

50. $(a - 2b)^2 - 12(a - 2b) + 36$

2.9 FACTORING BY GROUPING

Except for polynomials that can be factored by just removing a common factor, we have considered so far only the factorization of binomials and trinomials. Here we consider some special types of four- and six-term polynomials that are factorable by **grouping.** We present four examples.

Example 1. $ab + ad + cb + cd = a(b + d) + c(b + d)$
$$= (a + c)(b + d)$$

Example 2. $a^2 + 2ab + b^2 - c^2 = (a + b)^2 - c^2$
$$= (a + b + c)(a + b - c)$$

Example 3. $x^2 + 2xy + y^2 - u^2 + 2uv - v^2$
$$= (x^2 + 2xy + y^2) - (u^2 - 2uv + v^2)$$
$$= (x + y)^2 - (u - v)^2$$
$$= [(x + y) + (u - v)][(x + y) - (u - v)]$$
$$= (x + y + u - v)(x + y - u + v)$$

Example 4. $x^2 + 2xy + y^2 + 2x + 2y + 1$
$$= (x + y)^2 + 2(x + y) + 1$$
$$= [(x + y) + 1]^2 = (x + y + 1)^2$$

In each case the "trick" is to group certain terms of the polynomial together in such a way that it becomes a two- or three-term polynomial. Then we use our methods of factoring binomials and trinomials.

Exercise 2.9 /0-/0 _Test 2,10_

Factor by grouping. $a(x-y)$ b $(x-y)$

1. $ax - ay + bx - by$
2. $4p + 4q - ap - aq$
3. $5a^3 + 5a^2 - a - 1$ $(x-y)(a+b)$
4. $ax + 3x + 2ay + 6y$
5. $4xz + 3y - 4yz - 3x$
6. $3a - ab - 3c + cb$
7. $x^3 - 3x^2 + x - 3$
8. $2 + 4x - 10x^4 - 5x^3$
9. $a^3 - b^3 - a^2 + ab$
10. $6x^3 - 4x^2 + 9x - 6$
11. $x^2 - y^2 + 3x + 3y$
12. $x^2 + 2x + 1 + ax + a$
13. $x^3 - 3x^2 - 2x + 6$
14. $x^3 - y^3 - x + y$
15. $4a^2 - 4b^2 - c^2 - 4bc$
16. $y^2 + 2yz + z^2 - 4x^2$
17. $16a^2 - 1 - 9x^2 + 6x$
18. $ax + bx + ay + by - cx - cy$
19. $20x^2 + 15xy + 8xz + 6yz$
20. $x^2 - y^2 - 9z^2 - 6yz$
21. $4x^2 - 14x + 4xy - 7y + y^2$
22. $4x^2 - a^2 + y^2 - b^2 + 4xy + 2ab$
23. $x^3 + 3x^2 - 2xy - 6y$
24. $3x^3 + 3x^2 - x - 1$

$a(x-y)^2b$ $(x-y)$
$a(x-y)^2b$
$ab(x-y)$

$3(x^3 + x^2) - x - 1$
$3(x^5 - x) - 1$

2.10 CONCLUDING REMARKS ON FACTORING

We have emphasized throughout that our remarks on factoring are necessarily incomplete. There are many polynomials that are factorable but not by the methods given here. For example, $a^n + b^n$ is factorable for any odd positive integer n, but we have considered only the case when $n = 3$. Similarly, $a^4 + a^2b^2 + b^4$ is not factorable directly as a trinomial but, by writing it as $(a^4 + 2a^2b^2 + b^4) - a^2b^2$, we have

$$(a^4 + 2a^2b^2 + b^4) - a^2b^2 = (a^2 + b^2)^2 - (ab)^2$$
$$= (a^2 + b^2 + ab)(a^2 + b^2 - ab)$$

Furthermore, we have not defined exactly what it means for a polynomial not to be factorable since, as we have pointed out, $x^2 - 2y^2$ is factorable into $(x + \sqrt{2}y)(x - \sqrt{2}y)$ if we allow the use of irrational numbers and, of course, one could say that $x + y$ is factorable into $\frac{1}{2}(2x + 2y)$. For the purposes of this chapter, however, we will reject the use of irrational numbers in our factorizations and disregard "trivial" factorizations like $x + y = \frac{1}{2}(2x + 2y)$. We will not, in this book, use any polynomials which are factorable but which cannot be factored by one of the standard methods described in the previous four sections, or will we, until Chapter 6, use irrational numbers in our factoring.

Exercise 2.10 *10-10* *Test 10-10*

Factor completely by the methods of this chapter.

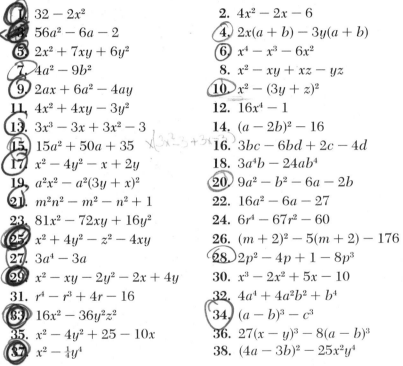

1. $32 - 2x^2$
2. $4x^2 - 2x - 6$
3. $56a^2 - 6a - 2$
4. $2x(a + b) - 3y(a + b)$
5. $2x^2 + 7xy + 6y^2$
6. $x^4 - x^3 - 6x^2$
7. $4a^2 - 9b^2$
8. $x^2 - xy + xz - yz$
9. $2ax + 6a^2 - 4ay$
10. $x^2 - (3y + z)^2$
11. $4x^2 + 4xy - 3y^2$
12. $16x^4 - 1$
13. $3x^3 - 3x + 3x^2 - 3$
14. $(a - 2b)^2 - 16$
15. $15a^2 + 50a + 35$
16. $3bc - 6bd + 2c - 4d$
17. $x^2 - 4y^2 - x + 2y$
18. $3a^4b - 24ab^4$
19. $a^2x^2 - a^2(3y + x)^2$
20. $9a^2 - b^2 - 6a - 2b$
21. $m^2n^2 - m^2 - n^2 + 1$
22. $16a^2 - 6a - 27$
23. $81x^2 - 72xy + 16y^2$
24. $6r^4 - 67r^2 - 60$
25. $x^2 + 4y^2 - z^2 - 4xy$
26. $(m + 2)^2 - 5(m + 2) - 176$
27. $3a^4 - 3a$
28. $2p^2 - 4p + 1 - 8p^3$
29. $x^2 - xy - 2y^2 - 2x + 4y$
30. $x^3 - 2x^2 + 5x - 10$
31. $r^4 - r^3 + 4r - 16$
32. $4a^4 + 4a^2b^2 + b^4$
33. $16x^2 - 36y^2z^2$
34. $(a - b)^3 - c^3$
35. $x^2 - 4y^2 + 25 - 10x$
36. $27(x - y)^3 - 8(a - b)^3$
37. $x^2 - \frac{1}{4}y^4$
38. $(4a - 3b)^2 - 25x^2y^4$

In problems 39–44, factor completely the following expressions, which occur in trigonometry problems.

39. $x^2 + 2xy$
40. $x^2(1 - y^2) - (1 - y^2)$
41. $2bc - b^2 - c^2 + a^2$
42. $x^2y + x^3 + xy^2 + y^3$
43. $3xy^2 + x(y^2 - x^2)$
44. $2x(1 - x^2) + x - 2x^3$

In problems 45–50, factor completely the following expressions, which occur in analytic geometry problems.

45. $2x^2 - 8x + 8$ 46. $y^2 - y + \frac{1}{4}$

47. $4x^2 + 4xy + y^2 - 2x - y - 20$

48. $4x^2 + 4xy + y^2 + 2x + y - 2$

49. $xy + 5x - 2y - 10$ 50. $10xy + 4x - 15y - 6$

In problems 51–56, factor completely the following expressions, which occur in calculus problems.

51. $3x^2 - 12x$ 52. $2x^3 - 16$

53. $(x - 1)^4 + 4(x - 1)^3(x - 6)$

54. $2x(1 - x^2)^2 + 4x(1 + x^2)(1 - x^2)$

55. $3(x^2 + 2)(x - 2)^2 - 2x(x - 2)^3$

56. $(x + 1)(x - 2)^2 - x[2(x + 1)(x - 2) + (x - 2)^2]$

RATIONAL ALGEBRAIC EXPRESSIONS

3

3.1 SIMPLIFICATION OF RATIONAL ALGEBRAIC EXPRESSIONS

A rational number was defined in Section 1.9 to be an integer divided by a nonzero integer. Similarly, a rational algebraic expression is simply a polynomial divided by a nonzero polynomial. Thus $\frac{1}{2}, \frac{x}{y}$, $\frac{a+b}{a^2+b^2}$, and $\frac{x^2-x+1}{x+2}$ are all rational algebraic expressions *provided* the denominators are different from zero. That is, $y \neq 0$ in the second example, a and b are not both zero in the third example, and $x \neq -2$ in the fourth example.

In Section 2.1, we saw that integers are polynomials. Thus an integer divided by an integer is a polynomial divided by a polynomial and, therefore, rational numbers are rational algebraic expressions. Moreover, any polynomial, $P(x)$, can be written as $\frac{P(x)}{1}$ and, since 1 is also a polynomial, we see that polynomials are rational algebraic expressions.

Just as rational numbers are sometimes called fractions, rational algebraic expressions are often called algebraic fractions or even, sometimes, just fractions.

We are now ready to discuss the arithmetic of algebraic fractions. First, we should note that when we are dealing with an expression such as $\frac{x+y}{x-y}$ we have no right to assume that x and y stand for inte-

66

gers. For example, it might turn out that, in some application, $x = \frac{2}{3}$ and $y = \sqrt{2}$. Then

$$\frac{\frac{2}{3} + \sqrt{2}}{\frac{2}{3} - \sqrt{2}}$$

is certainly not a rational number. However, as is shown in more advanced courses, perfectly consistent results are obtained if we treat all algebraic fractions as if they were rational numbers. Second, we must assume that $x \neq y$ for, if $x = y$, then $\dfrac{x + y}{x - y}$ is not a number since division by zero is impossible. In all work with algebraic fractions, then, we will make the tacit assumption that the letters involved never take on values which would make a denominator zero. Of course, in any application of these algebraic fractions to numerical problems, we must check to see that the denominators are actually nonzero.

The following four examples illustrate simplification of algebraic fractions. Notice that the key step lies in the factorization of the numerator and the denominator and that we again use the cancellation theorem (Theorem 1.5).

Example 1. $\dfrac{ax + ay}{bx + by} = \dfrac{a(x + y)}{b(x + y)} = \dfrac{a}{b}.$

Example 2. $\dfrac{ac - ad}{bd - bc} = \dfrac{a(c - d)}{b(d - c)} = \dfrac{a(c - d)}{(-b)(c - d)} = \dfrac{a}{-b} = -\left(\dfrac{a}{b}\right).$

Example 3. $\dfrac{x^2 - x - 2}{x^2 + 4x + 3} = \dfrac{(x - 2)(x + 1)}{(x + 3)(x + 1)} = \dfrac{x - 2}{x + 3}.$

Example 4.

$$\frac{a^2 - 3ab + 2b^2}{2b^2 + ab - a^2} = \frac{(a - 2b)(a - b)}{(2b - a)(b + a)} = \frac{[-(a - 2b)][+(a - b)]}{(2b - a)(b + a)}$$

$$= \frac{(2b - a)(b - a)}{(2b - a)(b + a)} = \frac{b - a}{b + a}$$

The student is again cautioned against the illegitimate use of canceling, as in

$$\frac{\not{a} + b}{\not{a}}, \frac{\not{2}a + b}{\not{2}}$$

The obvious way to avoid such errors is to refrain from the use of "canceling" altogether and compare $\dfrac{a+b}{a}$ with, for example, $\dfrac{a+ab}{a}$. In the second case we have

$$\frac{a+ab}{a} = \frac{a(1+b)}{a} = \frac{a}{a} \cdot \frac{1+b}{1} = 1+b$$

whereas, in the first, we certainly do not have

$$\frac{a+b}{a} = b \cdot \frac{a}{a} \qquad \text{or} \qquad \frac{1+b}{1} \cdot \frac{a}{a} \left(= \frac{a+ab}{a} \right)$$

An algebraic fraction whose numerator and denominator do not have any factors in common other than 1 and -1 is said to be **reduced to lowest terms**. Thus the answers in the four examples above are all in lowest terms.

Exercise 3.1 10-12

In problems 1–6, give the rational values of x for which the given algebraic expression is *not* a rational number.

1. $\dfrac{1}{x}$ 2. $\dfrac{1}{x-1}$ 3. $\dfrac{3}{1-x}$

4. $\dfrac{x}{x+2}$ 5. $\dfrac{x+3}{x(x-3)}$ 6. $\dfrac{5(x+3)}{x(x-5)(x+2)}$

In problems 7–20, reduce the given algebraic fraction to lowest terms.

7. $\dfrac{x^2y}{xy^2}$ 8. $\dfrac{2a^2b^3}{4ab^5}$ 9. $\dfrac{-9x^2y}{3xy^2}$

10. $\dfrac{-25x}{5x^3}$ 11. $\dfrac{-10a^7}{-2a^5}$ 12. $\dfrac{15a^3}{-30a^3}$

13. $\dfrac{8x^3y^5z^2}{-4x^2yz^3}$ 14. $\dfrac{a(x+y)}{3a}$ 15. $\dfrac{-4x^2(a+b)}{2x(a-b)}$

16. $\dfrac{18ax^2(x+y)(x-y)}{24a^2x(x-y)}$ 17. $\dfrac{3a+5ab}{2a}$

18. $\dfrac{a^2}{a^2-ab}$ 19. $\dfrac{x(x-y)}{y(y-x)}$ 20. $\dfrac{a(b-a)}{a^2(a-b)}$

21. $\dfrac{x(x+y)}{x^2-y^2}$ 22. $\dfrac{2x+2y}{x^2-y^2}$ 23. $\dfrac{5(x^2-2x)}{4-2x}$

24. $\dfrac{x^2 - x - 6}{x^2 + 4x + 4}$ 25. $\dfrac{2a^2 - 5a - 12}{16 - a^2}$ 26. $\dfrac{6 - x - x^2}{3x^2 - 3x - 6}$

27. $\dfrac{xy - xz}{xy + 3y - 3z - xz}$

28. $\dfrac{a^2 + ab + b^2}{a^3 - b^3}$

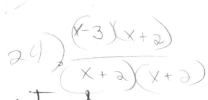

3.2 DIVISION OF POLYNOMIALS

A monomial divided by a monomial is a rational algebraic expression. Sometimes such expressions can be reduced to lower terms. For example, $9xy^2$ divided by $-3xy$ can be written

$$\frac{9xy^2}{-3x^2y} = -\frac{3y}{x}$$

Likewise, if a polynomial is divided by a polynomial, we have a rational algebraic expression, and it is a simple matter to reduce this to lowest terms if the polynomials are easily factored. However, as mentioned in Section 2.10, this is not always the case. In fact, even in a numerical example, such as $\frac{1767}{31}$, where the numerator is not easily factored, it is much easier to simply divide by "long division."

$$
\begin{array}{r}
57 \\
31 \overline{)1767} \\
155 \\
\hline
217 \\
217 \\
\hline
0
\end{array}
$$

Since the remainder is zero, we see that $\frac{1767}{31} = 57$.

The "long division" process for dividing a polynomial by a polynomial is essentially the same as for integers. Let us recall that division may be regarded as repeated subtraction. Thus $8 \div 2$ is equal to 4 because $8 - (2 \times 4) = 0$. In general, if we have $\dfrac{a}{b} = c$ where $a, b,$ and c are whole numbers, we are saying that $a - (bc) = 0$.

Now a problem such as $1767 \div 31$ could be done in "stages." We could first subtract fifty 31's from 1767 leaving 217; then we could subtract seven 31's from 217 leaving zero. Thus we find that we must subtract $(50 + 7)$ 31's from 1767 to get 0. Similarly, if we want to divide $x^2 + 5x + 6$ by $x + 2$, we can first subtract $x(x + 2)$'s from $x^2 + 5x + 6$, giving us $(x^2 + 5x + 6) - x(x + 2) = (x^2 + 5x + 6) - (x^2 + 2x) = 3x + 6$. Now we subtract three $(x + 2)$'s from $3x + 6$ and have $(3x + 6) - 3(x + 2) = 0$. Thus, if we subtract $(x + 3)(x + 2)$ from $x^2 + 5x + 6$ we get 0, and hence $(x^2 + 5x + 6) \div (x + 2) = x + 3$.

The process is made quite mechanical by writing down the problem as follows:

$$
\begin{array}{r}
x + 3 \\
x + 2 \overline{\smash{)}\, x^2 + 5x + 6} \\
\underline{x^2 + 2x } \\
+ 3x + 6 \\
\underline{+ 3x + 6} \\
0
\end{array}
$$

where we first divide x^2 by x to get x, multiply $x + 2$ by x, subtract the product from $x^2 + 5x + 6$, divide $3x$ by x to get 3, etc.

Two other examples are given to illustrate this method:

$$
\begin{array}{r}
2x^2 - x + 1 \\
x - 2 \overline{\smash{)}\, 2x^3 - 5x^2 + 3x - 1} \\
\underline{2x^3 + 4x^2 } \\
-x^2 + 3x - 1 \\
\underline{-x^2 + 2x } \\
x - 1 \\
\underline{x - 2} \\
1
\end{array}
\qquad
\begin{array}{r}
x^2 + xy + y^2 \\
x - y \overline{\smash{)}\, x^3 - y^3} \\
\underline{x^3 - x^2 y } \\
x^2 y \\
\underline{x^2 y - xy^2 } \\
xy^2 - y^3 \\
\underline{xy^2 - y^3} \\
0
\end{array}
$$

Note that in the first example we have a nonzero remainder. This means that $(2x^3 - 5x^2 + 3x - 1) \div (x - 2) = 2x^2 - x + 1 + \dfrac{1}{x - 2}$ just as $7 \div 2 = 3\frac{1}{2} = 3 + \frac{1}{2}$.

Sometimes we obtain fractional coefficients in performing the division, as in

$$
\begin{array}{r}
\frac{1}{2}x - \frac{3}{4} \\
2x - 1 \overline{\smash{)}\, x^2 - 2x + 2} \\
\underline{x^2 - \frac{1}{2}x } \\
-\frac{3}{2}x + 2 \\
\underline{-\frac{3}{2}x + \frac{3}{4}} \\
\frac{5}{4}
\end{array}
$$

Exercise 3.2    / 0-12

In problems 1–48, carry out the indicated divisions.

1. $\frac{5}{2}$ ⌐ᴢ

2. $\frac{10}{7}$

3. $\frac{15}{8}$

4. $\frac{6a + 8b}{2}$ $3a + 4b$

5. $\frac{15x^2 + 5x}{5}$ $3x + 1$

6. $\frac{-3x^2 + 15x}{-1}$

7. $\frac{-4a^3 + 6a^2}{-2a}$ $+ 2a - 3a$

8. $\frac{-7x^2 + 5x}{x}$

9. $\frac{4y^3 + 8y^5}{2y^2}$

10. $\frac{5a^3b^2 - ab^3 + 15ab}{ab}$

11. $\frac{8x^2y^4z + 4x^2yz^2 - 4x^2yz}{2x^2yz}$

12. $\frac{-15x^3y^3 + 20x^2y^2 - 10xy}{-5xy}$

13. $\frac{10s^3t^2 - 15s^4t^3 + 20t^6}{5t^2}$

14. $\frac{4x^4 - 8}{2x^3}$

15. $\frac{12x^3 - 9x - 4}{-6x^2}$

16. $\frac{6x^3 - 8x^2 + 5}{2x^2}$

17. $\frac{2x^3 - 3x^2 + 5x - 2}{-2x}$

18. $\frac{3a^3 - 5a^2 + 2}{3a^2}$

19. $\frac{15x^2y - 10x^2 - 3}{-5x^2}$

20. $\frac{9x^5y^2z - 12x^4yz^2 + 18x^2y^2z^2}{-3x^2yz}$

21. $\frac{x^2 - 5x + 6}{x - 2}$

22. $\frac{6x^2 + x - 2}{2x - 1}$

23. $\frac{7a^2 + 8a + 1}{7a + 1}$

24. $\frac{x^2 + 5x + 4}{x + 1}$

25. $\frac{3x^2 - x - 24}{x - 3}$

26. $\frac{x^3 + y^3}{x + y}$

27. $\frac{x^3 - 2x^2 + 3x - 2}{x - 1}$

28. $\frac{6a^2 - a - 15}{3a - 5}$

29. $\frac{x^3 - x^2 - 10x - 8}{x - 4}$

30. $\frac{6x^2 - xy - 2y^2}{2x + y}$

31. $\frac{2x^2 + 7x - 3}{x - 7}$

32. $\frac{x^2 + 10}{x - 5}$

33. $\frac{x^2 + x + 2}{x - 4}$

34. $\frac{x^3 + 7x^2 - 6x + 10}{x + 3}$

35. $\frac{2x^3 - 3x^2 + 7x - 3}{2x - 1}$

36. $\frac{6x^3 - 8x + 5x^2 + 7}{2x + 3}$

37. $\frac{4x^3 - 8x^2 - 9x + 7}{2x - 3}$

38. $\frac{2x^3 + 9x^2y + 12y^3 + 17xy^2}{2x + 3y}$

39. $\frac{10x^3 - 7x^2y - 16xy^2 + 12y^3}{5x - 6y}$

40. $\dfrac{x^4 - y^4}{x + y}$

41. $\dfrac{2x^3 + 3x^2 - x - 12}{x^2 + 3x + 4}$

42. $\dfrac{6x^3 + 12x - 19x^2 - 5}{3x^2 - 2x + 1}$

43. $\dfrac{5x^2 - 3x + 1}{2x - 1}$

44. $\dfrac{7x^2 + 5x - 2}{3x + 1}$

45. $\dfrac{3x^2 - 2x + 1}{2x + 3}$

46. $\dfrac{x^3 - 3x^2 + x - 1}{3x - 1}$

47. $\dfrac{2x^3 - 4x^2 + 2x + 1}{3x - 4}$

48. $\dfrac{3x^3 + 4x^2 - x + 5}{2x + 1}$

Perform the indicated divisions in problems 49–54. These are all taken from problems in the "integration of rational functions" in the calculus.

49. $\dfrac{x^4}{x^3 + 1}$

50. $\dfrac{x^4 + 2x + 1}{x(x + 1)(x + 2)}$

51. $\dfrac{x^3}{x^2 + x + 1}$

52. $\dfrac{x^4 + 3x^2 + 1}{x^2 - 1}$

53. $\dfrac{x^3 + 6x^2 + 5x - 12}{x^2 + 6x + 5}$

54. $\dfrac{x^4 + 2x^3 + x^2 + 3x + 1}{x^2 + 2x}$

3.3 SYNTHETIC DIVISION

The work involved in dividing one polynomial by another polynomial can be simplified in the special case where we are dividing by a linear polynomial of the form $x - a$. Consider, for example, the following division:

$$
\begin{array}{r}
2x^2 + x + 4 \\
x - 3{\overline{\smash{\big)}\,2x^3 - 5x^2 + x - 14}} \\
\underline{2x^3 - 6x^2} \\
x^2 + x - 14 \\
\underline{x^2 - 3x} \\
4x - 14 \\
\underline{4x - 12} \\
- 2
\end{array}
$$

We notice that all of the terms that are circled are duplicates of terms directly above them. Let us omit the circled terms and push the remaining terms up into a more compact form:

$$\begin{array}{r} 2x^2 + x + 4 \\ x - 3 \overline{)\; 2x^3 - 5x^2 + x - 14} \\ + 6x^2 - 3x - 12 \\ \hline x^2 + 4x - 2 \end{array}$$

Now the x's are only used in this compact form to keep the coefficients in order. If we are careful to write the terms of the polynomials in descending order of the powers of x, we need only write the coefficients. Moreover, the coefficient of x in the divisor is, by hypothesis, always 1 and, since it serves no purpose in this compact form, we do not write it down. Omitting the x's we now have:

$$\begin{array}{r} 2 1 4 \\ -3 \overline{)\; 2 {-5} 1 {-14}} \\ {-6} {-1} {-12} \\ \hline 1 4 {-\; 2} \end{array}$$

The numbers in the first row above (2, 1, 4) are the coefficients of the quotient polynomial that is always of degree one less than the degree of the dividend. The last number in the last row (-2) is the remainder that will always be a constant. (Remember that we are dividing by a linear polynomial, $x - a$.) We note that, except for the 2, the numbers in the first row are the same as the numbers beneath them in the last row. Let us, therefore, write the 2 in the last row and omit the first row:

$$\begin{array}{r} \underline{-3}   2 {-5} 1 {-14} \\ {-6} {-3} {-12} \\ \hline 2 1 4 \,\|{-\; 2} \end{array}$$

Now, the last row, except for the 2, is obtained by *subtracting* the second row from the first row. But the -6 is obtained by multiplying 2 by -3. Thus if we change the -3 in the box to 3, we will get 6 and we can obtain the last row by *adding* the second row to the first one:

$$\begin{array}{r} \underline{3}   2 {-5} 1 {-14} \\ 6 3 12 \\ \hline 2 1 4 \,\|{-\; 2} \end{array}$$

This process for dividing a polynomial by a polynomial of the form $x - a$ is called **synthetic division.** We have divided $x - 3$ (represented by the 3 in the box) into $2x^3 - 5x^2 + x - 14$ (whose coefficients make up the first row) and obtained the quotient polynomial $2x^2 + x + 4$

(whose coefficients are in the last row preceding the two vertical bars) and a remainder of −2 (the last number in the last row).

Let us divide $3x^4 - 10x^2 + x - 6$ by $x + 2$ by synthetic division. We set the problem up as follows:

$$\underline{-2|}\quad 3 \quad 0 \quad -10 \quad 1 \quad -6$$

When dividing by $x - a$ by synthetic division, we put a at the left. Here we are dividing by $x + 2 = x - (-2)$; thus $a = -2$.

Note also that if a term is "missing," such as the x^3 term in the polynomial $3x^4 - 10x^2 + x - 6$, its coefficient is 0, and this coefficient must be written in its proper place to keep the terms in descending order of the powers of x as we agreed to do. Now we begin by bringing down the leading coefficient 3. Then we multiply 3 by −2 and place the result under the 0. We add this column and multiply that result by −2 and place it under the −10, etc., as shown below.

$$
\begin{array}{r|rrrrr}
-2 & 3 & 0 & -10 & 1 & -6 \\
   &   & -6 & 12 & -4 & 6 \\
\hline
   & 3 & -6 & 2 & -3 \| & 0
\end{array}
$$

The quotient polynomial is $3x^3 - 6x^2 + 2x - 3$ and the remainder is 0.

Here is one final example with fractional coefficients. Divide $4x^3 + x^2 - 5$ by $x - \frac{1}{2}$.

$$
\begin{array}{r|rrrr}
\frac{1}{2} & 4 & 1 & 0 & -5 \\
   &   & 2 & \frac{3}{2} & \frac{3}{4} \\
\hline
   & 4 & 3 & \frac{3}{2} \| & -\frac{17}{4}
\end{array}
$$

The quotient polynomial is $4x^2 + 3x + \frac{3}{2}$ and the remainder is $-\frac{17}{4}$.

Exercise 3.3

Find the quotient polynomial and the remainder using synthetic division.

1. $\dfrac{x^3 + 2x^2 - 17x + 6}{x - 3}$

2. $\dfrac{2x^3 - 9x^2 + 13x - 12}{x + 3}$

3. $\dfrac{2x^3 + 5x^2 - 4x - 5}{x + \frac{1}{2}}$

4. $\dfrac{x^4 + 5x^2 - 4x - 2}{x - 1}$

5. $\dfrac{x^4 + 5x^2 - 4x - 2}{x + 4}$

6. $\dfrac{x^3 - 5x^2 + 3x + 7}{x + 2}$

7. $\dfrac{3x^4 - x^2 + 10}{x - 1}$

8. $\dfrac{3x^4 - x^2 + 10}{x - 5}$

9. $\dfrac{2x^3 + 3x^2 - 1}{x - \frac{1}{2}}$

10. $\dfrac{2x^3 + 3x^2 - 1}{x + \frac{1}{2}}$

11. $\dfrac{2x^3 + 3x^2 - 1}{x - \frac{1}{3}}$

12. $\dfrac{3x^4 - 5x}{x + \frac{2}{3}}$

13. $\dfrac{x^4 - 5}{x + 2}$

14. $\dfrac{x^3 + a^3}{x + a}$

15. $\dfrac{x^3 + a^3}{x - a}$

3.4 MULTIPLICATION AND DIVISION OF RATIONAL ALGEBRAIC EXPRESSIONS

Let us recall (Sections 1.9 and 1.10) the operations of multiplication and division for rational numbers. We have

$$\frac{a}{b} \cdot \frac{c}{d} = \frac{ac}{bd}$$

$$\frac{a}{b} \div \frac{c}{d} = \frac{a}{b} \cdot \frac{d}{c} = \frac{ad}{bc}$$

The following four examples illustrate the multiplication and division of algebraic fractions, using the definitions and theorems of Chapter 1 concerning rational numbers.

Example 1. $\dfrac{x^2 - y^2}{2x} \cdot \dfrac{4x^2}{x + y} = \dfrac{(x + y)(x - y)}{2x} \cdot \dfrac{4x^2}{x + y}$

$$= \frac{(x + y)(x - y) \cdot 2x \cdot 2x}{2x(x + y)}$$

$$= \frac{x + y}{x + y} \cdot \frac{2x}{2x} \cdot \frac{(x - y)2x}{1} = 2x(x - y)$$

Example 2.

$$\frac{x^3 - y^3}{x^2 - 5x + 6} \cdot \frac{x^2 - 4}{x^2 - 2xy + y^2} = \frac{(x - y)(x^2 + xy + y^2)}{(x + 2)(x - 3)} \cdot \frac{(x + 2)(x - 2)}{(x - y)^2}$$

$$= \frac{x - y}{x - y} \cdot \frac{x - 2}{x - 2} \cdot \frac{x^2 + xy + y^2}{x - 3} \cdot \frac{x + 2}{x - y}$$

$$= \frac{(x^2 + xy + y^2)(x + 2)}{(x - 3)(x - y)}$$

Example 3. $\dfrac{a^2 + ab}{3a + b} \div a = \dfrac{a(a + b)}{3a + b} \div \dfrac{a}{1} = \dfrac{a(a + b)}{3a + b} \cdot \dfrac{1}{a}$

$$= \dfrac{a}{a} \cdot \dfrac{a + b}{3a + b} = \dfrac{a + b}{3a + b}$$

Example 4.

$$\dfrac{3x^2 + 4x + 1}{x + 2} \div \dfrac{x^2 + 2x + 1}{x^2 + 5x + 6} = \dfrac{3x^2 + 4x + 1}{x + 2} \cdot \dfrac{x^2 + 5x + 6}{x^2 + 2x + 1}$$

$$= \dfrac{(3x + 1)(x + 1)}{x + 2} \cdot \dfrac{(x + 2)(x + 3)}{(x + 1)(x + 1)}$$

$$= \dfrac{x + 1}{x + 1} \cdot \dfrac{x + 2}{x + 2} \cdot \dfrac{(3x + 1)(x + 3)}{x + 1}$$

$$= \dfrac{(3x + 1)(x + 3)}{x + 1}$$

Notice that we have merely indicated the multiplications in the numerators and denominators of our answers since, for most purposes, the answer is more useful in this form.

Exercise 3.4 $10-19$

In problems 1–15, perform the indicated operations, reducing to lowest terms, but leaving the numerator and denominator in factored form.

1. $\dfrac{12a^2}{5b^2} \cdot \dfrac{5ab^3b}{24}$

2. $\dfrac{3x^2}{4y} \cdot \dfrac{16y^2}{27x}$

3. $\dfrac{9x^3y}{6xy^2} \div \dfrac{xy^4}{5x^3}$

4. $\dfrac{x + y}{2} \cdot \dfrac{4}{x^2 - y^2}$

5. $\dfrac{a^2 - 4}{x - y} \cdot \dfrac{3x - 3y}{a^2 + 6a + 8}$

6. $\dfrac{x + 3}{x - 4} \div \dfrac{x^2 - 9}{x^2 - 16}$

7. $\dfrac{x + 2}{x^2 - 8x + 12} \cdot \dfrac{x^2 - 9x + 18}{x^2 - 4}$

8. $\dfrac{x^2 + xy}{x - y} \div \dfrac{x^2 - y^2}{2x - 2y}$

9. $\dfrac{2a^2 + a - 1}{a^2 - 4} \div \dfrac{2a - 1}{a + 2}$

10. $\dfrac{x - 1}{x^2 - 4x} \cdot (x^2 - 16)$

11. $\dfrac{x^2 - x - 12}{x^2 + x - 2} \cdot \dfrac{1 - x^2}{x^2 - 2x - 8}$

12. $\dfrac{3x^2 + x - 2}{4x^2 - 4x - 3} \div \dfrac{2x^2 - x - 3}{6x^2 - x - 2}$

13. $\dfrac{ax^2 - ay^2}{ax^2 - 2axy + ay^2} \div \dfrac{x + y}{x - y}$

14. $\dfrac{ax + ay - bx - by}{y^2 - x^2} \cdot \dfrac{ax - ay + 2bx - 2by}{a^2 + ab - 2b^2}$

15. $\dfrac{9y^2 - 4x^2}{3x^2 + xy - 2y^2} \div (2x^2 - xy - 3y^2)$

3.5 THE LEAST COMMON MULTIPLE

The **least common multiple** (L.C.M.) of any number of given positive integers, $a, b, c, \ldots$, is simply the smallest positive integer M which is divisible by $a, b, c, \ldots$. Thus, for example, the L.C.M. of 6, 8, and 12 is 24, since 24 is the smallest positive integer divisible by 6, 8, and 12. When we seek to extend this definition to polynomials we are faced with the problem that, for example, x^2 is larger than x if $x = 2$ but smaller than x if $x = \frac{1}{2}$. In order to get around this difficulty, we rephrase the definition as follows:

A polynomial P is said to be a least common multiple of the polynomials $p, q, r, \ldots$ if (1) P is divisible by $p, q, r, \ldots$ and (2) any polynomial divisible by $p, q, r, \ldots$ is divisible by P.

Before we show how to find L.C.M.'s we need to clarify the notion of divisibility for polynomials by stating that if q and p are two polynomials, we say that q divides p if

$$p = rq$$

where r is a polynomial. Recalling (Section 2.1) that a number is a polynomial we see that, for example, not only does $(x + 1)^2$ divide $2(x + 1)^2$ but that, also, $2(x + 1)^2$ divides $(x + 1)^2$ since

$$(x + 1)^2 = \tfrac{1}{2}[2(x + 1)^2]$$

Least common multiples can be found by factoring. Thus, to find the L.C.M. of 24, 20, and 36, we write:

$$24 = 2^3 \cdot 3, \qquad 20 = 2^2 \cdot 5, \qquad \text{and} \qquad 36 = 2^2 \cdot 3^2$$

Then the L.C.M. of 24, 20, and 36 is

$$M = 2^3 \cdot 3^2 \cdot 5 = 360$$

where we have multiplied together the various prime factors occurring (2, 3, and 5) each to the highest power to which it occurs (3, 2, and 1, respectively). Similarly, to find a L.C.M. for $x^2 + 5x + 6$, $x^2 + 4x + 4$, and $x + 3$, we write

$$x^2 + 5x + 6 = (x + 2)(x + 3), \qquad x^2 + 4x + 4 = (x + 2)^2, \qquad x + 3$$

and have a L.C.M. given by

$$(x + 2)^2(x + 3)$$

(Customarily, we do not multiply out the L.C.M. but leave it in factored form.) In general, our procedure is to factor our given polynomials and then to take the product of the different factors occurring — each to the highest power to which it occurs. When, however, one of the factors is the negative of another we do not list it as a separate factor. Thus if we have given

$$3x^2 - 6xy + 3y^2 = 3(x - y)^2 \qquad \text{and} \qquad y^2 - x^2 = (y + x)(y - x)$$

it is true that $y - x = -(x - y)$, and so our L.C.M. is $3(x - y)^2(y + x)$ and not $3(x - y)^2(y - x)(y + x)$.

Since we take each different factor to the highest power to which it occurs in our factorizations, it should be rather evident that we do have a L.C.M. in the sense of our second definition. A complete proof of this fact, however, involves certain notions of divisibility and factorability that would take us somewhat far afield, and we conclude with the remark that, although we do have a unique L.C.M. for any given set of positive integers, there is more than one L.C.M. for a set of polynomials. Thus

$$(x + 2)^2(x + 3), \qquad 2(x + 2)^2(x + 3), \qquad 3(x + 2)^2(x + 3), \ldots$$

are all L.C.M.'s of $x^2 + 5x + 6$, $x^2 + 4x + 4$, and $x + 3$ according to our definition. If, however, we have given polynomials with integral coefficients and follow the procedure given above for finding a L.C.M. the answer will be unique (except possibly for sign). In any event, it would certainly seem that $(x + 2)^2(x + 3)$ is the "simplest" L.C.M. and, in fact, it is the natural one to use in problems involving the addition of fractions as discussed in the next section.

Exercise 3.5

Find the L.C.M. of the polynomials given in each problem.

1. 10, 30, 5
2. 12, 15, 20
3. 48, 18, 21
4. 18, 60, 72
5. $3bc$, $9ac$, $2ab$
6. $5xy^2$, $10xy$, $15x^2y^3$

7. $14a, 21b^2, 12ab^3$

8. $ab, b(a - b), a(b - a)$

9. $x^2y, xy(x - y), x^2y^2(y - x)$

10. $x + 2y, x^2 - 4y^2$

11. $a^2 + 2ab + b^2, a^2 - b^2$

12. $a + b, a(a - b), b^2 - a^2$

13. $2x - 1, x + 2, 2x^2 + 3x - 2$

14. $x^2 + x - 6, 2x^2 - x - 6, 9 - 4x^2$

15. $2x^2 + 3x - 5, 2x^2 + 7x + 5, 1 - x^2$

16. $x^2 + 5xy + 6y^2, x^2 + 6xy + 9y^2, x^2 - 9y^2$

3.6 ADDITION AND SUBTRACTION OF ALGEBRAIC FRACTIONS

The following three examples illustrate the addition and subtraction of algebraic fractions according to our definitions, theorems, and Fundamental Principle (page 25).

Example 1.

$$\frac{5}{2a + b} + \frac{7}{6a + 3b} = \frac{5}{2a + b} + \frac{7}{3(2a + b)} = \frac{5}{2a + b} \cdot \frac{3}{3} + \frac{7}{3(2a + b)}$$

$$= \frac{15}{3(2a + b)} + \frac{7}{3(2a + b)} = \frac{22}{3(2a + b)}$$

Example 2.

$$\frac{x}{x^2 - y^2} + \frac{2}{y - x} - 5 = \frac{x}{(x + y)(x - y)} + \frac{-2}{(-y + x)} - \frac{5}{1}$$

$$= \frac{x}{(x + y)(x - y)} - \frac{2}{x - y} \cdot \frac{x + y}{x + y}$$

$$- \frac{5}{1} \cdot \frac{(x + y)(x - y)}{(x + y)(x - y)}$$

$$= \frac{x}{(x + y)(x - y)} - \frac{2(x + y)}{(x + y)(x - y)}$$

$$- \frac{5(x + y)(x - y)}{(x + y)(x - y)}$$

$$= \frac{x - 2(x + y) - 5(x^2 - y^2)}{x^2 - y^2}$$

$$= \frac{x - 2x - 2y - 5x^2 + 5y^2}{x^2 - y^2}$$

$$= \frac{5y^2 - 5x^2 - x - 2y}{x^2 - y^2}$$

Example 3.

$$a + b - \frac{2ab}{a+b} = \frac{a+b}{1} \cdot \frac{a+b}{a+b} - \frac{2ab}{a+b} = \frac{(a+b)(a+b)}{a+b} - \frac{2ab}{a+b}$$

$$= \frac{(a+b)^2 - 2ab}{a+b} = \frac{a^2 + 2ab + b^2 - 2ab}{a+b} = \frac{a^2 + b^2}{a+b}$$

These examples have been written out in considerable detail to show exactly how our basic principles are applied. Various short cuts are perfectly possible, but the best way to the correct answer (especially for the beginning student) is to avoid short cuts until the fundamentals are mastered.

Notice that what we do in each example is to transform all of the fractions into equivalent fractions with the same denominator. Clearly this denominator is simply the L.C.M. of all the denominators — that is, the **least common denominator** (L.C.D.). Finally, note Example 2 where the denominator $y - x$ is changed to $x - y$.

Exercise 3.6

In problems 1–24, perform the indicated additions or subtractions. In each problem begin by using the Fundamental Principle to transform the given fractions into fractions with the same denominator and avoid short cuts at this stage. Reduce all answers to fractions in lowest terms.

1. $\dfrac{x}{3} + \dfrac{5x}{6} - \dfrac{3x}{4}$

2. $\dfrac{2x}{3} - \dfrac{5x}{4} + \dfrac{5x}{12}$

3. $\dfrac{4a}{3} + \dfrac{5a}{12} - \dfrac{7a}{15}$

4. $\dfrac{7y}{10} - \dfrac{y}{30} + \dfrac{4y}{5}$

5. $\dfrac{a-2}{5} - \dfrac{a-3}{3}$

6. $\dfrac{5x+3}{6} - \dfrac{4x-1}{9}$

7. $\dfrac{3}{2x-1} + \dfrac{1}{x+2} - \dfrac{5}{2x^2+3x-2}$

8. $\dfrac{1}{2x+1} - \dfrac{3}{3x-1} - \dfrac{5}{x+1}$

9. $\dfrac{x-2}{2x^2-7x+6} + \dfrac{3x+5}{2x^2-x-6} + \dfrac{2x-1}{9-4x^2}$

10. $\dfrac{3x-5}{2x^2+3x-5} + \dfrac{5x-2}{2x^2+7x+5} + \dfrac{4x}{1-x^2}$

11. $\dfrac{2x-5}{x^2+3x} - \dfrac{1}{x+3}$

12. $\dfrac{7x-1}{x^2-4} - \dfrac{5x-1}{x+2}$

13. $\dfrac{x}{x^2-3x+2} - \dfrac{x+1}{x^2-5x+6} - \dfrac{2}{x^2-4x+3}$

14. $\dfrac{4x-7}{x^2-7x} - \dfrac{4x+47}{3x^2-17x-28}$

15. $\dfrac{16}{x-2} + \dfrac{9}{x+3} - \dfrac{5}{x-1}$

16. $1 - \dfrac{6}{x} + \dfrac{1}{x^2-4x}$

17. $\dfrac{x-2}{x^2+x-2} + \dfrac{x+1}{x^2-x-6} - \dfrac{2}{(x-1)(x+2)(x-3)}$

18. $\dfrac{6a+2b}{a^2-b^2} + \dfrac{3}{b-a} - \dfrac{2}{a+b}$

19. $3 + \dfrac{2}{a+1}$

20. $-2 + \dfrac{5}{2x-1}$

21. $1 - \dfrac{4}{a+2}$

22. $x - 2y - \dfrac{(x+2y)^2}{x-2y}$

23. $\dfrac{x-a}{x^2-6ax+9a^2} - \dfrac{2x-a}{x^2-9a^2} + \dfrac{3}{x+3a}$

24. $4x^2 - 6x - \dfrac{27}{2x+3} + 9$

In problems 25–30, perform the indicated operations and obtain a fraction in lowest terms for your answer.

25. $\left(1 - \dfrac{x^2}{y^2}\right)\left(3 - \dfrac{3x}{x+y}\right)$

26. $\left(\dfrac{x}{y} - 1\right)\left(\dfrac{x}{y} + 1\right)$

27. $\dfrac{x}{2}\left(x^2 - \dfrac{3x}{2} - \dfrac{1}{3}\right) - \left(x^2 + \dfrac{3x}{4} - \dfrac{1}{6}\right)$

28. $\left(\dfrac{2}{3x} - \dfrac{5}{2y} + \dfrac{3}{z}\right) \div \left(\dfrac{3}{4x} + \dfrac{2}{3y} - \dfrac{5}{6z}\right)$

29. $\left(\dfrac{x-2}{x} - \dfrac{2}{x+3}\right)\left(\dfrac{2}{x+2} + \dfrac{3}{x-3}\right)$

30. $\left(x - \dfrac{2x}{2-x}\right)\left(x - 3 + \dfrac{2}{x}\right)$

Simplify the expressions in problems 31–38 to a single fraction in lowest terms. Problems 31–34 occur in analytic geometry; problems 36–38 occur in the calculus.

31. $\dfrac{(x+1)^2}{4} + \dfrac{(y+3)^2}{9}$

32. $\dfrac{(y+1)^2}{4} - \dfrac{(x+4)^2}{3}$

33. $\dfrac{(x-1)^2}{9} - \dfrac{(y-2)^2}{4}$

34. $\dfrac{(x+a)^2}{4a^2} + \dfrac{(y-2a)^2}{a^2}$

35. $\left[\dfrac{2x(x^2-1)-2x(x^2+1)}{(x^2-1)^2}\right] 2\left(\dfrac{x^2+1}{x^2-1}\right)$

36. $\left[\dfrac{(3+2x)(-2)-(3-2x)(2)}{(3+2x)^2}\right] 2\left(\dfrac{3-2x}{3+2x}\right)$

37. $\dfrac{3x^2}{x^3+2} + \dfrac{2x}{x^2+3}$

38. $\dfrac{4}{x} - \dfrac{6}{3x-4}$

3.7 COMPLEX FRACTIONS

A fraction in which one or more terms of its numerator or denominator are themselves fractions is called a **complex fraction**. The following three examples show how complex fractions may be reduced to simple fractions.

Example 1. $\dfrac{1+\dfrac{1}{x}}{1-\dfrac{1}{x}} = \dfrac{1+\dfrac{1}{x}}{1-\dfrac{1}{x}} \cdot \dfrac{x}{x} = \dfrac{\left(1+\dfrac{1}{x}\right)\cdot x}{\left(1-\dfrac{1}{x}\right)\cdot x} = \dfrac{x+1}{x-1}\,.$

Example 2.

$$\dfrac{\dfrac{x}{x+y}+\dfrac{y}{x-y}}{\dfrac{y}{x+y}-\dfrac{x}{x-y}} = \dfrac{\dfrac{x}{x+y}+\dfrac{y}{x-y}}{\dfrac{y}{x+y}-\dfrac{x}{x-y}} \cdot \dfrac{(x+y)(x-y)}{(x+y)(x-y)}$$

$$= \dfrac{x(x-y)+y(x+y)}{y(x-y)-x(x+y)} = \dfrac{x^2-xy+xy+y^2}{xy-y^2-x^2-xy}$$

$$= \dfrac{x^2+y^2}{-x^2-y^2} = \dfrac{x^2+y^2}{-(x^2+y^2)} = -1$$

Example 3.

$$1+\dfrac{1}{1+\dfrac{1}{1-x}} = 1+\dfrac{1}{\dfrac{1-x}{1-x}+\dfrac{1}{1-x}} = 1+\dfrac{1}{\dfrac{1-x+1}{1-x}}$$

$$= 1+\dfrac{1}{\dfrac{2-x}{1-x}} \cdot \dfrac{1-x}{1-x} = 1+\dfrac{1-x}{2-x}$$

$$= \dfrac{2-x}{2-x}+\dfrac{1-x}{2-x} = \dfrac{2-x+1-x}{2-x} = \dfrac{3-2x}{2-x}$$

Exercise 3.7

In problems 1–18, express the given fraction as a simple fraction in lowest terms.

1. $\dfrac{1}{1+\frac{3}{4}}$

2. $\dfrac{2}{1+\frac{2}{3}}$

3. $\dfrac{1+\frac{1}{2}}{4+\frac{3}{4}}$

4. $\dfrac{2-\frac{3}{5}}{3+\frac{7}{10}}$

5. $\dfrac{x-\dfrac{x^2}{y}}{1-\dfrac{x}{y}}$

6. $\dfrac{\dfrac{a^2}{6}-6}{1+\dfrac{a}{6}}$

7. $\dfrac{\dfrac{1}{2}-\dfrac{1}{x}}{\dfrac{1}{6x}-\dfrac{1}{3x^2}}$

8. $\dfrac{y+\dfrac{2x}{3}}{\dfrac{9y}{x}-\dfrac{4x}{y}}$

9. $\dfrac{x-2}{1-\dfrac{4}{x+2}}$

10. $\dfrac{x^2-xy-2y^2}{\dfrac{x-2y}{3x}}$

11. $\dfrac{x-y+\dfrac{1}{x}}{\dfrac{1}{x^2}+\dfrac{1}{x}+1}$

12. $\dfrac{x-\dfrac{16}{x}}{\dfrac{1}{x^2}+\dfrac{2}{x^3}-\dfrac{8}{x^4}}$

13. $\dfrac{1}{1-\dfrac{1}{1-\frac{1}{2}}}$

14. $\dfrac{1}{1+\dfrac{1}{1+\frac{1}{2}}}$

15. $\dfrac{1}{1-\dfrac{1}{1-\dfrac{1}{x}}}$

16. $\dfrac{1}{1+\dfrac{1}{1+\dfrac{1}{x}}}$

17. $\dfrac{\dfrac{x+2}{x}}{x+1-\dfrac{\dfrac{1}{x+1}}{1-\dfrac{x}{x+1}}}$

18. $\dfrac{\dfrac{x}{x+1}}{x+2-\dfrac{\dfrac{2}{x+1}}{1-\dfrac{x}{x+1}}}$

The complex fractions in problems 19–24 occur in trigonometry. Express each one as a simple fraction in lowest terms.

19. $\dfrac{2\,\dfrac{x}{y}}{1-\dfrac{x^2}{y^2}}$

20. $\dfrac{x+\dfrac{x}{y}}{1+\dfrac{1}{x}}$

21. $1-\dfrac{\dfrac{1}{x^2}}{\dfrac{1}{y^2}}$

22. $\dfrac{\dfrac{a}{x}+\dfrac{b}{y}}{1-\dfrac{a}{x}\cdot\dfrac{b}{y}}$

23. $\dfrac{\dfrac{y^2}{x^2}}{1+\dfrac{1}{x^2}}+1$

24. $\dfrac{\dfrac{x^2}{y^2}\left(\dfrac{1}{x}-1\right)}{\dfrac{1}{x}+1}$

The complex fractions in problems 25–32 occur in the calculus. Express each one as a simple fraction in lowest terms.

25. $\dfrac{\dfrac{1}{(x+h)^2}-\dfrac{1}{x^2}}{h}$

26. $\dfrac{\dfrac{(x+h)^2}{x+h+1}-\dfrac{x^2}{x-1}}{h}$

27. $\dfrac{y-\dfrac{x^2}{xy}}{y^2}$

28. $\left(\dfrac{1}{\dfrac{1-x}{1+x}}\right)\cdot\left[\dfrac{(1+x)-(1-x)}{(1+x)^2}\right]$

29. $\left(\dfrac{1}{\dfrac{1+x^2}{y}}\right)\left[\dfrac{2xy-(1+x^2)y}{y^2}\right]$

30. $\left[\dfrac{1}{1+\left(\dfrac{2x}{1-x^2}\right)^2}\right]\cdot\left[\dfrac{2(1-x^2)-(2x)(-2x)}{(1-x^2)^2}\right]$

31. $\dfrac{1}{x}+\dfrac{1}{1+\dfrac{x}{y}}\left(\dfrac{y-x}{y^2}\right)$

32. $\dfrac{\dfrac{2(1+t^2)-(2t)(2t)}{(1+t^2)^2}}{\dfrac{-2t(1+t^2)-(1-t^2)(2t)}{(1+t^2)^2}}$

WORD PROBLEMS

4

4.1 INTRODUCTION

As mentioned in Section 1.12, word problems of the type solved in elementary and intermediate algebra were considered by the Babylonians over 4,000 years ago. Many such problems have come down through the ages in only slightly altered form. For example, the problem of how long it takes to fill a reservoir into which several streams of water are flowing when we know how long it would take each separate stream to fill it appeared first in a book by a Greek author around A.D. 100 and is found in the algebra books of India of around 1150. The first printed version in English appeared in 1540.

By the sixteenth century many variants of the problem began to appear in which men built walls or houses — one man can build the wall in 3 days, another in 4; how long will it take them to do the job together, etc. One amusing variant was popular in the wine-drinking countries: If a man can drink a cask of wine alone in 20 days, and, when his wife drinks with him, in only 14 days, how long will it take his wife alone? An even more unusual version concerned a problem in which priests are praying for souls in purgatory!

In spite of the popularity of these problems in the past, and despite the present-day interest in puzzles, word problems in algebra have never been very popular with most students. Most of the lack of popularity is undoubtedly caused by the difficulty that most students have with such problems. In this chapter, suggestions will be given which should help in solving word problems.

For some students, however, it is not the difficulty of solution which is the primary stumbling block, but rather the feeling that many of the problems given are just not worth doing. Such a student, for example, is just not interested in finding out the answer to the question, "How old is Mary?" Now word problems do occur in engineering, physics, chemistry, etc., and, in fact, a common complaint of teachers of these subjects is that their students cannot handle such problems. Unfortunately, however, these applied problems cannot usually be done without a fairly extensive knowledge of the field in which they occur.

Faced with this difficulty, the only thing we can say to the students whose interest is primarily in useful applications is that a careful study of the *basic* methods underlying the solution of the problems given here will help them to solve many of the practical problems when they come to them. Note the italicized word *basic*, however. As long as formulas and special methods are relied upon, no real progress will be made.

Although "real life" problems are difficult to consider in intermediate algebra, we will give some examples of these in order to see how such things go. It will be advantageous, however, if the student can work up an interest in doing problems for their own sake, as a challenge to his ingenuity. In the long run, such an attitude will pay real dividends, as well as help make algebra more enjoyable.

4.2 THE BASIC TECHNIQUES

As every student will recall from elementary algebra, the central difficulty in word problems is to extract from a tangled mass of verbiage one or more equations in one or more unknowns which represent, algebraically, the verbal statement. Once this is done, the problem of solving the equation or equations is usually relatively simple. What we will do in this chapter is to concentrate on the setting up of equations. Methods for solving these equations will be considered in the next two chapters.

Our first suggestion in the solution of word problems is that the student separate the words of the problem into short sentences. We give two simple examples here and will present other, more complex, examples in the next section.

Example 1. $450 is to be divided between Brown and White so that White will receive $50 more than Brown. How much does each receive?

Our three sentences are

  (1) $450 is to be divided between Brown and White.
  (2) White will receive $50 more than Brown.
  (3) How much does each receive?

One of the sentences will certainly be a question about the size or amount of one or more quantities. We now suggest that the student adopt a letter or letters to represent the number or numbers desired, being sure to state exactly what the letters stand for, including the units of measurement (if any). Thus, for sentence 3 of Example 1, we may write:

Let $x =$ number of dollars that Brown will receive and $y =$ number of dollars that White will receive.

Next, we suggest translating each of the remaining sentences into a form involving the use of one or more of the symbols for the operations of arithmetic and the equality sign, but without using any letters. Thus the first sentence of Example 1 becomes

(amount Brown receives) + (amount White receives) = $450

and the second becomes:

(amount White receives) = (amount Brown receives) + $50

Finally, we replace the word phrases used by the appropriate symbols to obtain the desired equation or equations. By our definition of x and y this is easily done for Example 1 and we have

$$x + y = 450 \qquad \text{and} \qquad y = x + 50$$

as the two equations desired.

Example 2. One dimension of a rectangle is two-thirds of the other. Find the dimensions of the rectangle if its perimeter becomes 130 feet when each dimension is increased by 5 feet.

Our three sentences are

  (1) One dimension of a rectangle is two-thirds of the other.
  (2) The perimeter of the rectangle becomes 130 feet when each dimension is increased by 5 feet.

(3) What are the dimensions of the rectangle?

We then rewrite our sentences as

(1') (width of the rectangle) = $\frac{2}{3}$ (length).

(2') New perimeter = 2 (new width) + 2 (new length) = 130.

(3') Let x = number of feet in the width of the rectangle and y = number of feet in the length of the rectangle.

Then the number of feet in the new width is $x + 5$, the number of feet in the new length is $y + 5$, and our equations are

$$x = \tfrac{2}{3}y \qquad \text{and} \qquad 2(x + 5) + 2(y + 5) = 130$$

The student will notice that, in both examples, two equations in two unknowns have been obtained. It is easily possible to use only one letter in these problems, but the technique of using as many letters as there are unknowns is a useful one and often makes the analysis of the problem easier. There is no objection, of course, to the use of only one letter and writing, in the first example

$$x + (x + 50) = 450$$

and, in the second,

$$2(\tfrac{2}{3}y + 5) + 2(y + 5) = 130$$

Before proceeding with other examples, the student should try his skill on some simple problems. In every case, the search for the equations is really that of the search for the gist of the sentences. Ask yourself what the real point of each sentence is and note that the symbol "=" may stand for the words "equals," "is the same as," "is," "what is left is," "the result is," "gives," "leaving," and so on. Similarly, "+" may stand for "added to," "more than," "greater than," and so on. It is suggested that the student compile a list of words or phrases that may be represented by "=," "+," "−," "×," and "÷," and add to this list every time he finds a new expression.

Exercise 4.2

In the following problems set up one or more equations to represent the given verbal statement. Be sure to state exactly what your letter or letters represent, including the units of measurement, if any. You need not, at this time, solve the resulting equations.

1. If we add three to a certain number we get 15. Find the number.

2. Three times a certain number is equal to 15. Find the number.

3. A certain number is 3 more than another number and their sum is 25. Find the numbers.

4. A certain number is 3 less than another number and if we add the two numbers together we get 27. Find the numbers.

5. One number is 3 times another number and their sum is 28. Find the numbers.

6. The sum of a certain number and 3 less than that number is 27. Find the number.

7. The sum of a number and 3 times that number is 20. Find the number.

8. Three more than twice a certain number is 25. Find the number.

9. If twice a certain number is increased by 10 we obtain 44. Find the number.

10. If three times a certain number is decreased by 9 we obtain 30. Find the number.

11. If twice a certain number is increased by 3 we obtain 8 less than three times that number. Find the number.

12. Three times a certain number is 10 more than twice that number. Find the number.

13. A piece of string is 27 inches long. We cut the string into two pieces such that one is 3 inches less than the other. Find the length of the two parts.

14. A piece of string is 20 inches long. We cut the string into two pieces such that one piece is 3 times as long as the other. Find the length of the two parts.

15. John is three years younger than Jim and the sum of their ages is 27. Find their ages.

16. John has three more than twice as many marbles as Bill has. Jim has eight less than three times as many as Bill and John and Jim have the same number. Find the number of marbles that each has.

17. Jack and Bill earned $20 for taking care of a garden but Jack worked three times as long as Bill and received three times as much pay as Bill. Find what each received.

18. In a total of 10,000 votes, the Democrats won by 1,000 votes over all their opponents together. How many votes did the Democrats receive?

19. A town with a population of 11,425 is increasing at the rate of

750 per year. In how many years will it have a population of 16,000?

20. A farmer wrote in his will that he wished to divide his 600 acres between his two sons, leaving the younger son 150 acres more than half as much as the older son. How many acres did each receive?

21. A certain right triangle is such that one of the acute angles is 5 times the size of the other. Find the degree measure of each angle.

22. One side of a right triangle is one inch less than half the other side and the hypotenuse is 13 inches long. Find the length of the sides.

23. One side of a rectangular field is 5 yards less than the other side and the area of the field is 150 square yards. Find the length of the sides.

24. The perimeter of a rectangle is 8 times its width and its length is 6 feet. Find its width.

25. The length of a box is 2 inches greater than its width, whereas its height is one inch less than its width. Its volume is 30 cubic inches. Find the dimensions of the box.

26. A collection contains $6.70. There are twice as many dimes as quarters and two more nickels than dimes. Find the number of coins of each denomination.

27. A purse contains twice as many dimes as nickels and twice as many quarters as dimes. The total is $5.60. Find the number of coins of each denomination.

28. The receipts of a concert were $8,873.50. Adult tickets cost $1.25 and children's, $.50. How many adults and children bought tickets if a total of 8,372 tickets were sold?

29. A post office sold 1,000 stamps in denominations of 2 cents, 4 cents, 10 cents, and 15 cents. There were 4 times as many two-cent stamps sold as ten-cent and $\frac{1}{3}$ as many four-cent as two-cent stamps sold. The total receipts were $42.50. How many stamps of each denomination were sold?

30. A man invests $2,500 at one bank at a certain rate of interest and $3,000 at another bank at a rate of interest twice as high. His total annual income from these investments is $212.50. Find the rate of interest paid by each bank.

4.3 FURTHER EXAMPLES

We now continue with some more complicated examples.

Example 1. How long will it take Jones and Smith working together to plow a field which Jones can plow alone in 5 hours and Smith alone in 8 hours?

Here we have

  (1) How long will it take?
  (2) Jones and Smith working together plow a field.
  (3) Jones can plow it alone in 5 hours.
  (4) Smith can plow it alone in 8 hours.

These sentences become, in turn:

  (1') Let x = number of hours that it takes Jones and Smith to plow the field.
  (2') (Jones's fractional part of the work) + (Smith's fractional part of the work) = 1.
  (3') In one hour Jones does $\frac{1}{5}$ of the job, and in x hours, $\frac{x}{5}$ of the job.
  (4') In one hour Smith does $\frac{1}{8}$ of the job and, in x hours, $\frac{x}{8}$ of the job.

Hence we have

$$\frac{x}{5} + \frac{x}{8} = 1$$

Example 2. How many gallons of a mixture containing 80% alcohol should be added to 6 gallons of a 25% solution to give a 30% solution?

Here we have

  (1) How many gallons of the 80% mixture are needed?
  (2) (alcohol in 80% solution) + (alcohol in 25% solution)
       = (alcohol in 30% solution).

These sentences become, in turn,

  (1') Let x = number of gallons of the 80% solution;
  (2') $0.80x + (0.25)(6) = 0.30(x + 6)$;

since the amount of alcohol in any solution is obtained by taking the total amount of the solution times the percentage of alcohol contained in it.

Our last three examples are done in somewhat less formal a fashion. We still stress, however, the semi-algebraic equation (**basic equation**) which serves as a bridge between the purely verbal statement of the problem and the final, purely algebraic, statement.

Example 3. A plane can travel 100 miles per hour (mph) without any wind. Its fuel supply is 3 hours. (This means that it can fly 300 miles without a wind.) The pilot now proposes to fly east with a tail wind of 10 mph and to return with, of course, a head wind of 10 mph. How far out can he fly and return without running out of fuel?

(The student should first reflect on whether or not the answer is the same as it is without a wind—namely, 150 miles.)

Our basic equation is

$$\text{(time out)} + \text{(time back)} = 3$$

We now use the fact that time $= \dfrac{\text{(distance)}}{\text{(speed)}}$ so that if d represents the distance out (= distance back) in miles we have

$$\frac{d}{110} + \frac{d}{90} = 3$$

since the speed out is $(100 + 10)$ mph and the speed back is $(100 - 10)$ mph.

Example 4. A rectangle has its length 2 feet greater than its width. If the length is increased by 3 feet and the width by one foot, the area of the new rectangle will be twice the area of the old. What is the length and width of the original rectangle?

If w represents the number of feet in the width of the original rectangle, the number of feet in the length is $w + 2$. (This is the gist of the first and last sentence.) From the second sentence we gain the basic equation

$$\text{(area of new rectangle)} = 2(\text{area of original rectangle})$$

Since the area of a rectangle is the product of the width and the length we have

$$(w + 1)[(w + 2) + 3] = 2[w(w + 2)]$$

Our final example illustrates two points. First, the basic equation arises from a law of physics with which you may not be acquainted. Second, the resulting equation cannot (as you will learn later) be solved by any of the techniques commonly taught in intermediate algebra. Thus we reemphasize our point that applied problems exist but require knowledge of the subject of application and also point out that applied problems sometimes require for their solution more advanced algebraic techniques than those covered in intermediate algebra.

Example 5. How far does a wooden (spherical) ball of specific gravity 0.4 and radius 2 feet sink in water? (See Figure 4.1.)

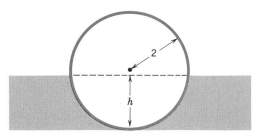

Figure 4.1

By "Archimedes' principle" we know that the ball will sink until it displaces a weight of water equal to the entire weight of the ball. Thus our basic equation is

(weight of ball) = (weight of displaced water)

Now the weight of the ball is its volume times its specific gravity times the density of water, w, in pounds per cubic foot. On the other hand, the volume of water displaced by the ball is the volume of a segment of a sphere and this is, by solid geometry, $\pi h^2 \left(2 - \dfrac{h}{3}\right)$ where h is the length in feet of the altitude of the segment. Hence, we have

$$\frac{4}{3} \pi \cdot 2^3 \cdot 0.4w = \pi h^2 \left(2 - \frac{h}{3}\right) w$$

Exercise 4.3

1. Find two consecutive positive integers the sum of which is 47.
2. Find two consecutive integers which are such that when we multiply them together we get 42.

3. Find two consecutive integers such that the difference of their squares is 9.

4. Find two consecutive even integers such that four times the first minus three times the second is half of the first.

5. Find two consecutive odd integers such that the sum of their squares is two more than 10 times the difference of their squares.

6. Find two consecutive integers such that the difference of their cubes is equal to 3.

7. One man can build a wall in 15 days. He and his partner working together can build it in 10 days. How long would it take his partner to build the same wall alone?

8. One man can do a job in 6 days and another can do it in 8 days. How long would it take if they worked together?

9. Three men can do a certain job in 5 days. If the first and the second working together can do it in 6 days while the first and the third can do it in 10 days, how long would it take each man alone to do the job?

10. A man starts a job and works alone for 5 days. He is joined by a second worker and they finish the job in 4 more days. If they could do the entire job together in 6 days, how long would it take each to do the job alone?

11. A swimming pool can be filled in 6 hours and emptied in 8 hours. If both the inlet and outlet are open, how long would it take to fill the pool?

12. A tank can be filled by one hose in 10 hours and by another hose in 15 hours. How long will it take to fill the tank using both hoses?

13. A confectioner has two kinds of candy worth 71 cents and 96 cents a pound, respectively. How many pounds of each kind of candy should be mixed together to obtain 100 pounds worth 76 cents a pound?

14. How many gallons of gasoline worth 24 cents a gallon should be mixed with 40 gallons of gasoline worth 36 cents a gallon to obtain a mixture worth 28 cents a gallon?

15. A wholesaler has 300 pounds of coffee worth 50 cents a pound. How many pounds of 60-cent coffee should he mix with this to obtain a blend worth 58 cents a pound?

16. An investor had $230 to invest. He bought 60 shares of which some were worth $3.00 a share and the others worth $5.00 a share. How many of each did he buy?

17. An investor buys 100 shares of stock for $320. If some of the

shares were worth $3.00 and the others worth $4.00, how many of each share did he buy?

18. An investor bought 30 shares of one stock worth $4.50 per share; then he bought a certain number of shares worth $2.00. Afterwards, he found that he had paid an average of $3.00 per share. How many shares of $2.00 stock did he buy?

19. How many gallons of a 60 per cent solution of alcohol must be added to 6 gallons of a 30 per cent solution to obtain a 40 per cent solution?

20. How many gallons of water must be added to 20 gallons of a 30 per cent solution of acid to obtain a 20 per cent solution?

21. How many quarts of pure acid must be added to 20 quarts of a 10 per cent solution to obtain a 15 per cent solution?

22. A car radiator with a 30-quart capacity has been prepared for fall driving with 3 quarts of antifreeze and 27 quarts of water, but winter driving requires a 20 per cent solution of antifreeze. How many quarts of the fall mixture must be withdrawn and replaced with pure antifreeze?

23. It is desired to obtain 10 tons of bronze which is 48 per cent copper from a mixture of bronze which is 60 per cent copper and another which is 40 per cent copper. Find the amounts of the two given bronzes that are needed.

24. How many gallons of cream which is 10 per cent butterfat must be mixed with 15 gallons which is 20 per cent butterfat to obtain cream which is 16 per cent butterfat?

25. A train leaves New York at noon and a second train leaves the same station in the same direction at 2:00 P.M. If the second train travels 20 mph faster than the first and overtakes the first train at 5:00 P.M., find the speed of the first train.

26. A bicyclist left a town going 10 mph at 8:00 A.M. Three hours later a car left in the same direction going 50 mph. When did the car overtake the bicycle?

27. Two airplanes leave an airport at the same time. One is flying north at 150 mph and the other is going south at 210 mph. How much later will they be 1,080 miles apart?

28. Two boats leave a port at the same time, traveling south. One can travel $1\frac{1}{2}$ times as fast as the other and, after 3 hours, they are 15 miles apart. How fast can each boat travel?

29. Two cars start traveling towards each other from cities 350 miles apart. One travels 20 mph less than twice as fast as the other and they meet after 5 hours. How fast are they traveling?

30. An airplane that can fly 250 mph in still air makes 675 miles in 3 hours against a wind. What is the speed of the wind?

31. A motorboat can go 20 miles up a river in 5 hours and can come down the same distance in $2\frac{1}{2}$ hours. What is the speed of the boat in still water and what is the speed of the current?

32. A boat went 15 miles down a tributary in one hour. Then it entered the main stream, which was flowing half as fast as the tributary and could only make 3 miles per hour against the current. What is the speed of the boat in still water and what is the speed of the main stream?

33. A wind is blowing at 40 mph. An airplane makes a round trip flight going from one town to another with the wind in one hour, and back in one hour and 10 minutes against the wind. How fast can this plane travel in still air and how far apart are the towns?

34. A car travels from one town to another in 6 hours. Coming back it increases its speed by 10 mph and makes the trip in 5 hours. How fast did the car travel and how far apart are the towns?

35. In commuting to his office, a man drives 50 mph to a river where he takes a ferry upstream. The ferry can travel 10 mph in still water and the river flows 2 mph. If it takes one hour to go to his office and 45 minutes to return, how far must he travel in his car and how far on the ferry?

FIRST-DEGREE EQUATIONS

5

5.1 DEFINITIONS

An **algebraic equation** is a mathematical statement that two algebraic expressions are equal. This equality is expressed symbolically by the equals sign (=), which divides the statement into two parts, the left-hand member or side, and the right-hand member or side. Thus, "$x^2 - 4 = 0$," "$x + y = 2$," "$(x + y)(x - y) = x^2 - y^2$" are all examples of algebraic equations.

There are two general types of equations. The **identical equation** or **identity** is a statement that is true for all **permissible**[1] values of the letters. Thus an identity is an equation which remains true no matter what permissible real numbers are substituted for the letters. We have already seen many identities such as

$$a + b = b + a$$

$$(a + b)(a + b) = a^2 + 2ab + b^2$$

$$5ax^2 - 15ax + 20a = 5a(x^2 - 3x + 4)$$

$$\frac{x^2 - y^2}{2x} \cdot \frac{4x^2}{x + y} = 2x(x - y)$$

[1] "Permissible" means any numerical value for which the algebraic expressions on both sides of the equality sign are defined. For example, $1/x = 1/x$ is a very trivial identity, but it would not be one if we did not restrict ourselves to *permissible* values. For to say that an identity holds for *all* values would be to insist that $1/x = 1/x$ when $x = 0$, whereas we know that $1/0$ is undefined. (See Section 1.9.)

100

The other type of equation is the **conditional equation** which is so-called because it is true only on the condition that the letters represent certain numbers, or it is never true no matter what numbers the letters represent. The process of finding the numbers (if any) which, when substituted for the letters, make the conditional equation true is known as "solving the equation." The resulting values are called **solutions** (or **roots**) of the equation, and the set of solutions is called the **solution set** of the equation.

Example 1. $x - 2 = 0$ has the single root 2. The solution set is

$$X = \{x|x - 2 = 0\}$$
$$= \{2\}$$

Example 2. $x^2 - 5x + 6 = 0$ has the two roots, 2 and 3. The solution set is

$$X = \{x|x^2 - 5x + 6 = 0\}$$
$$= \{2, 3\}$$

Example 3. $x + 2 = x + 3$ has no solution. The solution set is

$$X = \{x|x + 2 = x + 3\}$$
$$= \varnothing$$

Example 4. $x + y = 2$ has infinitely many solutions such as $(x = 1, y = 1)$, $(x = 2, y = 0)$, $(x = -1, y = 3)$, etc. Solution sets of equations in two variables will be studied in Section 5.7ff. Since there are infinitely many solutions, the solution set can only be written in set-builder notation.

or
$$X = \{(x, y) \mid x + y = 2\}$$
$$= \{(x, y) \mid y = 2 - x\}$$

Variables are sometimes used in equations without the idea of finding values of these variables which satisfy the equation. For example, the simple equation,

$$x - a = 0$$

would ordinarily be considered as an equation in which we wish to determine what x must be so that, for any value of a, the equation will be satisfied. Clearly, under these conditions, the solution is $x = a$ and not ($x = 1$, $a = 1$), ($x = 2$, $a = 2$), etc. Hence, when a conditional equation contains more than one variable we must specify in each case the letter or letters for which we wish to solve. Thus we might say that the equation $x - a = 0$ is to be "solved for x" or that it is an equation "in x."

A considerable part of algebra, both elementary and advanced, is concerned with solving conditional equations, which we shall simply call *equations* from now on. It will be clear from the context if an identity is intended. An equation in the form $P(x) = 0$ where $P(x)$ is a polynomial is called a polynomial equation. Polynomial equations are frequently named according to the degree of the polynomial. That is, first-degree equations are called **linear** equations, second-degree are called **quadratic**, third-degree are called **cubic**, etc.

> ***Examples.*** $x + 3y - 2 = 0$ is a linear equation (in x and y)
> $x^2 - 5x + 6 = 0$ is a quadratic equation in x
> $x^2y + y^2 = 0$ is a cubic equation in x and y, a quadratic equation in x and a quadratic equation in y.

In this chapter we shall study linear equations in one, two, three, and more variables.

Exercise 5.1

In problems 1–10, state whether the equation is an identity or a conditional equation and give a reason for your answer.

1. $x - 5 = 2$
2. $x + 10 = x + 5$
3. $\sqrt{x^2} = x$
4. $(x + 7)^2 = x^2 + 49$
5. $\dfrac{x}{x} = 1$
6. $(x + y)^2 = x^2 + 2xy + y^2$
7. $(x - y)^3 = x^3 - 3x^2y + 3xy^2 - y^3$
8. $(a + b)^2 = a^2 + b^2$
9. $\sqrt{a^2 + b^2} = a + b$
10. $(x - 1)(x^2 + x + 1) = x^3 - 1$

11. Write the general fifth-degree polynomial equation using the subscript notation.

12. Write the general seventh-degree polynomial equation using the subscript notation.

In problems 13–20, give the degree of the given equation. If more than one letter is involved give also the degree of the equation in each letter.

13. $2x^3 - 5x^2 - 3x = 0$ **14.** $5x^4 - 2x + 1 = 0$

15. $3x^2y^3 - 2xy^2 + x = 2$ **16.** $7x^3y - 4xy^2 + 3 = 0$

17. $3x^5 + 5x^3y^3 - 6x + 2 = 0$ **18.** $7x^2 + 3xy^3 - 3y^2 + 1 = 0$

19. $9x^4y^3 - 7x^5 + 15y^4 - 14 = 0$ **20.** $8x^3y^4 - 8x^6 - 13y^5 + 12 = 0$

In problems 21–26, assume the standard notation $a_nx^n + a_{n-1}x^{n-1} + \cdots + a_0 = 0$ and identify a_0, a_1, a_2, etc.

21. $2x^3 - 3x^2 + 5x - 1 = 0$ **22.** $3x^3 - 5x^2 + 2x - 7 = 0$

23. $x^4 - 5x^2 + x - 1 = 0$ **24.** $x^5 - 3x^3 + x^2 - x + 1 = 0$

25. $x^6 - 5x^2 + 2x = 0$ **26.** $x^7 - 3x^3 + 3x^2 = 0$

5.2 EQUIVALENT EQUATIONS

The solution of linear equations of the form $a_1x + a_0 = 0$ $(a_1 \neq 0)$ can easily be reduced to a formula. Before we do this, however, we will need a definition and two axioms concerning equations:

Definition. Two equations are said to be **equivalent** if they have the same solution set.

For example, the equations $x - 2 = 0$ and $x + 3 = 5$ are equivalent as they both have the single root 2. But $(x - 2)(x - 3) = 0$ is not equivalent to $x - 2 = 0$ since the equation $(x - 2)(x - 3) = 0$ also has the root $x = 3$.

Axiom 1. Adding (or subtracting) the same number to (from) both members of an equation yields an equivalent equation.

For example, $x - 2 = 0$ is equivalent to $(x - 2) + 2 = 0 + 2$.

Axiom 2. Multiplying (or dividing) both members of an equation by any number not equal to zero yields an equivalent equation.

For example, $\frac{1}{3}x = 2$ is equivalent to $3(\frac{1}{3}x) = 3(2)$; $5x = 15$ is equivalent to $5x \div 5 = 15 \div 5$.

Now let us find the solution of the general linear equation

$$a_1x + a_0 = 0 \qquad (a_1 \neq 0) \tag{1}$$

By Axiom 1 we will obtain an equivalent equation if we subtract a_0 from both sides to get

$$(a_1x + a_0) - a_0 = 0 - a_0 \quad \text{or} \quad a_1x = -a_0 \tag{2}$$

Now by Axiom 2 we will again obtain an equivalent equation if we divide both members by a_1 since $a_1 \neq 0$. Thus

$$\frac{a_1x}{a_1} = \frac{-a_0}{a_1} \quad \text{or} \quad x = \frac{-a_0}{a_1} = -\frac{a_0}{a_1} \tag{3}$$

Now equation (3) is equivalent to equation (2) which in turn is equivalent to equation (1). Therefore any root of equation (3) is also a root of equation (2) and likewise a root of equation (1). But equation (3) is in a form that expresses its root explicitly, and this root is also a root of equation (1). To check that $-\dfrac{a_0}{a_1}$ is actually a solution of equation (1) we substitute in $a_1x + a_0 = 0$ and the left-hand member becomes

$$a_1\left(-\frac{a_0}{a_1}\right) + a_0 = -a_0 + a_0 = 0$$

Thus the left member equals the right member and the equation is said to be **satisfied.**

Of course, the work just done is simply to show the generality of the method, and the student is not advised to treat the result as a "formula." Rather, a problem such as $6x + 2 = 5$ is to be done directly by subtracting 2 from both sides and then dividing by 6.

Exercise 5.2

In problems 1–10, solve the given equations.

1. $7x + 2 = 16$
2. $4x + 8 = 24$
3. $5x + 9 = 34$
4. $3x - 2 = 5$
5. $9x - 15 = 27$
6. $4x - 6 = 2$
7. $-5x + 1 = -6$
8. $-3x + 7 = -8$
9. $-2x - 5 = -7$
10. $-3x - 8 = -15$

In problems 11–20, determine which of the given pairs of equations are equivalent and state the reason for your answer.

11. $5x + 3 = 0$, $x + \frac{3}{5} = 0$ 12. $3x + 2 = 0$, $x + \frac{2}{3} = 0$

13. $7x - 4 = 0$, $x - \frac{4}{7} = 0$ 14. $2x - 3 = 0$, $x - \frac{3}{2} = 0$

15. $x - 4 = 0$, $(x - 4)^2 = 0$ yes   16. $2x - 1 = 0$, $(2x - 1)^2 = 0$

17. $x - 2 = 0$, $x^2 - 4 = 0$ 18. $x - 5 = 0$, $x^2 - 25 = 0$

19. $x - 1 = 0$, $x(x - 1) = 0$ 20. $x + 2 = 0$, $x(x + 2) = 0$

5.3 SOLUTION OF A LINEAR EQUATION

Linear equations do not always appear in as simple a form as equation (1) of the preceding section but all linear equations may be solved by simplifying the algebraic expressions involved, if necessary, and then applying the axioms of the preceding section.

Example 1. Solve the equation $2(\frac{2}{3}y + 5) + 2(y + 5) = 130$. (See Example 2 of Chapter 4.)

SOLUTION:

Step 1.	$\frac{4}{3}y + 10 + 2y + 10 = 130$	Distributive property
Step 2.	$\frac{4}{3}y + 2y = 110$	Subtracting 20 from both sides
Step 3.	$\frac{10}{3}y = 110$	Combining like terms
Step 4.	$y = 110 \cdot \frac{3}{10} = 33$	Dividing by $\frac{10}{3}$

Step 5. The left-hand member of our equation becomes, when $y = 33$,

$$2[\tfrac{2}{3} \cdot 33 + 5] + 2(33 + 5) = 2(22 + 5) + 2(38)$$
$$= 2(27) + 76 = 54 + 76 = 130.$$

Since the right-hand side is also equal to 130, we have checked our result.

(*Note:* Step 5, the check, is worthwhile only if carefully done. If carelessly done as a routine it will not reveal any errors that you might have made.)

Example 2. Solve the equation $a(x + b) = bx + c$ for x if $a \neq b$.

SOLUTION:

Step 1.	$ax + ab = bx + c$	Distributive property
Step 2.	$ax - bx = c - ab$	Subtracting bx and ab from both sides
Step 3.	$(a - b)x = c - ab$	Combining like terms
Step 4.	$x = \dfrac{c - ab}{a - b}$ if $a \neq b$	Dividing by $a - b$

(What is the situation if $a = b$?)

Step 5. When $x = \dfrac{c - ab}{a - b}$ the left-hand member of our equation becomes

$$a\left(\frac{c - ab}{a - b} + b\right) = a\left(\frac{c - ab}{a - b} + \frac{b}{1} \cdot \frac{a - b}{a - b}\right)$$

$$= a\left[\frac{c - ab + b(a - b)}{a - b}\right] = a\left(\frac{c - ab + ba - b^2}{a - b}\right)$$

$$= a\left(\frac{c - b^2}{a - b}\right) = \frac{a(c - b^2)}{a - b}$$

On the other hand, when $x = \dfrac{c - ab}{a - b}$ the right-hand member of our equation becomes

$$b \cdot \frac{c - ab}{a - b} + c = \frac{b}{1} \cdot \frac{c - ab}{a - b} + \frac{c}{1} \cdot \frac{a - b}{a - b}$$

$$= \frac{b(c - ab) + c(a - b)}{a - b} = \frac{bc - ab^2 + ca - cb}{a - b}$$

$$= \frac{ca - ab^2}{a - b} = \frac{a(c - b^2)}{a - b}$$

Since, then, the left-hand side of the equation is equal to the right-hand side of the equation when $x = \dfrac{c - ab}{a - b}$, we have checked our result.

Exercise 5.3

In problems 1–20, find the value of x which satisfies the given equation and check.

1. $5x - 3 = -2x + 10$ 2. $3x + 7 = -5x + 8$

3. $5(x - 3) - 2 = 2x - 6(2 - x)$

4. $2(x - 5) - 1 = 5x - 6(1 - x)$

5. $-(x - 1) - 4 = -2(x - 7) - 4x$

6. $-(6x + 2) - 3 = -3(x - 4) - 5x$

7. $-2(-x + 1) - 3x + 7 = -5[(7 - x) - 3(x + 2)]$

8. $-3(1 - x) - 4x - 2 = -3[(2x - 1) - 2(x + 3)]$

9. $4(\frac{3}{4}x + 1) + 3(x + 5) = 20$ 10. $3(\frac{5}{3}x - 1) + 2(x - 1) = 5$

11. $7(\frac{2}{7}x + \frac{3}{7}) - 3(x - 2) = 11$ 12. $2(\frac{3}{2}x - 2) - 5(x - 1) = 10$

13. $-5(\frac{2}{5}x - \frac{1}{5}) + 2x = 12 - x$ 14. $5(x - a) + 2(x - b) = 7$

15. $3(x - b) + 5(x + a) = 2$ 16. $qx + r(x - a) = q(x - r)$

17. $rx + p(x - a) = r(x - a)$ 18. $(a + b)x = c$

19. $(2a - 1)x = a + b$ 20. $(3a - b)x = a - b$

21. Work problem 1 of Exercise 4.2.

22. Work problem 2 of Exercise 4.2.

23. Work problem 6 of Exercise 4.2.

24. Work problem 7 of Exercise 4.2.

25. Work problem 8 of Exercise 4.2.

26. Work problem 9 of Exercise 4.2.

27. Work problem 10 of Exercise 4.2.

28. Work problem 11 of Exercise 4.2.

29. Work problem 12 of Exercise 4.2.

30. Work problem 13 of Exercise 4.2.

31. Work problem 14 of Exercise 4.2.

32. Work problem 19 of Exercise 4.2.

33. Work problem 21 of Exercise 4.2.

34. Work problem 24 of Exercise 4.2.

35. Work problem 1 of Exercise 4.3.

36. Work problem 3 of Exercise 4.3.

37. Work problem 14 of Exercise 4.3.

38. Work problem 15 of Exercise 4.3.

39. Work problem 25 of Exercise 4.3.

40. Work problem 26 of Exercise 4.3.

41. Work problem 27 of Exercise 4.3.

42. Work problem 29 of Exercise 4.3.

5.4 THE RECTANGULAR COORDINATE SYSTEM

In setting up the word problems of Chapter 4, linear equations were often encountered with more than one variable. If we had two variables we found two equations and three variables implied three equations. In general, n variables require n equations if the solution set is to be finite.

Thus if we have a single equation, say $x + y = 5$, infinitely many pairs of numbers satisfy the equation. For example, $(x = 1, y = 4)$, $(x = 2, y = 3)$, $(x = -5, y = 10)$, $(x = 4\frac{1}{2}, y = \frac{1}{2})$, etc. Each solution consists of two numbers, one value for x and one value for y. If we agree to write the two numbers always in the same order, the solutions can be given as **ordered pairs**, (x, y), where the first number is the value for x and the second number the value for y. Ordered pairs are always written in parenthesis and separated by a comma. Thus the solutions of $x + y = 5$ given above would be written $(1, 4)$, $(2, 3)$, $(-5, 10)$, $(\frac{7}{2}, \frac{1}{2})$, etc. Since, by subtracting x from both sides of the equation, we have $y = 5 - x$, we may take x to be any number and determine y to be 5 minus that number. The solution set can be given conveniently in set-builder notation:

$$X = \{(x, y) \mid y = 5 - x\}$$

This notation should be read, "X equals the set of ordered pairs x, y such that y equals five minus x." It is understood that x and y are real numbers.

If, however, in addition to $x + y = 5$ we have the equation $2x - y = 4$ and wish to find the pairs of numbers which satisfy both equations, we find that only one pair will: $x = 3, y = 2$.

Although problems such as these can be handled in a purely algebraic fashion, it will help to present them first graphically. We thus introduce what is called a **rectangular coordinate system** by first drawing two perpendicular lines in a plane, as shown in Figure 5.1.

The horizontal line is called the **x-axis** and the vertical line the **y-axis**, while the intersection, O, of the two lines is called the **origin**. Then, on each of the lines, we establish a **scale**, positive to the right of O on the x-axis and above O on the y-axis, and negative to the left of O on the x-axis and below O on the y-axis.

We now have a device which enables us to describe geometrically

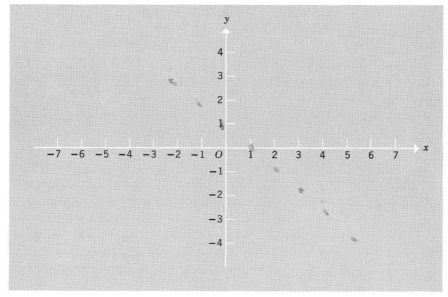

Figure 5.1

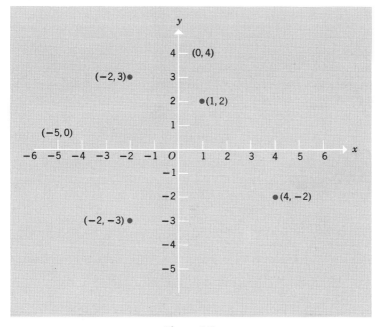

Figure 5.2

pairs of numbers (just as the single line of Chapter 1 enabled us to describe geometrically single numbers). For example, the point corresponding to the pair of numbers (1, 2) is one unit to the right of the *y*-axis and two units above the *x*-axis as shown in Figure 5.2. Similarly, (−2, −3) is two units to the left of the *y*-axis and three units below the *x*-axis; (4, −2) is four units to the right of the *y*-axis and two units below the *x*-axis; (−2, 3) is two units to the left of the *y*-axis and three units above the *x*-axis; (0, 4) lies on the *y*-axis four units above the *x*-axis; and (−5, 0) lies on the *x*-axis five units to the left of the *y*-axis.

The first number of the pair is called the **x-coordinate** or **abscissa;** the second number of the pair is called the **y-coordinate** or **ordinate.** Locating a point in a rectangular coordinate system is called **plotting** the point.

Exercise 5.4

1. Plot the points with coordinates: (1, 3), (1, −3), (−2, −5), (0, 5), (6, 0), (0, 0), (−1, −3), ($\sqrt{2}$, $\frac{1}{2}$), ($\frac{1}{3}$, −$\frac{1}{6}$), (10, −1), ($\sqrt{5}$, $\sqrt{3}$).

2. Plot the points with coordinates: (2, 1), (−2, 3), (−3, 1), (−4, −1), (0, 2), (3, 0), (−3, 0), (0, 0), ($\sqrt{3}$, $\frac{1}{2}$), ($\frac{1}{4}$, −$\frac{1}{5}$), (9, −2), ($\sqrt{3}$, $\sqrt{5}$).

3. Plot all points whose distance from the *x*-axis is 1.

4. Plot all points whose distance from the point (0, 0) is 2.

5. Plot all points whose coordinates are equal.

6. Plot all points whose first coordinate is the negative of the second coordinate.

7. Plot all points whose second coordinate is the square of the first coordinate.

8. Plot all points whose second coordinate is twice the first coordinate.

9. Plot all points whose second coordinate is one less than the first coordinate.

5.5   THE GRAPH OF A LINEAR EQUATION

With the aid of this rectangular coordinate system we may draw a "picture" of equations such as

$$x + y = 5$$

For, as pointed out in Section 5.4, there are infinitely many pairs of numbers satisfying this equation as, for example, $(1, 4), (2, 3), (-5, 10)$, $(4\frac{1}{2}, \frac{1}{2})$. If we plot the four points corresponding to these pairs of numbers on a rectangular coordinate system as in Figure 5.3, it is apparent that they all seem to lie on a straight line. Trials with further pairs of numbers which satisfy the equation in question such as $(-1, 6), (0, 5)$, and $(5, 0)$ strengthen this hypothesis.

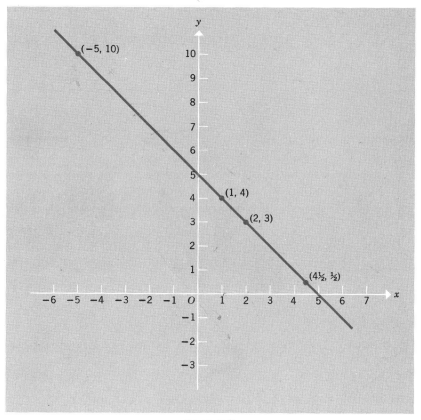

Figure 5.3

In the next section, it is shown that the set of all points whose coordinates satisfy any given linear equation $Ax + By + C = 0$ do lie on a straight line and, conversely, the coordinates of any point lying on a given straight line satisfy a linear equation in x and y. For the time being we will take this for granted. Thus we are assuming that to each linear equation there corresponds a straight line called the **graph** of the equation. In fact, we call a first-degree equation **linear** because its graph is a straight line.

To obtain the coordinates of points lying on a line defined by an equation $Ax + By + C = 0$ we usually solve for y, then choose any convenient values of x and obtain the corresponding values of y. Or, conversely, we may solve for x and choose values for y. Thus, if we have given $2x + 3y = 5$ we may write

$$y = \frac{5 - 2x}{3}$$

and obtain for $x = 1$, $y = 1$; for $x = 4$, $y = -1$, etc. Or we may write

$$x = \frac{5 - 3y}{2}$$

and obtain for $y = 1$, $x = 1$; for $y = -1$, $x = 4$, etc. Notice, too, that if our equation is of the form $By + C = 0$ $(B \neq 0)$ our straight line is parallel to the x-axis. Thus if $y - 2 = 0$, we have $y = 2$ and for *any* value of x, we have $y = 2$. Hence the points with coordinates $(0, 2)$, $(1, 2)$, $(-1, 2)$, etc., are all on the line with equation $y = 2$. Similarly, an equation such as $x = 5$ is the equation of a line parallel to the y-axis.

Clearly only two points need be plotted to determine any line. It is a wise precaution, however, to plot at least one other point as a check.

Exercise 5.5

Graph each of the following equations.

1. $x + y = 1$	2. $x + y = 3$	3. $x + y = 5$
4. $x + y = -1$	5. $x + y = -2$	6. $x + y = 7$
7. $x + y = 0$	8. $x + 2y = 0$	9. $x - y = 0$
10. $x - 2y = 0$	11. $2x + 3y = 1$	12. $3x + 4y = 2$
13. $5x + 11y = 4$	14. $2x - 3y = 1$	15. $3x + 7y = 1$
16. $3x - 5y = 0$	17. $3x + 2 = 5y$	18. $4x + 5 = 2y$
19. $3y + 5 = 4x$	20. $-2y + 5 = -5x$	21. $x - 5 = 0$
22. $2x + 3 = 0$	23. $2y - 5 = 0$	24. $y + 7 = 0$

5.6   THE SLOPE OF A LINE

The concept of the **slope** of a line is a very useful one. It is, roughly, a measure of the "slant" of a line relative to some horizontal line — the more nearly vertical a line is, the larger slope it has. A horizontal line has slope 0, and a line making an angle of 45° with the positive portion of the x-axis has a slope of 1. In general, we make:

Definition 5.1. Consider any nonvertical line L. Let P_1 and P_2 be two distinct points on L and let the coordinates of P_1 and P_2 be (x_1, y_1) and (x_2, y_2), respectively. Then the **slope,** m, of L (relative to our coordinate system) is defined to be

$$m = \frac{y_2 - y_1}{x_2 - x_1} \left(= \frac{y_1 - y_2}{x_1 - x_2} \right)$$

Example 1. Let L be the line passing through the points with coordinates $(3, -2)$ and $(5, -1)$. Its slope is given by

$$m = \frac{-1 - (-2)}{5 - 3} = \frac{1}{2}$$

Example 2. Let L be the line passing through the points with coordinates $(-\frac{9}{2}, 2)$ and $(-4, -1)$. Its slope is given by

$$m = \frac{-1 - 2}{-4 - (-\frac{9}{2})} = \frac{-3}{\frac{1}{2}} = -6$$

Example 3. Let L be the line passing through the points with coordinates $(5, 1)$ and $(2, 1)$. Its slope is given by

$$m = \frac{1 - 1}{2 - 5} = 0$$

Example 4. Let L be the line passing through the points with coordinates $(1, 5)$ and $(1, 3)$. It is a vertical line for which we do not define a slope. (Notice that $x_2 - x_1 = 0$ so that our formula for the slope of nonvertical lines certainly cannot be used.)

One should, of course, ask whether or not the calculation for the slope of a line would result in the same answer if two different points

P_3 and P_4 with coordinates (x_3, y_3) and (x_4, y_4) respectively were used. With the help of Figure 5.4 and the properties of similar triangles, the student should be able to answer this question in the affirmative.

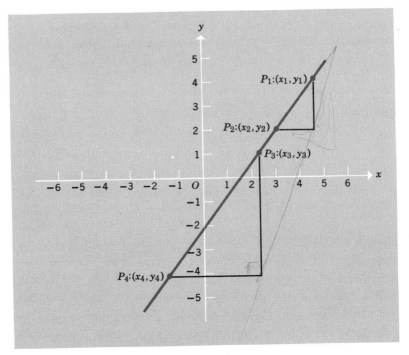

Figure 5.4

We are now ready to prove the assertion made in Section 5.5.

Theorem 5.1. (1) The coordinates (x, y) of any point lying on a given straight line satisfy a linear (first-degree) equation in x and y. (2) The set of points whose coordinates satisfy a given linear equation $Ax + By + C = 0$ (A and B not both zero) all lie on a straight line.

To prove (1) we first observe that a vertical line always has an equation of the form $x = k$ (why?) or $1 \cdot x + 0 \cdot y + (-k) = 0$. If our line L is not a vertical line, we consider two distinct points P_1 and P_2 on it with respective coordinates (x_1, y_1) and (x_2, y_2). Then, if P with coordinates (x, y) is any other point on L, we have

$$\frac{y - y_1}{x - x_1} = \frac{y_2 - y_1}{x_2 - x_1} = m$$

since we have observed that the slope of a line is independent of the choice of points on it. Thus

$$y - y_1 = m(x - x_1) = mx - mx_1$$

or

$$mx + (-1)y + (y_1 - mx_1) = 0$$

which is a linear equation in x and y. (Why can we be sure that $x - x_1 \neq 0$ and $x_2 - x_1 \neq 0$?)

To prove (2), on the other hand, we first observe that the equation $y = mx + b$ is certainly the equation of a straight line. For if (x_1, y_1) and (x_2, y_2) $(x_2 \neq x_1)$ satisfy the equation, we have

$$y_1 = mx_1 + b \qquad \text{and} \qquad y_2 = mx_2 + b$$

Subtraction then gives us

$$y_2 - y_1 = mx_2 - mx_1 = m(x_2 - x_1)$$

and hence

$$\frac{y_2 - y_1}{x_2 - x_1} = m$$

Thus a line joining two points on the curve whose equation is $y = mx + b$ has the same slope, m, as the line joining any other two points on the curve and hence this curve is a straight line.

Now if $B \neq 0$, the equation $Ax + By + C = 0$ may be written as

$$y = -\frac{A}{B}x + \left(-\frac{C}{A}\right) = mx + b$$

and hence the set of all points satisfying $Ax + By + C = 0$ is a straight line.

Finally, if $B = 0$, we have $A \neq 0$ (why?) so that $x = -\frac{C}{A}$ which is clearly the equation of a vertical line.

Theorem 5.2. Two lines L_1 and L_2 with the same slope m are either parallel or they coincide.

In the proof of Theorem 5.1 we have shown that L_1 has an equation of the form $y = mx + b_1$ and L_2 has an equation of the form $y = mx + b_2$. Suppose now that the two lines are not parallel and, instead, intersect at a point with coordinates (x_0, y_0). Then

$$y_0 = mx_0 + b_1 \quad \text{and} \quad y_0 = mx_0 + b_2$$

Thus $b_1 = b_2$ and L_1 coincides with L_2.

Exercise 5.6

In problems 1–12, find the slope (if it exists) of the line joining the two points whose coordinates are given and draw the line.

1. (1, 3) and (2, 5)
3. (−1, −3) and (2, 5)
5. (−1, −4) and (−3, −2)
7. (0, 0) and (−3, −4)
9. (5, 4) and (−1, 4)
11. (4, −2) and (4, −7)

2. (1, 7) and (0, 1)
4. (5, 4) and (0, 0)
6. (−3, −2) and (−7, −2)
8. (−5, 2) and (7, −3)
10. (−3, 2) and (−7, 2)
12. (2, −3) and (2, 4)

In problems 13–28, find the slope (if it exists) of the line with the given equation and draw the line.

13. $y = 3x + 5$
15. $y = x + 4$
17. $y - 3 = 4(x - 5)$
19. $y = 5(x - 3)$
21. $4x + 3y = 2$
23. $4x + 5y = 0$
25. $y = 4$
27. $x = -1$

14. $y = -2x + 7$
16. $y - 1 = 2(x - 3)$
18. $y - 2 = x - 4$
20. $-2y + 5 = x - 1$
22. $2x - 3y = 0$
24. $y = -1$
26. $x = 3$
28. $y = x$

29. Which of the lines in problems 1–12 are parallel?
30. Which of the lines in problems 13–28 are parallel?

5.7 GRAPHICAL SOLUTION OF A PAIR OF LINEAR EQUATIONS

Returning now to pairs of equations, let us consider the pair

$$x + y = 5$$

and

$$2x - y = 4$$

We arrange the necessary material in tabular form:

For the equation $x + y = 5$ For the equation $2x - y = 4$

x	y
0	5
5	0
2	3

x	y
2	0
0	-4
4	4

Drawing the two straight lines as shown in Figure 5.5, we see that they intersect at the point with coordinates (3, 2). This, then, gives the solution of the pair of equations, since this point lies on *both* graphs and therefore the coordinates of this point satisfy *both* equations.

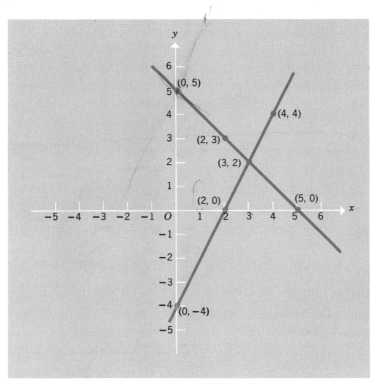

Figure 5.5

This method of solution is not a practical one. If, for example, the solution of a pair of equations was $x = \frac{7}{8}$, $y = \frac{15}{16}$, a graph, unless drawn on an enormously large scale, would not give the exact solution. Furthermore, the algebraic methods to be described are much easier. The purpose of giving the graphical method was simply to give a geo-metrical interpretation of the solution of a pair of linear equations.

It may happen, of course, that when the two lines are graphed they are coincident or parallel. We shall not consider these possibilities at present, however, but we shall investigate them in Section 5.12.

Exercise 5.7

Solve the following systems of equations graphically.

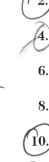

1. $x + 2y = 1$
   $2x + 5y = 3$

2. $3x + 2y = 4$
   $7x + 5y = 7$

3. $3x + 4y = 0$
   $5x + 7y = 1$

4. $2x + 3y = 1$
   $x + y = 0$

5. $3x - 4y = -6$
   $5x - y = 7$

6. $x - 2y = -4$
   $3x + y = -5$

7. $x + 4y = 3$
   $3x - 4y = -23$

8. $2x + y = 4$
   $x - y = 5$

9. $2x + y = 0$
   $3x + y = -2$

10. $x - 4y = 13$
    $2x + y = -1$

11. $x + y = 2$
    $x - y = -1$

12. $y = 2x$
    $2x = 10 - y$

5.8   ALGEBRAIC SOLUTION OF A PAIR OF LINEAR EQUATIONS

Geometrically, the solution of a pair of linear equations is represented by the intersection of their graphs. Algebraically, the solution set of a pair of linear equations is the intersection of the solution sets of the two equations. Thus the set X_1 of all ordered pairs that satisfy the equation $x + y = 5$ is

$$X_1 = \{(x, y) \mid x + y = 5\}$$

and the set X_2 of all ordered pairs satisfying the equation $2x - y = 4$ is

$$X_2 = \{(x, y) \mid 2x - y = 4\}$$

The intersection of X_1 and X_2 is the set consisting of the single ordered pair, (3, 2), that satisfies both equations:

$$X_1 \cap X_2 = \{(x, y) \mid x + y = 5 \text{ and } 2x - y = 4\}$$
$$= \{(3, 2)\}.$$

We shall give four algebraic methods of solution, two in this section, one in Section 5.9, and one in Section 5.12. The first is called the method of **substitution.** In this method we choose either one of the two equations and solve for either one of the two unknowns in that equation. Thus in the example previously considered there are four variations. We might choose the first equation, $x + y = 5$ and either solve for x to get $x = 5 - y$ or for y to get $y = 5 - x$. Or we might choose the second equation, $2x - y = 4$, and either solve for x to get $x = \dfrac{4 + y}{2}$ or for y to get $y = 2x - 4$. Our choice is either arbitrary or, in this case, determined by a desire to avoid fractions. Thus the variation which leads to $x = \dfrac{4 + y}{2}$ would be rejected.

Suppose we take the first variation which gives $x = 5 - y$. As we are looking for values of x which satisfy both equations, it follows that the value for $x(= 5 - y)$ must satisfy the second equation too. Thus we substitute $5 - y$ for x in the second equation to obtain

$$2(5 - y) - y = 4$$

which is a linear equation in one variable. Solving for y we find $y = 2$. Going back, then, to the equation $x = 5 - y$ and substituting 2 for y, we find $x = 3$. The student should work through the other variations and verify that the same values of x and y are obtained in all four cases.

The second method of solving linear systems is known as **Gauss'
elimination method.**[2] This method is chosen because of its complete generality and its adaptability later to matrix representation. The basic concept used is that of *equivalent* systems of equations — systems that have the same solution set. Gauss' elimination method, then, consists of replacing the given system with a sequence of equivalent systems until we obtain a system where the solution set is obvious.

We first illustrate the method by solving the system

$$2x + y = -2 \quad (1)$$
$$-3x + 2y = 4 \quad (2)$$

---

[2] Named for Karl Friedrich Gauss, German mathematician, 1777–1855.

First, we divide both sides of equation (1) by 2 to get equation (1′). Equation (2′) is the same as equation (2).

$$x + \tfrac{1}{2}y = -1 \quad (1')$$
$$-3x + 2y = 4 \quad (2')$$

Now we need to multiply both sides of equation (1′) by 3 to get

$$3x + \tfrac{3}{2}y = -3$$

Then, from this equation and equation (2′), we obtain

$$(-3x + 2y) + (3x + \tfrac{3}{2}y) = 4 + (-3)$$

or

$$\tfrac{7}{2}y = 1$$

As an abbreviation to describe this process we shall simply say "add 3 times equation (1′) to equation (2′) to get equation (2″)."

$$x + \tfrac{1}{2}y = -1 \quad (1'')$$
$$\tfrac{7}{2}y = 1 \quad (2'')$$

Now, we multiply equation (2″) by $\tfrac{2}{7}$ to get equation (2‴)

$$x + \tfrac{1}{2}y = -1 \quad (1''')$$
$$y = \tfrac{2}{7} \quad (2''')$$

Finally (using our abbreviated language again), we subtract $\tfrac{1}{2}$ times equation (2‴) from equation (1‴) to get equation (1^{IV}).

$$x = -\tfrac{8}{7} \quad (1^{\text{IV}})$$
$$y = \tfrac{2}{7} \quad (2^{\text{IV}})$$

Each system, as we shall show later, is equivalent to the preceding one and, therefore, the solution set, X, of the original system is the same as the solution set of the last system which is obvious. That is,

$$X = \{(x, y) \mid 2x + y = -2 \text{ and } -3x + 2y = 4\}$$
$$= \{(x, y) \mid x = -\tfrac{8}{7} \text{ and } y = \tfrac{2}{7}\}$$
$$= \{(-\tfrac{8}{7}, \tfrac{2}{7})\}$$

The method used above to solve the system can be used to find the solution set of any system of two equations in two variables as follows:

(1) By an appropriate multiplication make the coefficient of x equal to 1 in the first equation. (This is always possible, for the coefficient of x must be different from zero in at least one of the equations.

Otherwise, we would not have two equations in two unknowns. If the coefficient of x is zero in the first equation, we simply interchange the two equations.)

(2) Eliminate x from the second equation by adding an appropriate multiple of the first equation to the second equation.

(3) Solve the second equation for y.

(4) Eliminate y from the first equation by adding an appropriate multiple of the second equation to the first equation. (Unless, of course, y does not occur in the first equation.)

Example 1. Find the solution set of the system

$$2x - 12y = 3 \quad (1)$$
$$3x + 9y = 4 \quad (2)$$

Step 1. Divide equation (1) by 2.

$$x - 6y = \tfrac{3}{2} \quad (1')$$
$$3x + 9y = 4 \quad (2')$$

Step 2. Add -3 times equation (1') to equation (2').

$$x - 6y = \tfrac{3}{2} \quad (1'')$$
$$27y = -\tfrac{1}{2} \quad (2'')$$

Step 3. Solve equation (2'') for y.

$$x - 6y = \tfrac{3}{2} \quad (1''')$$
$$y = -\tfrac{1}{54} \quad (2''')$$

Step 4. Add 6 times equation (2''') to equation (1''').

$$x = \tfrac{25}{18} \quad (1^{\text{IV}})$$
$$y = -\tfrac{1}{54} \quad (2^{\text{IV}})$$

The solution set is now obviously

$$X = \{(x, y) \mid 2x - 12y = 3 \text{ and } 3x + 9y = 4\}$$
$$= \{(\tfrac{25}{18}, -\tfrac{1}{54})\}$$

Example 2. Find the solution set of the system

$$2x - 3y = 12 \quad (1)$$
$$4x - 6y = 13 \quad (2)$$

We first divide equation (1) by 2 to obtain

$$x - \tfrac{3}{2}y = 6 \quad (1')$$
$$4x - 6y = 13 \quad (2')$$

and then subtract 4 times equation (1′) from equation (2′).

$$x - \tfrac{3}{2}y = 6 \quad (1'')$$
$$0 \cdot y = -11 \quad (2'')$$

This system is equivalent to the original system. Thus the solution set is

$$X = \{(x, y) \mid 2x - 3y = 12 \text{ and } 4x - 6y = 13\}$$
$$= \{(x, y) \mid x - \tfrac{3}{2}y = 6 \text{ and } 0 \cdot y = -11\}$$

Since there is no ordered pair of numbers (x, y) such that $0 \cdot y = -11$, we conclude that the solution set is empty:

$$X = \varnothing$$

An advantage of the Gauss' elimination method is that it not only always leads to solutions if there are any but also reveals the impossibility of solutions when the system has none.

Example 3. Let us solve

$$x + \tfrac{3}{2}y = \tfrac{5}{2} \quad (1)$$
$$2x + 3y = 5 \quad (2)$$

Subtracting 2 times equation (1) from equation (2), we have

$$x + \tfrac{3}{2}y = \tfrac{5}{2} \quad (1')$$
$$0 \cdot x + 0 \cdot y = 0 \quad (2')$$

Hence the solution set is given by

$$X = \{(x, y) \mid x + \tfrac{3}{2}y = \tfrac{5}{2} \text{ and } 2x + 3y = 5\}$$
$$= \{(x, y) \mid x + \tfrac{3}{2}y = \tfrac{5}{2} \text{ and } 0 \cdot x + 0 \cdot y = 0\}$$

Since $0 \cdot x + 0 \cdot y = 0$ for all real numbers x and y this statement places no restrictions on the solution set; thus

$$X = \{(x, y) \mid x + \tfrac{3}{2}y = \tfrac{5}{2}\}$$

is the solution set of the given system. This set contains an infinite number of ordered pairs and any ordered pair that satisfies equation (1) will also satisfy equation (2).

The steps we have used to solve the above problems by Gauss' elimination method yield equivalent systems. Multiplication of any equation of the system by a nonzero number is, of course, justified by the remarks made in Section 5.2. Indeed, the only procedure that might look suspect to the student is that of adding a nonzero multiple of one equation to the other equation. But this procedure does yield an equivalent system according to the following theorem.

Theorem 5.3.

$$a_1x + b_1y + c_1 = 0 \quad (1)$$
$$a_2x + b_2y + c_2 = 0 \quad (2)$$

and

$$a_1x + b_1y = c_1 = 0 \quad (1')$$
$$k(a_1x + b_1y + c_1) + a_2x + b_2y + c_2 = 0 \quad (2')$$

with $k \neq 0$ are equivalent systems.

PROOF: If (x_0, y_0) satisfies the first system, it obviously satisfies the second system since $k \cdot 0 + 0 = 0$. Conversely, if (x_0, y_0) satisfies the second system, then $a_1x_0 + b_1y_0 + c_1 = 0$ by $(1')$ and equation $(2')$ reduces to

$$a_2x_0 + b_2y_0 + c_2 = 0$$

Therefore (x_0, y_0) also satisfies the first system.

Exercise 5.8

In problems 1–10, solve the system by the method of substitution.

1. $x + 2y = 1$
    $2x + 5y = 3$

2. $2x + 3y = 1$
    $x + y = 0$

3. $x - 2y = -4$
    $3x + y = -5$

4. $x + 4y = 3$
    $3x - 4y = -23$

5. $2x + y = 4$
    $x - y = 5$

6. $2x + y = 0$
    $3x + y = -2$

7. $x - 4y = 13$
  $2x + y = -1$

9. $x - 3y = 1$
  $4x + 7y = 0$

8. $x + y = 2$
  $x - y = -1$

10. $3x + y = 5$
  $2x + 3y = 7$

In problems 11–20, solve the system by Gauss' elimination method.

11. $2x - 3y = 0$
  $-2x + 5y = 1$

13. $2x + y = 4$
  $4x + 5y = 11$

15. $7x - 2y = 0$
  $2x + y = 0$

17. $7x - 6y = 2$
  $5x + 2y = 3$

19. $3x - 2y = 4$
  $2x - 7y = 11$

12. $3x + y = 1$
  $2x + y = 1$

14. $4x + 3y = 0$
  $5x - 2y = 0$

16. $3x + 6y = 1$
  $2x + 5y = 1$

18. $2x + 3y = 1$
  $5x + 7y = 6$

20. $5x + 11y = 17$
  $7x - 25y = 11$

Some of the following word problems can be worked quite easily by setting up only one equation. For the sake of the practice, however, the student should solve the problems by setting up a system of two equations in two unknowns.

21. Work problem 3 of Exercise 4.2.
22. Work problem 4 of Exercise 4.2.
23. Work problem 15 of Exercise 4.2.
24. Work problem 16 of Exercise 4.2.
25. Work problem 17 of Exercise 4.2.
26. Work problem 18 of Exercise 4.2.
27. Work problem 20 of Exercise 4.2.
28. Work problem 28 of Exercise 4.2.
29. Work problem 13 of Exercise 4.3.
30. Work problem 16 of Exercise 4.3.
31. Work problem 17 of Exercise 4.3.
32. Work problem 23 of Exercise 4.3.
33. Work problem 31 of Exercise 4.3.
34. Work problem 35 of Exercise 4.3.
35. The perimeter of a triangle is 200 feet; two of the sides of the triangle are equal and the sum of the length of these two equal sides is 50 feet more than the length of the other side. Find the dimensions of the sides of the triangle.

36. The perimeter of a rectangle is 13 feet while the length is 7 feet less than three times the width. What is the area of the rectangle?

37. A man has $10,000 in two banks paying 3 and 5 per cent interest, respectively. The total interest he obtains from the two banks is $400. How much does he have in each bank?

38. The sum of the digits of a two-digit number is 10 and twice the first digit plus three times the second is 25. What is the number?

39. The sum of the digits of a two-digit number is 11 and the first digit is one less than twice the second. Find the number.

40. The second digit of a two-digit number is one more than twice the first. When the digits are reversed, the number obtained is 45 more than the original number. What was the original number?

41. The first digit of a two-digit number is three times the second. When the digits are reversed, the number obtained is 36 less than the original number. What was the original number?

42. $S = -16t^2 + V_0 t + S_0$ is the equation of motion of a falling body where $V_0 =$ initial velocity in feet per second, $S_0 =$ initial height in feet, $t =$ time in seconds, $S =$ height in feet after t seconds. If $S = 5$ feet when $t = 1$ second, and $S = 3$ feet when $t = 2$ seconds, find V_0 and S_0.

43. In analytic geometry it is shown that the graph of an equation of the form $y = ax^2 + b$ is a parabola. Find an equation whose graph is the parabola passing through the points with coordinates (1, 2) and (2, 4).

5.9 SOLUTION BY MATRIX REPRESENTATION

The system

$$2x - 3y = 0$$
$$x + 5y = 3$$

is given in a standard notation. That is, each equation is written with the x-term first, then the y-term, then the equals sign and, finally, the constant on the right. If we agree always to write equations in this order, then we need only write the coefficients. The above system, for example, can then be given by the array

$$\begin{pmatrix} 2 & -3 & 0 \\ 1 & 5 & 3 \end{pmatrix}$$

Similarly, the array

$$\begin{pmatrix} 5 & 0 & -2 \\ 2 & -1 & -1 \end{pmatrix}$$

represents the system

$$\begin{aligned} 5x \quad &= -2 \\ 2x - y &= -1 \end{aligned}$$

An array of numbers such as the above is known as a **two by three** (2×3) **matrix.** It has two **rows** and three **columns.**

Let us agree to let the rows of a matrix represent equations of a system. Then we may perform the same operations on the rows of a matrix as were performed on the equations of a system and arrive at a matrix representing an equivalent system of equations. That is, we may

(1) Interchange two rows.

(2) Multiply or divide the elements of a row by any (nonzero) constant.

(3) Add (or subtract) k times any row to any other row.

Let us consider the following system and its corresponding matrix:

$$\begin{aligned} 2x - 3y &= 0 \\ x + 5y &= 3 \end{aligned} \qquad \begin{pmatrix} 2 & -3 & 0 \\ 1 & 5 & 3 \end{pmatrix}$$

To use Gauss' elimination method we wish to make the coefficient of x equal to 1 in the first equation. This can be most easily accomplished by interchanging the equations (rows of the matrix) to obtain

$$\begin{aligned} x + 5y &= 3 \\ 2x - 3y &= 0 \end{aligned} \qquad \begin{pmatrix} 1 & 5 & 3 \\ 2 & -3 & 0 \end{pmatrix}$$

Now we add -2 times the first equation (row of the matrix) to the second equation (row).

$$\begin{aligned} x + 5y &= 3 \\ -13y &= -6 \end{aligned} \qquad \begin{pmatrix} 1 & 5 & 3 \\ 0 & -13 & -6 \end{pmatrix}$$

Now we solve the second equation for y by dividing both sides by -13. (In the matrix we divide each element of the second row by -13.)

$$x + 5y = 3 \qquad \begin{pmatrix} 1 & 5 & 3 \\ 0 & 1 & \frac{6}{13} \end{pmatrix}$$
$$y = \frac{6}{13}$$

Finally, we add -5 times the second equation (row) to the first equation (row).

$$x \qquad = -\tfrac{9}{13} \qquad \begin{pmatrix} 1 & 0 & -\frac{9}{13} \\ 0 & 1 & \frac{6}{13} \end{pmatrix}$$
$$y = \qquad \tfrac{6}{13}$$

Since a matrix in the form

$$\begin{pmatrix} 1 & 0 & a \\ 0 & 1 & b \end{pmatrix}$$

corresponds to the system

$$x \qquad = a$$
$$y = b$$

and that is what we wish to obtain when we solve a linear system, it now becomes apparent how we can use matrices to advantage in solving systems of equations. We write the system in matrix representation and then use the three operations listed above to reduce the matrix to the desired form. Then we write the system of equations corresponding to the "reduced form" of the matrix.

Let us use this matrix method to solve the system

$$3x - 5y = 1$$
$$x = 2y = 3$$

The corresponding matrix is

$$\begin{pmatrix} 3 & -5 & 1 \\ 1 & 2 & 3 \end{pmatrix}$$

The following steps will solve this system.

Interchange the rows:

$$\begin{pmatrix} 1 & 2 & 3 \\ 3 & -5 & 1 \end{pmatrix}$$

Add -3 times row 1 to row 2:

$$\begin{pmatrix} 1 & 2 & 3 \\ 0 & -11 & -8 \end{pmatrix}$$

Divide row 2 by -11:

$$\begin{pmatrix} 1 & 2 & 3 \\ 0 & 1 & \frac{8}{11} \end{pmatrix}$$

Add -2 times row 2 to row 1:

$$\begin{pmatrix} 1 & 0 & \frac{17}{11} \\ 0 & 1 & \frac{8}{11} \end{pmatrix}$$

This matrix is now in the desired form. Its corresponding system of equations is

$$x = \tfrac{17}{11}$$
$$y = \tfrac{8}{11}$$

Thus the solution set of the given system is $\{(\tfrac{17}{11}, \tfrac{8}{11})\}$.

Exercise 5.9

Use matrix representation to solve the systems of equations in problems 1–20 of Exercise 5.8.

5.10 THREE EQUATIONS IN THREE VARIABLES

Gauss' method and the corresponding matrix notation can be used to solve any number of equations in any number of variables. In matrix notation, a system of three equations in three variables is represented by a 3 × 4 (three by four) matrix and the desired solution matrix is of the form

$$\begin{pmatrix} 1 & 0 & 0 & a \\ 0 & 1 & 0 & b \\ 0 & 0 & 1 & c \end{pmatrix}$$

For example, the system

$$x + y - z = 1$$
$$x + 2y \quad\;\; = 3$$
$$2x \quad\;\; + z = 2$$

is represented by the matrix

$$\begin{pmatrix} 1 & 1 & -1 & 1 \\ 1 & 2 & 0 & 3 \\ 2 & 0 & 1 & 2 \end{pmatrix}$$

We wish to use the permissible operations on the rows of this matrix to reduce it to the desired form. The following steps will solve this system.

Subtract row 1 from row 2 and also subtract 2 times row 1 from row 3 to obtain

$$\begin{pmatrix} 1 & 1 & -1 & 1 \\ 0 & 1 & 1 & 2 \\ 0 & -2 & 3 & 0 \end{pmatrix}$$

Now subtract row 2 from row 1 and add 2 times row 2 to row 3 to obtain

$$\begin{pmatrix} 1 & 0 & -2 & -1 \\ 0 & 1 & 1 & 2 \\ 0 & 0 & 5 & 4 \end{pmatrix}$$

Divide row 3 by 5 to obtain

$$\begin{pmatrix} 1 & 0 & -2 & -1 \\ 0 & 1 & 1 & 2 \\ 0 & 0 & 1 & \frac{4}{5} \end{pmatrix}$$

Finally, add 2 times row 3 to row 1 and subtract row 3 from row 2 to obtain

$$\begin{pmatrix} 1 & 0 & 0 & \frac{3}{5} \\ 0 & 1 & 0 & \frac{6}{5} \\ 0 & 0 & 1 & \frac{4}{5} \end{pmatrix}$$

This is the solution matrix and the corresponding system is

$$x = \tfrac{3}{5}$$
$$y = \tfrac{6}{5}$$
$$z = \tfrac{4}{5}$$

Let us attempt to solve the following system by matrix representation:

$$x + 2y - 2x = 5$$
$$3x - y - z = -2$$
$$2x - 3y + z = 1$$

The student should determine what operations are performed on the rows of the matrix in each of the following steps:

$$\begin{pmatrix} 1 & 2 & -2 & 5 \\ 3 & -1 & -1 & -2 \\ 2 & -3 & 1 & 1 \end{pmatrix}$$

$$\begin{pmatrix} 1 & 2 & -2 & 5 \\ 0 & -7 & 5 & -17 \\ 0 & -7 & 5 & -9 \end{pmatrix}$$

$$\begin{pmatrix} 1 & 2 & -2 & 5 \\ 0 & 1 & -\frac{5}{7} & \frac{17}{7} \\ 0 & -7 & 5 & -9 \end{pmatrix}$$

$$\begin{pmatrix} 1 & 0 & -\frac{4}{7} & \frac{1}{7} \\ 0 & 1 & -\frac{5}{7} & \frac{17}{7} \\ 0 & 0 & 0 & 8 \end{pmatrix}$$

We did not, and cannot, obtain the desired solution matrix. However, let us write the system corresponding to the last matrix.

$$x \qquad - \tfrac{4}{7}z = \tfrac{1}{7}$$
$$y - \tfrac{5}{7}z = \tfrac{17}{7}$$
$$0 = 8$$

Since $0 \neq 8$, we see that the solution set, X, of this system is empty, i.e., $X = \varnothing$.

Finally, let us attempt to solve by matrix representation the system

$$3x - 2y + 5z = 0$$
$$x + 3y + 4z = 0$$
$$2x - 5y + z = 0$$

Again, the student should determine what operations are used in each step.

$$\begin{pmatrix} 3 & -2 & 5 & 0 \\ 1 & 3 & 4 & 0 \\ 2 & -5 & 1 & 0 \end{pmatrix}$$

$$\begin{pmatrix} 1 & 3 & 4 & 0 \\ 3 & -2 & 5 & 0 \\ 2 & -5 & 1 & 0 \end{pmatrix}$$

$$\begin{pmatrix} 1 & 3 & 4 & 0 \\ 0 & -11 & -7 & 0 \\ 0 & -11 & -7 & 0 \end{pmatrix}$$

$$\begin{pmatrix} 1 & 3 & 4 & 0 \\ 0 & 1 & \frac{7}{11} & 0 \\ 0 & -11 & -7 & 0 \end{pmatrix}$$

$$\begin{pmatrix} 1 & 0 & \frac{23}{11} & 0 \\ 0 & 1 & \frac{7}{11} & 0 \\ 0 & 0 & 0 & 0 \end{pmatrix}$$

Again, we are not led to the desired solution matrix, but the corresponding system of equations is

$$x \qquad + \tfrac{23}{11}z = 0$$
$$y + \tfrac{7}{11}z = 0$$
$$0 = 0$$

The system of equations in this example is called a **homogeneous system,** one in which all of the constants are zero. Such a system always has the solution $x = 0$, $y = 0$, $z = 0$ and, therefore, the solution set is never empty. The solution $x = 0$, $y = 0$, $z = 0$ is called the **trivial solution,** but some homogeneous systems, as, for example, this one, have nontrivial solutions. In this case, the matrix method led to one equation involving x and z, a second equation involving y and z, and a universally true statement, $0 = 0$. Since $0 = 0$ imposes no restrictions on x, y, and z, the solutions of the system are given by the other equations. In such cases it is customary to solve for x and y in terms of z and give the solutions as, in this case,

$$x = -\tfrac{23}{11}z$$
$$y = \tfrac{7}{11}z$$

Now any value assigned to z will give a particular solution. For example, if we let $z = 0$, then $x = 0$, and $y = 0$. This is the trivial solution. If $z = 1$, then $x = -\tfrac{23}{11}$ and $y = \tfrac{7}{11}$. If $z = 11$, then $x = -23$ and $y = 7$, etc. There are, of course, an infinite number of solutions for this system since z can be any real number.

The method of solving linear systems of equations using matrix representation can be applied to any number of equations in any number of variables. It is applicable, as we have seen, to homogeneous systems, to systems where there are more equations than variables or *vice versa*, and even to systems where the solution set turns out to be empty. Because of this versatility and also because of its step-by-step mechanical process, this method is especially suitable for programming in computing machines.

Exercise 5.10

In problems 1–9, solve each of the systems by using matrix representation.

1. $\begin{aligned} x + 2y + 3z &= 0 \\ 2x + 5y + 7z &= 3 \\ 3x + 6y + 10z &= 0 \end{aligned}$

2. $\begin{aligned} 4x + 8y + 9z &= 1 \\ 2x + 5y + 5z &= 0 \\ 3x + 6y + 7z &= -2 \end{aligned}$

3. $\begin{aligned} 4x + 8y + 9z &= 1 \\ -5x - 9y - 10z &= 3 \\ 3x + 6y + 7z &= 1 \end{aligned}$

4. $\begin{aligned} a + b \quad\;\; &= 6 \\ 5a + 7b + c &= 0 \\ 3a + 4b + c &= 1 \end{aligned}$

5. $\begin{aligned} 5x_1 + 2x_2 &= 1 \\ 3x_2 - x_3 &= 3 \\ 4x_1 - x_3 &= 5 \end{aligned}$

6. $\begin{aligned} 3r + 2s - t &= 0 \\ r + s \quad\;\; &= -5 \\ r + t + 1 &= -2t \end{aligned}$

7. $\begin{aligned} 3x + 2y - z &= 7 \\ 4x + 3y + 2z &= 1 \\ 5x - 3y - 6z &= -7 \end{aligned}$

8. $\begin{aligned} 3x_1 + 2x_2 &= 1 \\ 4x_2 - 3x_3 &= 23 \\ 5x_1 - 2x_3 &= -13 \end{aligned}$

9. $\begin{aligned} x + y + z + 2w &= 1 \\ 2x + 3y + 3z + w &= 0 \\ 4x + y + 5z - w &= 2 \\ x - y - z - w &= 0 \end{aligned}$

In problems 10 and 11, solve for x, y, and z by first solving for $\dfrac{1}{x}, \dfrac{1}{y}$, and $\dfrac{1}{z}$.

10. $\begin{aligned} \frac{2}{x} - \frac{1}{y} - \frac{1}{z} &= -1 \\[4pt] \frac{1}{x} - \frac{2}{y} + \frac{1}{z} &= -8 \\[4pt] -\frac{1}{x} + \frac{1}{y} - \frac{2}{z} &= 9 \end{aligned}$

11. $\begin{aligned} \frac{1}{x} + \frac{2}{y} + \frac{3}{z} &= 5 \\[4pt] \frac{2}{x} - \frac{4}{y} + \frac{1}{z} &= -4 \\[4pt] \frac{2}{x} + \frac{1}{y} + \frac{2}{z} &= 9 \end{aligned}$

Some of the following word problems can be worked quite easily by setting up only one or two equations. For the sake of the practice, however, the student should solve the problem by setting up a system of three equations in three unknowns.

12. Do problem 26 of Exercise 4.2.

13. Do problem 29 of Exercise 4.2.

14. Do problem 9 of Exercise 4.3.

15. If the length of a rectangle is decreased by one foot and the width is increased by 2 feet, then the area is increased by 2 square feet. If the length is increased by 2 feet and the width decreased by 1 foot, then the area is decreased by one square foot. What is the area of the rectangle?

16. A three-digit number has the sum of its digits equal to 10, while the ten's digit is equal to three times the unit's digit. If the digits are reversed, the number remains the same. What is the number?

17. A three-digit number has the second digit equal to one more than the sum of the first and third digits and the third digit is twice the first. If the digits are reversed, the new number is 198 more than the original number. What is the number?

18. A man divides $10,000 among three investments at 3, 4, and 6 per cent per year, respectively. His annual income from the first two investments is $80 less than his income from the third investment, while his total income is $460 per year. Find the amount invested at each rate.

19. A man has $30,000 divided among three investments at 4, 5, and 6 per cent per year, respectively. He has two thirds as much invested at 5 as at 4 per cent, while his annual income from his investments is $1,480. Find the amount invested at each rate.

20. If the length of a certain rectangle is increased by 4 feet and the width is decreased by 3 feet, the area will be decreased by 13 square feet. On the other hand, if the length is decreased by one foot and the width is increased by 2 feet, the area will be increased by 37 square feet. Find the dimensions of the rectangle.

21. In analytic geometry it is shown that the graph of an equation of the form $x^2 + y^2 + Dx + Ey + F = 0$ is a circle. Find an equation whose graph is the circle passing through the points with coordinates $(1, 2)$, $(0, 5)$, and $(4, 1)$, respectively.

22. In analytic geometry it is shown that the graph of an equation of the form $y = ax^2 + bx + c$ is a parabola if $a \neq 0$. Find an equation whose graph is the parabola passing through the points with coordinates $(2, 1)$, $(1, 2)$, and $(-1, 1)$, respectively.

23. In analytic geometry it is shown that the graph of an equation of the form $x^2 + Ay^2 + Bx + Cy = 0$ is an ellipse, provided $A \neq 0$. Find an equation for the ellipse passing through the points with coordinates $(1, 1)$, $(1, -1)$, and $(3, 1)$, respectively.

5.11   THE DOUBLE-SUBSCRIPT NOTATION

A very notable characteristic of mathematicians is their desire to
treat general cases rather than special ones. In this way, they obtain
formulas covering all cases. Students, also, are usually only too glad
to have formulas so that they can solve problems by merely substituting in values and "turning the crank." The mathematician's attitude,
however, is quite different, since he is interested mainly in *obtaining*
the formulas and not so much in the use of the formulas. In the present
situation, the general problem is that of solving the system of equations

$$ax + by = c$$
$$dx + ey = f \tag{1}$$

A change in notation is desirable, however, not so much because it
is needed in this simple case, but because it is used extensively in
more advanced work. If it is learned now it will be available for your
use when needed.

To see a little more clearly the desirability for a different notation,
let us consider the solution of three equations in three variables. Our
general problem in a notation similar to that used above for the case of
two equations in two variables would be

$$ax + by + cz = d$$
$$ex + fy + gz = h \tag{2}$$
$$ix + jy + kz = l$$

It is easy now to see the disadvantage of this notation. For example,
if in a formula for the solution of the system, the student sees a "j,"
can he recall offhand what j stands for? Further work with four equations in four unknowns would be still more troublesome and, in fact,
we would soon exhaust the alphabet!

A notation intermediate in usefulness between that shown in (1)
and (2) above and the one we will use is given by

$$a_1x + b_1y = c_1 \qquad a_1x + b_1y + c_1z = d_1$$
$$a_2x + b_2y = c_2 \qquad a_2x + b_2y + c_2z = d_2 \tag{1'}$$
$$a_3x + b_3y + c_3z = d_3$$

where the subscript notation discussed in Section 2.1 is employed. However, for still larger numbers of variables and equations this notation would again become awkward. So our perfected notation is

$$a_{11}x_1 + a_{12}x_2 = b_1 \qquad a_{11}x_1 + a_{12}x_2 + a_{13}x_3 = b_1$$
$$a_{21}x_1 + a_{22}x_2 = b_2 \qquad a_{21}x_1 + a_{22}x_2 + a_{23}x_3 = b_2 \qquad (1'')$$
$$a_{31}x_1 + a_{32}x_2 + a_{33}x_3 = b_3$$

Here "a_{13}" is read "a one three"; "a_{22}" is read "a two two," etc. This **double-subscript** notation, complicated looking at first sight, is actually very simple and rewarding and is used extensively in advanced mathematics.

In explanation, note first that the variables are called x_1, x_2, x_3, etc., so that any number of them may be written down. Then, the first subscript variable tells us in which equation the coefficient occurs while the second subscript tells us which of the coefficients it precedes. For example, the coefficient a_{11} is the coefficient of x_1 in the first equation; a_{12} is the coefficient of x_2 in the first equation; a_{21} is the coefficient of x_1 in the second equation. Now, if, for example, we had six equations in six variables, we could immediately identify a_{46} as the coefficient of x_6 in the fourth equation.

Using this notation, let us now solve the general problem of two linear equations in two unknowns:

We shall use the matrix representation for Gauss' elimination method:

$$\begin{pmatrix} a_{11} & a_{12} & b_1 \\ a_{21} & a_{22} & b_2 \end{pmatrix}$$

We first assume $a_{11} \neq 0$ and divide the first row by a_{11} to obtain

$$\begin{pmatrix} 1 & \dfrac{a_{12}}{a_{11}} & \dfrac{b_1}{a_{11}} \\ a_{21} & a_{22} & b_2 \end{pmatrix}$$

Now we subtract a_{21} times the first row from the second row to obtain

$$\begin{pmatrix} 1 & \dfrac{a_{12}}{a_{11}} & \dfrac{b_1}{a_{11}} \\ 0 & \dfrac{a_{11}a_{22} - a_{21}a_{12}}{a_{11}} & \dfrac{a_{11}b_2 - a_{21}b_1}{a_{11}} \end{pmatrix}$$

Now we assume $a_{11}a_{22} - a_{21}a_{12} \neq 0$ and divide the second row by $\dfrac{a_{11}a_{22} - a_{21}a_{12}}{a_{11}}$ to obtain

$$\begin{pmatrix} 1 & \dfrac{a_{12}}{a_{11}} & \dfrac{b_1}{a_{11}} \\ 0 & 1 & \dfrac{a_{11}b_2 - a_{21}b_1}{a_{11}a_{22} - a_{21}a_{12}} \end{pmatrix}$$

Finally, we subtract $\dfrac{a_{12}}{a_{11}}$ times the second row from the first row to obtain

$$\begin{pmatrix} 1 & 0 & \dfrac{a_{22}b_1 - a_{12}b_2}{a_{11}a_{22} - a_{21}a_{12}} \\ 0 & 1 & \dfrac{a_{11}b_2 - a_{21}b_1}{a_{11}a_{22} - a_{21}a_{12}} \end{pmatrix}$$

Therefore, the solution is given by

$$x_1 = \frac{a_{22}b_1 - a_{12}b_2}{a_{11}a_{21} - a_{21}a_{12}} \tag{3}$$

$$x_2 = \frac{a_{11}b_2 - a_{21}b_1}{a_{11}a_{12} - a_{21}a_{12}} \tag{4}$$

Now in obtaining these solutions we made two assumptions: (1) $a_{11} \neq 0$ and (2) $a_{11}a_{22} - a_{21}a_{12} \neq 0$. The first assumption is permissible as was discussed in Section 5.8. The legitimacy of the second assumption will be discussed in Section 5.12.

At this point, however, it is important to note that we have only shown that *if* there are solutions x_1 and x_2 they must be given by (3) and (4), respectively. (Compare Section 5.2.) To prove that x_1 and x_2 are actually solutions, we must substitute these values into the given equations and see that they actually satisfy these equations. For our first equation this amounts to showing that

$$a_{11} \cdot \frac{a_{22}b_1 - a_{12}b_2}{a_{11}a_{22} - a_{12}a_{21}} + a_{12} \cdot \frac{a_{11}b_2 - a_{21}b_1}{a_{11}a_{22} - a_{12}a_{21}} = b_1$$

Simplifying the left-hand side, we have

$$\frac{a_{11}(a_{22}b_1 - a_{12}b_2) + a_{12}(a_{11}b_2 - a_{21}b_1)}{a_{11}a_{22} - a_{12}a_{21}}$$

$$= \frac{a_{11}a_{22}b_1 - a_{11}a_{12}b_2 + a_{12}a_{11}b_2 - a_{12}a_{21}b_1}{a_{11}a_{22} - a_{12}a_{21}}$$

$$= \frac{b_1(a_{11}a_{22} - a_{12}a_{21})}{a_{11}a_{22} - a_{12}a_{21}} = b_1$$

It is left to the student to perform a similar verification for the second equation.

Exercise 5.11

In problems 1–10, compare the given system with the double subscript notation and identify a_{11}, a_{12}, a_{21}, a_{22}, b_1, and b_2. Then use the formulas (3) and (4) in the text to find the solution.

1. $x + 2y = 1$
   $2x + 5y = 3$

2. $2x + y = 0$
   $3x + y = -2$

3. $2x - y = 4$
   $-5x + 2y = 11$

4. $x - \frac{1}{2}y = -2$
   $x + \frac{3}{4}y = -\frac{1}{2}$

5. $2x = 3y$
   $1 - 2x = 5y$

6. $y = m_1 x + b_1$
   $y = m_2 x + b_2$

7. $\dfrac{x}{a} + \dfrac{y}{b} = c$

   $\dfrac{x}{b} - \dfrac{y}{a} = c$

8. $x + y = -3a$
   $3x + y = -a$

9. $2x + 3y = 0$
   $y = 5 - a$

10. $x + y + a = 0$
    $3x - y - 2a = 0$

11. Verify that the formulas (3) and (4) in the text give solutions of $a_{21}x_1 + a_{22}x_2 = b_2$.

5.12 DETERMINANTS OF SECOND ORDER

The formulas (3) and (4) just given are difficult to remember without the aid of a device called a **determinant**, first used in 1693 in Europe by Leibnitz (co-inventor with Newton of the calculus) but used in Japan by Seki Kōwa some ten years earlier.

In the determinant notation, the formulas (3) and (4) are written as

$$x_1 = \frac{\begin{vmatrix} b_1 & a_{12} \\ b_2 & a_{22} \end{vmatrix}}{\begin{vmatrix} a_{11} & a_{12} \\ a_{21} & a_{22} \end{vmatrix}} \quad \text{and} \quad x_2 = \frac{\begin{vmatrix} a_{11} & b_1 \\ a_{21} & b_2 \end{vmatrix}}{\begin{vmatrix} a_{11} & a_{12} \\ a_{21} & a_{22} \end{vmatrix}}$$

To **evaluate** a second-order determinant, we multiply along diagonals as shown below, and subtract the second product from the first:

$$\begin{vmatrix} a_{11} & a_{12} \\ a_{12} & a_{22} \end{vmatrix} = a_{11}a_{22} - a_{12}a_{21}$$

For example,

$$\begin{vmatrix} 3 & 5 \\ -2 & 1 \end{vmatrix} = 3 \cdot 1 - 5 \cdot (-2) = 3 + 10 = 13$$

In this form our formulas are easy to remember. The denominators for both x_1 and x_2 have the coefficients in the order in which they appear in the equations. The determinant numerator for x_1 is obtained by replacing the first "column" by the constant terms and the determinant numerator for x_2 is obtained by replacing the second "column" by the constant terms. For example, the determinant solution of the first example of Section 5.8

$$\begin{aligned} x + y &= 5 \\ 2x - y &= 4 \end{aligned} \quad \text{or} \quad \begin{aligned} 1 \cdot x + 1 \cdot y &= 5 \\ 2 \cdot x + (-1) \cdot y &= 4, \end{aligned}$$

is

$$x = \frac{\begin{vmatrix} 5 & 1 \\ 4 & -1 \end{vmatrix}}{\begin{vmatrix} 1 & 1 \\ 2 & -1 \end{vmatrix}} = \frac{-5 - 4}{-1 - 2} = \frac{-9}{-3} = 3 \qquad y = \frac{\begin{vmatrix} 1 & 5 \\ 2 & 4 \end{vmatrix}}{\begin{vmatrix} 1 & 1 \\ 2 & -1 \end{vmatrix}} = \frac{4 - 10}{-1 - 2} = \frac{-6}{-3} = 2$$

For the sake of understanding the double-subscript notation, the student should verify that $a_{11} = 1, a_{12} = 1, a_{21} = 2, a_{22} = -1, b_1 = 5, b_2 = 4$. But he should not get the idea that the method of solution is to write down a_{11}, a_{12}, etc., and substitute in the formulas. Rather, he should think in terms of the coefficients of x, the coefficients of y, and the constant terms as explained above.

With the aid of our determinant notation, we are now ready to see what happens when

$$a_{11}a_{22} - a_{12}a_{21} = \begin{vmatrix} a_{11} & a_{12} \\ a_{21} & a_{22} \end{vmatrix} = 0 \tag{1}$$

Our interpretation will be a geometrical one, so we rewrite our equations as

$$\begin{aligned} a_{11}x + a_{12}y &= b_1 \\ a_{21}x + a_{22}y &= b_2 \end{aligned} \tag{2}$$

Then each of our equations is that of a straight line. If $a_{12} = 0$, (1) implies that $a_{11}a_{22} = 0$ so that either $a_{11} = 0$ or $a_{22} = 0$. But if $a_{11} = 0$, our first equation becomes $0 + 0 = b_1$, whereas if $a_{22} = 0$, we no longer have two variables. A similar argument shows that $a_{22} \neq 0$, and hence we may rewrite (2) as

$$y = -\frac{a_{11}}{a_{12}} x + \frac{b_1}{a_{12}}$$

$$y = -\frac{a_{21}}{a_{22}} x + \frac{b_2}{a_{22}}$$

(3)

But from (1) we have

$$\frac{a_{11}}{a_{12}} = \frac{a_{21}}{a_{22}}$$

Thus the slopes of the lines whose equations are given in (3) are equal, and hence the lines are coincident or parallel by Theorem 5.2.

Furthermore, the lines are coincident if and only if they are parallel and also

$$\frac{b_1}{a_{12}} = \frac{b_2}{a_{22}}$$

or $b_1 a_{22} - a_{12}b_2 = 0$. But

$$b_1 a_{22} - a_{12}b_2 = \begin{vmatrix} b_1 & a_{12} \\ b_2 & a_{22} \end{vmatrix}$$

Now, clearly, if the lines are parallel we have no solution, and if they are coincident we have an unlimited number of solutions. Thus we have

Theorem 5.4. The system of equations

$$a_{11}x_1 + a_{12}x_2 = b_1$$
$$a_{21}x_1 + a_{22}y_2 = b_2$$

has a unique solution, no solution, or an unlimited number of solutions, according as:

(1)
$$\begin{vmatrix} a_{11} & a_{12} \\ a_{21} & a_{22} \end{vmatrix} \neq 0$$

(2)
$$\begin{vmatrix} a_{11} & a_{12} \\ a_{21} & a_{22} \end{vmatrix} = 0 \quad \text{but} \quad \begin{vmatrix} b_1 & a_{12} \\ b_2 & a_{22} \end{vmatrix} \neq 0$$

or

(3)
$$\begin{vmatrix} a_{11} & a_{12} \\ a_{21} & a_{22} \end{vmatrix} = 0 \quad \text{and} \quad \begin{vmatrix} b_1 & a_{12} \\ b_2 & a_{22} \end{vmatrix} = 0$$

If (1) holds, we call the system of equations **consistent** and **independent**; if (2) holds, we call the system of equations **inconsistent**; and if (3) holds, we call the system of equations **dependent.**

Example 1. We have already seen that the system of equations $x + y = 5$ and $2x - y = 4$ is a consistent and independent system.

Example 2. The system of equations $x + y = 5$ and $x + y = 10$ is an inconsistent system, since

$$\begin{vmatrix} a_{11} & a_{12} \\ a_{21} & a_{22} \end{vmatrix} = \begin{vmatrix} 1 & 1 \\ 1 & 1 \end{vmatrix} = 0$$

but

$$\begin{vmatrix} b_1 & a_{12} \\ b_2 & a_{22} \end{vmatrix} = \begin{vmatrix} 5 & 1 \\ 10 & 1 \end{vmatrix} = -5 \neq 0$$

Example 3. The system of equations $x + y = 5$ and $2x + 2y = 10$ is a dependent system, since

$$\begin{vmatrix} a_{11} & a_{12} \\ a_{21} & a_{22} \end{vmatrix} = \begin{vmatrix} 1 & 1 \\ 2 & 2 \end{vmatrix} = 0$$

and

$$\begin{vmatrix} b_1 & a_{12} \\ b_2 & a_{22} \end{vmatrix} = \begin{vmatrix} 5 & 1 \\ 10 & 2 \end{vmatrix} = 0$$

If we try to solve an inconsistent system by the method of addition or subtraction we arrive at a contradiction. Thus in Example 2 suppose there are numbers x_0 and y_0 such that $x_0 + y_0 = 5$ and $x_0 + y_0 = 10$. Then we have

$$\begin{array}{r} x_0 + y_0 = 10 \\ x_0 + y_0 = 5 \\ \hline 0 = 5 \end{array}$$

Since $0 \neq 5$, there do not exist such numbers x_0 and y_0, and hence the system $x + y = 5$, $x + y = 10$ has no solution. On the other hand, the method of addition or subtraction when applied to a dependent system (Example 3) produces the statement "$0 = 0$":

$$2x + 2y = 10$$
$$\underline{2x + 2y = 10}$$
$$0 = 0$$

There is one solution for an independent and consistent system, no solution for an inconsistent system, and an infinite number of solutions for a dependent system. For example, from $x + y = 5$ and $2x + 2y = 10$ we have $y = 5 - x$ and hence the solutions $x = 0$, $y = 5$; $x = 1$, $y = 4$; $x = -1$, $y = 6$; etc.

Exercise 5.12

In problems 1–10, evaluate the given determinants.

1. $\begin{vmatrix} 1 & 2 \\ 3 & 4 \end{vmatrix}$

2. $\begin{vmatrix} 2 & 1 \\ 4 & 3 \end{vmatrix}$

3. $\begin{vmatrix} 2 & -3 \\ 1 & -5 \end{vmatrix}$

4. $\begin{vmatrix} 3 & -1 \\ 4 & -3 \end{vmatrix}$

5. $\begin{vmatrix} 0 & 1 \\ 5 & 0 \end{vmatrix}$

6. $\begin{vmatrix} 2 & 0 \\ 1 & 4 \end{vmatrix}$

7. $\begin{vmatrix} -5 & -7 \\ -6 & -1 \end{vmatrix}$

8. $\begin{vmatrix} -8 & -1 \\ -2 & -3 \end{vmatrix}$

9. $\begin{vmatrix} 0 & 1 \\ 0 & 2 \end{vmatrix}$

10. $\begin{vmatrix} 5 & 0 \\ 2 & 0 \end{vmatrix}$

In problems 11–20, solve the given system by use of determinants.

11. $2x - 3y = 0$
    $-2x + 5y = 1$

12. $3x + y = 1$
    $2x + y = 1$

13. $2x + y = 4$
    $4x + 5y = 11$

14. $3x + 6y = 1$
    $2x + 5y = 1$

15. $7x - 6y = 2$
    $5x + 2y = 3$

16. $2x + 3y = 1$
    $5x + 7y = 6$

17. $x + 2y = 1$
    $2x + 5y = 3$

18. $2x + 3y = 1$
    $x + y = 0$

19. $3x - 2y = 4$
    $2x - 7y = 11$

20. $5x + 11y = 17$
    $7x - 25y = 11$

In problems 21–30, try to solve each of the given systems by three methods: (1) graphically, (2) by determinants, and (3) by matrix representation. State for each problem whether the system is independent and consistent, dependent, or inconsistent. In each case, give the solution set.

21. $2x + y = 1$
$3x + 2y = 2$

22. $x + 3y = 1$
$2x + 6y = 2$

23. $3x + 2y = 1$
$4x + 3y = 6$

24. $5x - 4y = 1$
$-15x + 12y = 5$

25. $x - y = 4$
$y = x + 4$

26. $2x - 4y = 5$
$x - 2y = 3$

27. $x - y = 3$
$2y = 2x - 6$

28. $3x + 2y = 1$
$3x + y = 1$

29. $5x - 3y = 2$
$10x - 6y = 5$

30. $7x - 2y = 1$
$14x - 4y = 2$

5.13 DETERMINANTS OF HIGHER ORDER

In order to see how useful the determinant notation is, let us write, without proof, the determinant formulas for the solution of a system of three equations in three variables

$$a_{11}x_1 + a_{12}x_2 + a_{13}x_3 = b_1$$
$$a_{21}x_1 + a_{22}x_2 + a_{23}x_3 = b_2$$
$$a_{31}x_1 + a_{32}x_2 + a_{33}x_3 = b_3$$

These are

$$x_1 = \frac{\begin{vmatrix} b_1 & a_{12} & a_{13} \\ b_2 & a_{22} & a_{23} \\ b_3 & a_{32} & a_{33} \end{vmatrix}}{\begin{vmatrix} a_{11} & a_{12} & a_{13} \\ a_{21} & a_{22} & a_{23} \\ a_{31} & a_{32} & a_{33} \end{vmatrix}}, \quad x_2 = \frac{\begin{vmatrix} a_{11} & b_1 & a_{13} \\ a_{21} & b_2 & a_{23} \\ a_{31} & b_3 & a_{33} \end{vmatrix}}{\begin{vmatrix} a_{11} & a_{12} & a_{13} \\ a_{21} & a_{22} & a_{23} \\ a_{31} & a_{32} & a_{33} \end{vmatrix}}, \text{ and } x_3 = \frac{\begin{vmatrix} a_{11} & a_{12} & b_1 \\ a_{21} & a_{22} & b_2 \\ a_{31} & a_{32} & b_3 \end{vmatrix}}{\begin{vmatrix} a_{11} & a_{12} & a_{13} \\ a_{21} & a_{22} & a_{23} \\ a_{31} & a_{32} & a_{33} \end{vmatrix}}$$

providing, of course, that:

$$\begin{vmatrix} a_{11} & a_{12} & a_{13} \\ a_{21} & a_{22} & a_{23} \\ a_{31} & a_{32} & a_{33} \end{vmatrix} \neq 0 \tag{1}$$

The student should try to write down the general "four by four" case and the solution by determinants.

Of course, so far we have only a formal solution since we have not said how to evaluate a third-order determinant. This evaluation is accomplished by breaking the third-order determinant up into second-order determinants and can be done in several ways.

Let us consider the symbol for a third-order determinant as a "3 by 3 array" of numbers. There are 3 rows (horizontal) and 3 columns (vertical) with a total of 9 members. Each member or **element** can be associated with a second-order determinant known as its **minor**, which is formed by crossing out the row and column containing the element in question. For example, the minor of a_{11} is formed by crossing out the first row and the first column,

$$\begin{vmatrix} a_{11} & a_{12} & a_{13} \\ a_{21} & a_{22} & a_{23} \\ a_{31} & a_{32} & a_{33} \end{vmatrix}$$

to obtain the second-order determinant

$$\begin{vmatrix} a_{22} & a_{23} \\ a_{32} & a_{33} \end{vmatrix}$$

The minor of a_{23} is obtained by crossing out the row and column containing a_{23},

$$\begin{vmatrix} a_{11} & a_{12} & a_{13} \\ a_{21} & a_{22} & a_{23} \\ a_{31} & a_{32} & a_{33} \end{vmatrix}$$

and we get

$$\begin{vmatrix} a_{11} & a_{12} \\ a_{31} & a_{32} \end{vmatrix}$$

Now the position of each element of the determinant can be classified as either odd or even depending upon whether its two subscripts add up to an odd number or an even number. Thus the position of a_{11} is even (since $1 + 1 = 2$, which is even), the position of a_{31} is even (why?), while the position of a_{23} is odd (why?), etc.

A third-order determinant can be evaluated as follows: first, choose any row (or column) and take each element or the negative of the element of the row (column) according as to whether the position of the element is even or odd. Then form the product of each of these elements or negatives of elements of the row (column) with its minor and, finally, take the sum of these products. For example, let us choose the second row of our determinant (1). (We then say we are "expanding by the second row.") We have

$$-a_{21}\begin{vmatrix} a_{12} & a_{13} \\ a_{32} & a_{33} \end{vmatrix} + a_{22}\begin{vmatrix} a_{11} & a_{13} \\ a_{31} & a_{33} \end{vmatrix} - a_{23}\begin{vmatrix} a_{11} & a_{12} \\ a_{31} & a_{32} \end{vmatrix}$$

where each second-order determinant may then be expanded to give

$$-a_{21}(a_{12}a_{33} - a_{13}a_{32}) + a_{22}(a_{11}a_{33} - a_{13}a_{31}) - a_{23}(a_{11}a_{32} - a_{12}a_{31})$$

for the value of this third-order determinant.

Or, if we chose to expand by the third column we would have

$$a_{13}\begin{vmatrix} a_{21} & a_{22} \\ a_{31} & a_{32} \end{vmatrix} - a_{23}\begin{vmatrix} a_{11} & a_{12} \\ a_{31} & a_{32} \end{vmatrix} + a_{33}\begin{vmatrix} a_{11} & a_{12} \\ a_{21} & a_{22} \end{vmatrix}$$

In problems where the double-script notation is not in use there is another method of deciding whether to use an element or the negative of an element in forming a product of the expansion. This method is as follows: Consider an element in the mth row and n^{th} column. If $m + n$ is even, use the element itself in the product whereas if $m + n$ is odd, use the negative of the element. For example, in

$$\begin{vmatrix} 1 & 3 & -2 \\ 1 & 0 & -4 \\ 2 & -1 & 0 \end{vmatrix} \tag{2}$$

2 is in the third row and first column, $3 + 1 = 4$, which is even, so we have

$$2\begin{vmatrix} 3 & -2 \\ 0 & -4 \end{vmatrix}$$

-1 is in the third row and second column, $3 + 2 = 5$, and we have

$$-(-1)\begin{vmatrix} 1 & -2 \\ 1 & -4 \end{vmatrix}$$

Let us now evaluate the determinant (2) just given. If a 0 occurs it is usually best to evaluate along a row or column containing it for reasons which will be immediately apparent. Thus let us choose the second column to obtain

$$-3\begin{vmatrix}1 & -4\\2 & 0\end{vmatrix} + 0\begin{vmatrix}1 & -2\\2 & 0\end{vmatrix} - (-1)\begin{vmatrix}1 & -2\\1 & -4\end{vmatrix}$$
$$= -3(8) + 0(4) + 1(-4 + 2) = -24 - 2 = -26$$

On the other hand, evaluation by the third row yields

$$2\begin{vmatrix}3 & -2\\0 & -4\end{vmatrix} - (-1)\begin{vmatrix}1 & -2\\1 & -4\end{vmatrix} + 0\begin{vmatrix}1 & 3\\1 & 0\end{vmatrix}$$
$$= 2(-12) + 1(-2) + 0(-3) = -26$$

The student should evaluate this determinant along still other rows and columns to verify further the fact that the choice is arbitrary.

A fourth-order determinant can be evaluated by breaking it up into the sum of third-order determinants in exactly the same way, except that the minors of each element of a fourth-order determinant are third-order determinants, which in their turn will require breaking down into second-order determinants to evaluate them. Thus it can be seen that a determinant of any order can be evaluated, although it rapidly becomes a rather formidable task. This difficulty can be partially overcome, however, by use of various devices discussed in more advanced books on algebra.

Before leaving the topic of determinants, however, we should notice that we have not proved the validity of our determinant formulas for the solution of a system of three equations in three variables (although we did for two equations in two variables). These formulas may be proved by the methods described in Section 5.10, but we shall not do so here. For four or more variables with the corresponding number of equations the proof that the determinant formulas are valid is even more lengthy, unless more powerful methods of analysis are available. We would also like to point out that there is a corresponding geometric interpretation for the solution of a system of three equations in three variables based on the fact that an equation of the form $Ax + By + Cz = D$ is the equation of a plane. Thus an independent and consistent system represents three planes which intersect at a point. The remaining cases, however, are not so simple, since we may have three coincident planes, two coincident planes with the third plane intersecting, two coincident planes with the third plane parallel, etc. For details on these topics the student is referred to the books on the theory of equations listed in the bibliography.

We conclude this section with an example of the use of determinants in solving three equations in three variables.

Example. Solve by determinants the system of equations

$$3x - 5y - 2z = 20$$
$$x + y + z = 3$$
$$8x + 4y + 3z = 27$$

SOLUTION: We have, for the denominator,

$$\begin{vmatrix} 3 & -5 & -2 \\ 1 & 1 & 1 \\ 8 & 4 & 3 \end{vmatrix} = 3\begin{vmatrix} 1 & 1 \\ 4 & 3 \end{vmatrix} - \begin{vmatrix} -5 & -2 \\ 4 & 3 \end{vmatrix} + 8\begin{vmatrix} -5 & -2 \\ 1 & 1 \end{vmatrix}$$

$$= 3(3-4) - (-15+8) + 8(-5+2) = -3 + 7 - 24 = -20$$

Hence

$$x = \frac{\begin{vmatrix} 20 & -5 & -2 \\ 3 & 1 & 1 \\ 27 & 4 & 3 \end{vmatrix}}{-20} = \frac{20\begin{vmatrix} 1 & 1 \\ 4 & 3 \end{vmatrix} - 3\begin{vmatrix} -5 & -2 \\ 4 & 3 \end{vmatrix} + 27\begin{vmatrix} -5 & -2 \\ 1 & 1 \end{vmatrix}}{-20}$$

$$= \frac{20(3-4) - 3(-15+8) + 27(-5+2)}{-20} = \frac{-20+21-81}{-20}$$

$$= \frac{-80}{-20} = 4;$$

$$y = \frac{\begin{vmatrix} 3 & 20 & -2 \\ 1 & 3 & 1 \\ 8 & 27 & 3 \end{vmatrix}}{-20} = \frac{3\begin{vmatrix} 3 & 1 \\ 27 & 3 \end{vmatrix} - \begin{vmatrix} 20 & -2 \\ 27 & 3 \end{vmatrix} + 8\begin{vmatrix} 20 & -2 \\ 3 & 1 \end{vmatrix}}{-20}$$

$$= \frac{3(9-27) - (60+54) + 8(20+6)}{-20} = \frac{-54 - 114 + 208}{-20}$$

$$= \frac{40}{-20} = -2;$$

$$z = \frac{\begin{vmatrix} 3 & -5 & 20 \\ 1 & 1 & 3 \\ 8 & 4 & 27 \end{vmatrix}}{-20} = \frac{3\begin{vmatrix} 1 & 3 \\ 4 & 27 \end{vmatrix} - \begin{vmatrix} -5 & 20 \\ 4 & 27 \end{vmatrix} + 8\begin{vmatrix} -5 & 20 \\ 1 & 3 \end{vmatrix}}{-20}$$

$$= \frac{3(27-12) - (-135-80) + 8(-15-20)}{-20}$$

$$= \frac{45 + 215 - 280}{-20} = \frac{-20}{-20} = 1$$

Exercise 5.13

In problems 1–8, evaluate the given determinants.

1. $\begin{vmatrix} 1 & 3 & 7 \\ 2 & 5 & 6 \\ 1 & 0 & 4 \end{vmatrix}$

2. $\begin{vmatrix} 2 & 5 & 7 \\ -4 & 6 & 3 \\ 0 & 6 & 1 \end{vmatrix}$

3. $\begin{vmatrix} -1 & 2 & 4 \\ 3 & -1 & 5 \\ 0 & 4 & -2 \end{vmatrix}$

4. $\begin{vmatrix} 3 & -1 & 2 \\ 0 & -5 & 4 \\ 1 & 2 & 7 \end{vmatrix}$

5. $\begin{vmatrix} 1 & 3 & 5 \\ 0 & 2 & 1 \\ 5 & 1 & 6 \end{vmatrix}$

6. $\begin{vmatrix} 0 & 1 & 1 \\ 2 & 0 & -1 \\ 1 & 5 & 2 \end{vmatrix}$

7. $\begin{vmatrix} 1 & 7 & 9 \\ 2 & -1 & 4 \\ 7 & -2 & 6 \end{vmatrix}$

8. $\begin{vmatrix} 4 & -2 & 7 \\ 5 & -1 & 2 \\ -3 & 4 & 9 \end{vmatrix}$

9. Show that $\begin{vmatrix} a & b & c \\ a & b & c \\ d & e & f \end{vmatrix} = 0$

10. Show that $\begin{vmatrix} a & b & c \\ d & e & f \\ g & h & i \end{vmatrix} = -\begin{vmatrix} a & b & c \\ g & h & i \\ d & e & f \end{vmatrix}$

*11. Show that $\begin{vmatrix} a+kd & b+ke & c+kf \\ d & e & f \\ g & h & i \end{vmatrix} = \begin{vmatrix} a & b & c \\ d & e & f \\ g & h & i \end{vmatrix}$

12–20. Solve problems 1–9 of Exercise 5.10 by using determinants.

EQUATIONS OF HIGHER DEGREE

6

6.1 INTRODUCTION

We have seen in the previous chapter how one line of generalization leads from the simple linear equation in one variable to the study of systems of two or more linear equations in a corresponding number of variables. In this chapter our concern is with the generalization which takes us from linear (first-degree) equations to quadratic (second-degree) equations and beyond.

It would be fine, of course, if we could give a general method for solving all polynomial equations, $a_n x^n + a_{n-1} x^{n-1} + \cdots + a_1 x + a_0 = 0$, but such a project is beyond the scope of a course in intermediate algebra.[1] Further work along this line may be found in the books on college algebra and the theory of equations given in the bibliography. Here we shall discuss only the quadratic equation in detail and shall confine ourselves to some miscellaneous comments on equations of higher degree.

6.2 SOLUTION OF EQUATIONS BY FACTORING

Now we will apply the types of factorizations discussed in Chapter 2 to the solution of some equations of degree greater than one. However, we must be clear first about two things:

[1] In fact, in a certain sense, no such general method exists. But it is at least possible to show how to obtain approximations (to any desired degree of accuracy) to the roots of any polynomial equation with real number coefficients.

1. Just as with linear equations, a solution is a number that "satisfies" the equation, and the solution set is the set of all solutions.

2. Equations such as $x^2 + 10x + 24 = (x + 6)(x + 4)$, $x^2 + x - 6 = (x + 3)(x - 2)$, etc., are universally valid, that is, yield a true statement whenever any number is substituted for x and are called identities (Section 5.1).

In this chapter we shall restrict solutions to real numbers. Thus, the solution set of an identity is the set, R^*, of all real numbers.

Example 1. Solve the equation $x^2 + 8x + 15 = 0$.

SOLUTION: Since $x^2 + 8x + 15 = (x + 5)(x + 3)$, if $x^2 + 8x + 15 = 0$, then $(x + 5)(x + 3) = 0$. Hence

$$x + 5 = 0 \qquad \text{or} \qquad x + 3 = 0$$

since if the product of two numbers is zero, one of the numbers must be zero. Thus $x = -5$ or $x = -3$, and the solution set is $X = \{-5, -3\}$.

The student should note carefully that $x = -5$ *or* $x = -3$. We are certainly not making the ridiculous statement that $x = -5$ *and* $x = -3$. Also, the student should check that both these numbers do actually satisfy the given equation and hence are solutions. Thus,

$$(-5)^2 + 8(-5) + 15 = 0, \qquad (-3)^2 + 8(-3) + 15 = 0$$

Sometimes, in order to emphasize the two different possibilities we use x_1 and x_2 as symbols for the two roots and write, for example, $x_1 = -5$, $x_2 = -3$ without using the word "or."

Example 2. Find all solutions of the equation $x^3 - 3x^2 - 10x = 0$.

SOLUTION: Since $x^3 - 3x^2 - 10x = x(x^2 - 3x - 10) = x(x - 5)(x + 2)$, if $x^3 - 3x^2 - 10x = 0$, then

$$x(x - 5)(x + 2) = 0$$

Therefore $x = 0$ or $x - 5 = 0$ or $x + 2 = 0$. Thus $x = 0$ or $x = 5$ or $x = -2$. The solution set is $X = \{0, 5, -2\}$.

We have shown that, if there is a number x such that $x^3 - 3x^2 - 10x = 0$, then $x = 0$ or $x = 5$ or $x = -2$. Finally, to see that these three numbers are actually solutions, we substitute each of them in turn in the original equation to see whether or not it satisfies the equation $x^3 - 3x^2 - 10x = 0$. This check is left as an exercise for the student.

Exercise 6.2

Find the solution set of the following equations by factoring.

1. $x^2 - x - 6 = 0$
2. $y^2 - 2y - 15 = 0$
3. $y^2 + 7y + 10 = 0$
4. $t^2 - 6t - 55 = 0$
5. $2x^2 - 5x - 3 = 0$
6. $6t^2 - 7t - 20 = 0$
7. $8t^2 + 6t - 9 = 0$
8. $-6x^2 + 19x - 15 = 0$
9. $y^2 - y = 0$
10. $x^2 + 3x = 0$
11. $9z^2 - 4z = 0$
12. $z^3 - z^2 = 0$
13. $x^3 - 2x^2 = 0$
14. $25y^3 - 16y = 0$
15. $y^3 - y^2 - 6y = 0$
16. $x^3 - 6x^2 - 16x = 0$
17. $x^3 - 8x^2 + 15x = 0$
18. $6z^3 - z^2 - 2z = 0$
19. $3z^3 - 14z^2 - 5z = 0$
20. $24x^3 - 38x^2 - 7x = 0$
21. $x^2 = ax + bx - ab$ (solve for x)
22. $x^2 + bx = x + b$ (solve for x, $b \neq 0$)
23. $(y + c)^2 - 25a^2 = 0$ (solve for y)
24. $(x - 4a)^2 - 9b^2 = 0$ (solve for x)
25. $y^2 - 6by + 9b^2 - k^2 = 0$ (solve for y)
26. $a^2x^2 - 2abx + b^2 - k^4 = 0$ (solve for x)

6.3 SQUARE ROOTS

Not all quadratic equations can be handled by factoring of the simple type discussed so far. Even such a simple equation as

$$x^2 - 2 = 0$$

does not yield to this approach. Before discussing a more general method which will eventually yield a formula for solving all quadratic equations, we need to discuss square roots of numbers.

First, the student should note that there are two numbers whose square is a given positive number. Thus both 2 and −2, when squared, yield 4. But there is only one square root of 4: $\sqrt{4} = 2$; $\sqrt{4}$ does not equal −2 nor ±2. If we wish to designate −2, we must write $-\sqrt{4} = -2$. The following definition guarantees that the square root of any positive number is positive.

Definition 6.1. For any nonnegative number a, $\sqrt{a} = |b|$ where $b^2 = a$.

For example, noting that $2^2 = 4$, we have $\sqrt{4} = |2| = 2$. Or, noting that $(-2)^2 = 4$, we could write $\sqrt{4} = |-2| = 2$. Thus, in any case, $\sqrt{4} = 2$. From another point of view, note that if we have $b^2 = a$ we cannot conclude that $b = \sqrt{a}$ but only that $|b| = \sqrt{a}$. Thus if $b^2 = a$, we have $b = \sqrt{a}$ or $-b = \sqrt{a}$, from which we conclude that $b = \sqrt{a}$ or $b = -\sqrt{a}$. For example, if $b^2 = 4$, then $|b| = 2$. Thus $b = 2$ or $-b = 2$. That is, $b = 2$ or $b = -2$. We sometimes write "$b = \pm 2$" as an abbreviation for "$b = 2$ or $b = -2$".

We then have the two basic properties of square roots:

$$\sqrt{ab} = \sqrt{a}\sqrt{b} \quad \text{and} \quad \sqrt{\frac{a}{b}} = \frac{\sqrt{a}}{\sqrt{b}} \ (b \neq 0)$$

for nonnegative numbers a and b. Thus, for example,

$$\sqrt{72} = \sqrt{36 \cdot 2} = \sqrt{36}\sqrt{2} = 6\sqrt{2},$$
$$\sqrt{48} = \sqrt{16 \cdot 3} = \sqrt{16}\sqrt{3} = 4\sqrt{3}$$

and

$$\sqrt{\frac{1}{3}} = \sqrt{\frac{1}{3} \cdot \frac{3}{3}} = \sqrt{\frac{3}{9}} = \frac{\sqrt{3}}{\sqrt{9}} = \frac{\sqrt{3}}{3} = \frac{1}{3}\sqrt{3},$$
$$\sqrt{\frac{7}{12}} = \sqrt{\frac{7}{12} \cdot \frac{3}{3}} = \sqrt{\frac{21}{36}} = \frac{\sqrt{21}}{\sqrt{36}} = \frac{\sqrt{21}}{6} = \frac{1}{6}\sqrt{21}$$

The proofs of these properties are quite simple. By Definition 6.1, $\sqrt{ab}$ is that positive number whose square is ab. But a and b are both positive numbers and $(\sqrt{a}\sqrt{b})^2 = (\sqrt{a}\sqrt{b})(\sqrt{a}\sqrt{b}) = (\sqrt{a}\sqrt{a})(\sqrt{b}\sqrt{b}) = ab$. (What axioms are used here?) Thus $\sqrt{ab} = \sqrt{a}\sqrt{b}$. The proof of the second property is similar and is left as an exercise for the student.

Exercise 6.3

In problems 1–12, use the basic properties of square roots to write the given square root as $a\sqrt{b}$ where a is a rational number and b is an integer without perfect square factors.

1. $\sqrt{27}$ 2. $\sqrt{48}$ 3. $\sqrt{18}$ 4. $\sqrt{45}$

5. $\sqrt{96}$ 6. $\sqrt{75}$ 7. $\sqrt{\frac{1}{5}}$ 8. $\sqrt{\frac{2}{7}}$

9. $\sqrt{\dfrac{5}{12}}$ 10. $\sqrt{\dfrac{3}{8}}$ 11. $\sqrt{\dfrac{2}{27}}$ 12. $\sqrt{\dfrac{5}{18}}$

In problems 13–20, simplify the square root as in problems 1–12, and then simplify the resulting fraction by removing common factors from numerator and denominator.

13. $\dfrac{2 + \sqrt{8}}{2}$ 14. $\dfrac{3 + \sqrt{27}}{3}$ 15. $\dfrac{5 + \sqrt{75}}{5}$

16. $\dfrac{7 + \sqrt{343}}{14}$ 17. $\dfrac{5 + \sqrt{125}}{15}$ 18. $\dfrac{-4 - \sqrt{24}}{6}$

19. $\dfrac{-22 - \sqrt{363}}{22}$ 20. $\dfrac{-10 + \sqrt{200}}{15}$

*21. Prove that $\sqrt{3}$ is irrational.

*22. Prove that there are infinitely many irrational numbers.

*23. Prove that $\sqrt{P}$ is irrational if P is a prime number.

*24. Prove that $\sqrt[3]{5}$ is irrational.

6.4 SOLUTION BY COMPLETING THE SQUARE

With the use of irrational numbers we can certainly solve any quadratic equation of the form $x^2 + c = 0$ for any negative number c. Thus the solutions of $x^2 - 4 = 0$ are 2 and -2, whereas the solutions of $x^2 - 2 = 0$ are $\sqrt{2}$ and $-\sqrt{2}$. The three following examples show how we may reduce the solution of any [2] quadratic equation to the solution of the simple type just mentioned by **completion of the square.** We shall describe the method in steps and illustrate each step by three examples. We have given, then, three special cases of the general quadratic equation

$$ax^2 + bx + c = 0 \qquad (a \neq 0)$$

Example 1	*Example 2*	*Example 3*
$x^2 + 5x + 6 = 0$,	$3x^2 - 6x - 2 = 0$,	$6x^2 + x - 12 = 0$
$(a = 1, b = 5, c = 6)$	$(a = 3, b = -6, c = -2)$	$(a = 6, b = 1, c = -12)$

[2] Some quadratic equations, such as $x^2 + 1 = 0$, do not have real numbers as roots. Our method may still be used in such a situation, but we will avoid such types until discussion of complex numbers in Chapter 11.

Step 1: Subtract the constant term, c, from both sides of the equation.

$$x^2 + 5x = -6 \qquad 3x^2 - 6x = 2 \qquad 6x^2 + x = 12$$

Step 2: Divide both sides of the equation by the coefficient, a, of the first term.

$$x^2 + 5x = -6 \qquad x^2 - 2x = \tfrac{2}{3} \qquad x^2 + \tfrac{1}{6}x = 2$$

Step 3: Add the square of $\tfrac{1}{2}$ the coefficient of x in the new equation to both sides of the new equation.

$$x^2 + 5x + (\tfrac{5}{2})^2 = -6 + (\tfrac{5}{2})^2 \qquad x^2 - 2x + (-\tfrac{2}{2})^2 = \tfrac{2}{3} + (-\tfrac{2}{2})^2$$

$$x^2 + \tfrac{1}{6}x + (\tfrac{1}{12})^2 = 2 + (\tfrac{1}{12})^2$$

Step 4: Express the left side of the equation as the square of a binomial.

$$(x + \tfrac{5}{2})^2 = \tfrac{25}{4} - 6 \qquad (x - 1)^2 = \tfrac{2}{3} + 1 \qquad (x + \tfrac{1}{12})^2 = 2 + \tfrac{1}{144}$$

Step 5: From the remarks on page 153, we know that

$$\left| x + \frac{5}{2} \right| = \sqrt{\frac{25}{4} - 6} \qquad\qquad |x - 1| = \sqrt{\frac{2}{3} + 1}$$

$$\left| x + \frac{1}{12} \right| = \sqrt{2 + \frac{1}{144}}$$

and we conclude that

$$x + \frac{5}{2} = \pm \sqrt{\frac{25}{4} - 6} \qquad\qquad x - 1 = \pm \sqrt{\frac{2}{3} + 1}$$

$$x + \frac{1}{12} = \pm \sqrt{2 + \frac{1}{144}}$$

Step 6: Solve for x and simplify the answer.

$$x = -\frac{5}{2} \pm \sqrt{\frac{25}{4} - 6} \qquad\qquad x = 1 \pm \sqrt{\frac{2}{3} + 1}$$

$$= -\frac{5}{2} \pm \sqrt{\frac{25}{4} - \frac{24}{4}} \qquad\qquad = 1 \pm \sqrt{\frac{2}{3} + \frac{3}{3}}$$

$$= -\frac{5}{2} \pm \sqrt{\frac{1}{4}} = -\frac{5}{2} \pm \frac{1}{2} \qquad\qquad = 1 \pm \sqrt{\frac{5}{3} \cdot \frac{3}{3}} = 1 \pm \sqrt{\frac{15}{9}}$$

$$x_1 = -\frac{5}{2} + \frac{1}{2} = -2 \qquad\qquad = \frac{3}{3} \pm \frac{\sqrt{15}}{3} = \frac{3 \pm \sqrt{15}}{3}$$

$$x_2 = -\frac{5}{2} - \frac{1}{2} = -3 \qquad\qquad x_1 = \frac{3 + \sqrt{15}}{3}$$

$$x_2 = \frac{3 - \sqrt{15}}{3}$$

$$x = -\frac{1}{12} \pm \sqrt{\frac{288}{144} + \frac{1}{144}}$$

$$= -\frac{1}{12} \pm \sqrt{\frac{289}{144}} = -\frac{1}{12} \pm \frac{17}{12}$$

$$= \frac{-1 \pm 17}{12}$$

$$x_1 = \frac{-1 + 17}{12} = \frac{16}{12} = \frac{4}{3}$$

$$x_2 = \frac{-1 - 17}{12} = -\frac{18}{12} = -\frac{3}{2}$$

Step 7: Check the answers obtained.

We will leave this step as an exercise for the student in Examples 1 and 3. For Example 2 we have

$$3\left(\frac{3 + \sqrt{15}}{3}\right)^2 - 6\left(\frac{3 + \sqrt{15}}{3}\right) - 2 = 3\left(\frac{9 + 6\sqrt{15} + 15}{9}\right)$$

$$- 6\left(\frac{3 + \sqrt{15}}{3}\right) - \frac{6}{3} = \frac{24 + 6\sqrt{15} - 18 - 6\sqrt{15} - 6}{3} = \frac{0}{3} = 0$$

and similarly for the other root.

Exercise 6.4

Solve the following equations by completing the square and check your answers.

1. $x^2 - 2x - 3 = 0$ 2. $x^2 - x - 2 = 0$

3. $x^2 - 7x + 12 = 0$ 4. $2x^2 + 4x - 1 = 0$

5. $3x^2 + 6x + 2 = 0$ 6. $5x^2 + 7x + 1 = 0$

7. $x^2 - x - 1 = 0$ 8. $5x^2 + 4x - 4 = 0$

9. $3x^2 - 3\sqrt{3}x + 1 = 0$ 10. $x^2 + 2\sqrt{2}x - 2 = 0$

(handwritten annotations in top margin) $\frac{6}{3}$ 2 $\frac{12}{3}$ 4

11. $5x^2 + 2\sqrt{3}x - 1 = 0$ **12.** $x^2 + px + q = 0$

13. $x^2 - px + q = 0$ **14.** $x^2 - px - q = 0$

6.5   COMPLETING THE SQUARE IN GENERAL

Actually the method of completing the square is infrequently used in the solution of quadratic equations. If the quadratic polynomial can be factored easily the solution is immediate, whereas if it cannot be factored easily one usually uses the formula developed in the next section. However, the general procedure of completing the square is used extensively in both analytic geometry and calculus, and we now give two examples of procedures actually used in these subjects.

Example 1. Complete the square in $x^2 + x - 1$.

SOLUTION: We proceed as in Section 6.4 by adding the square of half the coefficient of x and, in this case, also subtracting it. That is, we write

$$x^2 + x - 1 = x^2 + x + (\tfrac{1}{2})^2 - 1 - (\tfrac{1}{2})^2 = (x + \tfrac{1}{2})^2 - \tfrac{5}{4}$$

Example 2. Complete the square in both x and y in $x^2 + 2x + y^2 - 3y$.

SOLUTION:

$$x^2 + 2x + y^2 - 3y = x^2 + 2x + 1^2 + y^2 - 3y + (\tfrac{3}{2})^2 - 1^2 - (\tfrac{3}{2})^2$$
$$= (x + 1)^2 + (y - \tfrac{3}{2})^2 - \tfrac{13}{4}$$

Exercise 6.5

It is shown in analytic geometry that the graph of an equation of the form $(x - h)^2 + (y - k)^2 = r^2$ is a circle of radius r and center with coordinates (h, k). In problems 1–10, find, by completing the square in both x and y, the center and radius of the circle whose equation is given.

1. $x^2 - 2x + y^2 - 4y = 11$ **2.** $x^2 - 6x + y^2 - 8y = 11$

3. $x^2 - 4x + y^2 - 6y = 12$ **4.** $x^2 + y^2 + 4x + 6y = 3$

5. $x^2 + y^2 + 2x + 4y = 20$ **6.** $x^2 + y^2 + 6x + 8y = 11$

7. $2x^2 + 2y^2 - 5x - 4y = 0$ 8. $3x^2 + 3y^2 - 4x - 6y = 0$

9. $2x^2 + 2y^2 + 5x = 0$ 10. $4x^2 + 4y^2 + 7x = 0$

The polynomials in problems 11–18 occur in the calculus where it is necessary to complete the square as in Example 1 of this section.

11. $x^2 + 10x + 30$ 12. $x^2 - 4x + 8$

13. $x^2 + 2x + 5$ 14. $x^2 - 12x + 8$

15. $20 + 8x - x^2$ 16. $28 - 12x - x^2$

17. $5 - 4x - x^2$ 18. $4x - x^2$

6.6 THE QUADRATIC FORMULA

The following theorem gives the promised formula for the solution of the general quadratic equation.

Theorem 6.1. Let a, b, and c be real numbers with $a \neq 0$ and $b^2 - 4ac$ positive or zero.[3] Then the equation $ax^2 + bx + c = 0$ has the solutions

$$x_1 = \frac{-b + \sqrt{b^2 - 4ac}}{2a} \quad \text{and} \quad x_2 = \frac{-b - \sqrt{b^2 - 4ac}}{2a}$$

PROOF: We follow the same procedure as in the numerical examples of Section 6.4. Thus

Step 1: $ax^2 + bx = -c$.

Step 2: $x^2 + \dfrac{b}{a}x = -\dfrac{c}{a}$

Step 3: $x^2 + \dfrac{b}{a}x + \left(\dfrac{b}{2a}\right)^2 = -\dfrac{c}{a} + \left(\dfrac{b}{2a}\right)^2$

Step 4: $\left(x + \dfrac{b}{2a}\right)^2 = -\dfrac{c}{a} + \dfrac{b^2}{4a^2} = \dfrac{b^2 - 4ac}{4a^2}$

Step 5: $\left|x + \dfrac{b}{2a}\right| = \sqrt{\dfrac{b^2 - 4ac}{4a^2}} = \dfrac{\sqrt{b^2 - 4ac}}{\sqrt{4a^2}} = \dfrac{\sqrt{b^2 - 4ac}}{2|a|}$

[3] The restriction that $b^2 - 4ac$ be positive or zero is necessary because of the fact that $b^2 - 4ac$ will appear under the square root sign and, at this stage, we have not yet discussed square roots of negative numbers. Later, when we discuss complex numbers, in Chapter 11, we will remove this restriction and our formula will then apply to any quadratic equation. (See also the footnote in Section 6.4.)

so that

$$x + \frac{b}{2a} = \frac{\sqrt{b^2 - 4ac}}{2|a|} \quad \text{or} \quad x + \frac{b}{2a} = -\frac{\sqrt{b^2 - 4ac}}{2|a|} \qquad (1)$$

At this point let us consider $|a|$ in the denominator of $\dfrac{\sqrt{b^2 - 4ac}}{2|a|}$.
It is there, of course, because $\sqrt{a^2} = |a|$ (Definition 6.1). Now if a were positive we would have $|a| = a$ and (1) would become

$$x + \frac{b}{2a} = \frac{\sqrt{b^2 - 4ac}}{2a} \quad \text{or} \quad x + \frac{b}{2a} = -\frac{\sqrt{b^2 - 4ac}}{2a} \qquad (2)$$

On the other hand, if a were negative we would have $|a| = -a$ and (1) would become

$$x + \frac{b}{2a} = \frac{\sqrt{b^2 - 4ac}}{2(-a)} = -\frac{\sqrt{b^2 - 4ac}}{2a} \quad \text{or}$$

$$x + \frac{b}{2a} = -\frac{\sqrt{b^2 - 4ac}}{2(-a)} = \frac{\sqrt{b^2 - 4ac}}{2a} \qquad (3)$$

which, we see, is exactly the same statement as (2) except for the order in which the two equations are written. This means that (1) can always be replaced by

$$x + \frac{b}{2a} = \frac{\sqrt{b^2 - 4ac}}{2a} \quad \text{or} \quad x + \frac{b}{2a} = -\frac{\sqrt{b^2 - 4ac}}{2a} \qquad (1')$$

Step 6: $x_1 = -\dfrac{b}{2a} + \dfrac{\sqrt{b^2 - 4ac}}{2a} = \dfrac{-b + \sqrt{b^2 - 4ac}}{2a},$

$\qquad x_2 = -\dfrac{b}{2a} - \dfrac{\sqrt{b^2 - 4ac}}{2a} = \dfrac{-b - \sqrt{b^2 - 4ac}}{2a}$

i.e., the solution set of $ax^2 + bx + c = 0$ is

$$X = \left\{ \frac{-b + \sqrt{b^2 - 4ac}}{2a}, \ \frac{-b - \sqrt{b^2 - 4ac}}{2a} \right\}$$

We often write

$$x = \frac{-b \pm \sqrt{b^2 - 4ac}}{2a}$$

as a shorthand for the answers given in Step 6.

Note once again that we have not actually shown that either

$$\frac{-b + \sqrt{b^2 - 4ac}}{2a} \quad \text{or} \quad \frac{-b - \sqrt{b^2 - 4ac}}{2a}$$

is a solution of the quadratic equation $ax^2 + bx + c = 0$. Rather, we have shown that *if* the equation $ax^2 + bx + c = 0$ has solutions, they are

$$\frac{-b + \sqrt{b^2 - 4ac}}{2a} \quad \text{and} \quad \frac{-b - \sqrt{b^2 - 4ac}}{2a}.$$

To prove that they are solutions, it remains to substitute, in turn,

$$\frac{-b + \sqrt{b^2 - 4ac}}{2a} \quad \text{and} \quad \frac{-b - \sqrt{b^2 - 4ac}}{2a}$$

for x in $ax^2 + bx + c = 0$ and show that the equation is satisfied by both of these two numbers. We leave this as an exercise for the student.

Example 1. Solve the equation $x^2 + 5x + 6 = 0$ by the quadratic formula.

SOLUTION: For this equation, $a = 1$, $b = 5$, $c = 6$. Therefore the solutions are

$$x = \frac{-5 \pm \sqrt{25 - 24}}{2}$$

or

$$x_1 = \frac{-5 + 1}{2} = -2, \quad x_2 = \frac{-5 - 1}{2} = -3$$

Example 2. $3x^2 + 2x - 4 = 0$.

SOLUTION: In this case, $a = 3$, $b = 2$, $c = -4$. Therefore the solutions are

$$x = \frac{-2 \pm \sqrt{4 - (4)(3)(-4)}}{6} = \frac{-2 \pm \sqrt{52}}{6}$$

$$= \frac{-2 \pm 2\sqrt{13}}{6} = \frac{-1 \pm \sqrt{13}}{3}$$

or

$$x_1 = \frac{-1 + \sqrt{13}}{3}, \quad x_2 = \frac{-1 - \sqrt{13}}{3}$$

Example 3. Solve for y if $6x^2 + 9y^2 + x - 6y = 0$.

SOLUTION: We first put the equation in standard form,

$$9y^2 - 6y + (6x^2 + x) = 0$$

where, then, $a = 9$, $b = -6$, and $c = 6x^2 + x$. Therefore the solutions are

$$y = \frac{6 \pm \sqrt{36 - 36(6x^2 + x)}}{18}$$

$$= \frac{6 \pm \sqrt{36[1 - (6x^2 + x)]}}{18}$$

$$= \frac{6 \pm 6\sqrt{1 - 6x^2 - x}}{18} = \frac{1 \pm \sqrt{1 - 6x^2 - x}}{3}$$

or

$$y_1 = \frac{1 + \sqrt{1 - 6x^2 - x}}{3}, \quad y_2 = \frac{1 - \sqrt{1 - 6x^2 - x}}{3}$$

Because problems such as the one just given will occur later in discussing graphing, as well as in other courses in mathematics, and because students generally find them difficult, we give another example.

Example 4. Solve for y if $2x^2 + y^2 + 2xy - 2x = 0$.

SOLUTION: In standard form, the equation becomes

$$y^2 + (2x)y + (2x^2 - 2x) = 0$$

where, then, $a = 1$, $b = 2x$, and $c = 2x^2 - 2x$. Therefore the solutions are

$$y = \frac{-2x \pm \sqrt{4x^2 - 4(2x^2 - 2x)}}{2}$$

$$= \frac{-2x \pm \sqrt{4[x^2 - (2x^2 - 2x)]}}{2}$$

$$= \frac{-2x \pm 2\sqrt{x^2 - 2x^2 + 2x}}{2} = -x \pm \sqrt{2x - x^2}$$

or

$$y_1 = -x + \sqrt{2x - x^2}, \quad y_2 = -x - \sqrt{2x - x^2}$$

Exercise 6.6

In problems 1–18, solve the given equation by using the quadratic formula.

1. $x^2 - 2x - 3 = 0$
2. $x^2 - x - 2 = 0$
3. $x^2 - 7x + 12 = 0$
4. $z^2 + 2z - 2 = 0$
5. $w^2 + 2w - 5 = 0$
6. $3t^2 + 6t - 4 = 0$
7. $5u^2 - 2u - 2 = 0$
8. $3x^2 + 7x - 5 = 0$
9. $5v^2 + 8v + 2 = 0$
10. $y^2 + 5y - 41 = 0$
11. $x^2 + x - 1 = 0$
12. $\sqrt{2}\, t^2 + t - \sqrt{2} = 0$
13. $p^2 + \sqrt{3}\, p - 1 = 0$
14. $0.3a^2 - a - 1 = 0$
15. $0.2r^2 - r + 1 = 0$
16. $\frac{2}{3}b^2 - \frac{1}{6}b - 3 = 0$
17. $\frac{4}{3}x^2 + \frac{1}{2}x - 5 = 0$
18. $\frac{11}{4}x^2 - x - 1 = 0$

In problems 19–24, solve for y in terms of x.

19. $x^2 + 2xy - 4y^2 = 0$
20. $2x^2 + 3xy - 4y^2 = 0$
21. $x^2 + (3 + x)y - 2y^2 = 0$
22. $x^2 + x + 3y + y^2 - 1 = 0$
23. $2x - x^2 + xy + 2y^2 - 2 = 0$
24. $3x^2 + 4 + 4y - 2y^2 = 0$

25. Work problem 22 of Exercise 4.2.

26. Work problem 23 of Exercise 4.2.

27. Work problem 5 of Exercise 4.3.

28. The length of a rectangle is 6 feet more than the width and the area is 112 square feet. What is the perimeter of the rectangle?

29. Find two consecutive integers whose product is 272.

30. The sum of the integers $1, 2, \ldots, n$ is given by the formula $S = \frac{1}{2}n(n + 1)$. Find n if $S = 435$.

31. Each edge of a cube is increased in length by one inch. As a result the volume is increased by 37 cubic inches. What is the surface area of the cube?

32. A fenced-in rectangular field is bordered on one side by a river. The area of the field is 1,250 square yards. The total length of the fence is 100 yards. What are the dimensions of the field?

33. When the square of a number and the square of its double are added together we have 6 more than the number. What is the number?

34. The sum of two numbers is 13 and their product is 42. Find the numbers.

35. The perimeter of a rectangle is 20 inches and its area is 24 square inches. Find its dimensions.

36. A number plus its square is equal to 132. Find the number.

37. The sum of the squares of three consecutive integers is 50. Find the integers.

38. Two objects are 5,200 miles apart and are traveling at constant speed in straight-line paths which intersect at right angles. One object travels 700 mph faster than the other. If the objects collide after 4 hours, how fast is each traveling?

39. The equation $s = -16t^2 + v_0 t$ gives the distance s, in feet, that a body will be above the earth at time t, in seconds, if projected upward with an initial velocity of v_0 feet per second. If $s = 12$ feet and $v_0 = 32$ feet per second, find t.

Skip

6.7 ALTERNATIVE METHOD OF DERIVING THE QUADRATIC FORMULA

It may be worthwhile to give another proof of the important quadratic formula, especially since the method demonstrated here is similar to that employed in the derivation of formulas for the solution of the cubic (third-degree) and quartic (fourth-degree) equations. In this method, we ask if we can choose m so that the substitution of $y + m$ for x will give us a simple equation of the form $y^2 = r$.

Thus $ax^2 + bx + c = 0$ becomes $a(y + m)^2 + b(y + m) + c = 0$. Multiplying out, we have

$$ay^2 + 2amy + am^2 + by + bm + c = 0$$

Collecting terms in y^2 and y we have

$$ay^2 + (2am + b)y + am^2 + bm + c = 0$$

Now the gist of the procedure lies in the fact that if the coefficient of y can be made zero, the resulting quadratic will be easy to solve. But this is possible since m is at our disposal. Thus we wish to choose m so that $2am + b = 0$ or $m = -\dfrac{b}{2a}$. With this value of m our quadratic equation in y becomes

$$ay^2 + 0y + a\left(\frac{-b}{2a}\right)^2 + b\left(\frac{-b}{2a}\right) + c = 0$$

so that

$$ay^2 = -a\left(\frac{b^2}{4a^2}\right) + \frac{b^2}{2a} - c$$

Thus

$$y^2 = -\frac{b^2}{4a^2} + \frac{b^2}{2a^2} - \frac{c}{a} = \frac{b^2}{4a^2} - \frac{c}{a} = \frac{b^2 - 4ac}{4a^2}$$

Hence

$$|y| = \sqrt{\frac{b^2 - 4ac}{4a^2}} \qquad \text{or} \qquad y = \pm \frac{\sqrt{b^2 - 4ac}}{2a}$$

where, for the same reason as given in Section 6.6, we do not need to write $\sqrt{4a^2} = 2|a|$. Now since

$$y + m = x$$

it follows that

$$x = -\frac{b}{2a} \pm \frac{\sqrt{b^2 - 4ac}}{2a}$$

or

$$x = \frac{-b \pm \sqrt{b^2 - 4ac}}{2a}$$

6.8 EQUATIONS INVOLVING RADICALS

A technique of solving equations involving radicals is illustrated by the examples below. In these examples we follow a procedure of squaring to obtain an equation which can be solved by methods already developed. Certain precautions must be observed, however, since it is possible that $a \neq b$ while $a^2 = b^2$. For example, $-1 \neq 1$, but $(-1)^2 = 1^2$. Thus the equation $\sqrt{x} = -1$ obviously has no solution since $\sqrt{x}$ denotes that *positive* number whose square is x (Definition 6.1). However, if both sides of this equation are squared, then

$$(\sqrt{x})^2 = (-1)^2 \qquad \text{or} \qquad x = 1$$

is obtained and the number 1 is a solution to this final equation, although the original equation $\sqrt{x} = -1$ has no solution. Of course, if we check, it is clear that 1 is not a solution, since $\sqrt{1} = 1 \neq -1$.

Thus, although a check is always a desirable feature in the solution of any equation, it is a vital *necessity* in the solution of certain equations involving radicals.

Example 1. Find the solution set of the equation $\sqrt{x+7} = 2x - 1$.

SOLUTION: Assume that there is a number x such that $\sqrt{x+7} = 2x - 1$. Squaring both sides, we have

$$(\sqrt{x+7})^2 = (2x-1)^2$$

or

$$x + 7 = 4x^2 - 4x + 1$$

Thus

$$4x^2 - 5x - 6 = (4x+3)(x-2) = 0$$

and $x = 2$ or $x = -\frac{3}{4}$.

Note that at this point we have *not* proved that either $x = 2$ or $x = -\frac{3}{4}$ is a solution of our equation, but simply that *if* there is any solution, it must be either 2 or $-\frac{3}{4}$. Thus we now try 2, to find that

$$\sqrt{2+7} = 3 = 2 \cdot 2 - 1 = 3$$

so that 2 is indeed a solution of our equation. On the other hand, $-\frac{3}{4}$ is not a solution since

$$\sqrt{-\tfrac{3}{4}+7} = \sqrt{\tfrac{25}{4}} = \tfrac{5}{2} \neq 2(-\tfrac{3}{4}) - 1 = -\tfrac{5}{2}$$

Thus the solution set is $\{2\}$.

A number such as $-\frac{3}{4}$ obtained in this way is sometimes called an *extraneous* root—a term we prefer not to use since it implies that we do have a root of some kind or another.

Note also that if the equation had been given in the form

$$\sqrt{x+7} + 1 = 2x$$

and we had squared both sides, we would have obtained

$$x + 7 + 2\sqrt{x+7} + 1 = 4x^2$$

and would not have eliminated the radical. For this reason we always "isolate" a radical on one side of the equation before squaring.

Example 2. Find the solution set of the equation $\sqrt{x+5} = \sqrt{2x-1}$.

SOLUTION: Assume that there is a number x such that $\sqrt{x+5} = \sqrt{2x-1}$. Squaring both sides, we have

$$x + 5 = 2x - 1$$

and hence $x = 6$. This is a root since $\sqrt{6+5} = \sqrt{11} = \sqrt{2 \cdot 6 - 1}$. Therefore the solution set is $\{6\}$.

Sometimes a single squaring is not sufficient to reduce the problem to one that does not involve radicals.

Example 3. Find the solution set of the equation

$$\sqrt{3 - 2x} = 3 - \sqrt{2x + 2}$$

SOLUTION: Assume that there is a number x such that $\sqrt{3 - 2x} = 3 - \sqrt{2x + 2}$. Squaring both sides, we have

$$3 - 2x = 9 - 6\sqrt{2x + 2} + 2x + 2$$

Thus

$$6\sqrt{2x + 2} = 4x + 8$$

or

$$3\sqrt{2x + 2} = 2x + 4$$

Squaring both sides of this new equation, we have

$$9(2x + 2) = 4x^2 + 16x + 16$$

Hence $4x^2 - 2x - 2 = 0$ and therefore

$$2x^2 - x - 1 = (2x + 1)(x - 1) = 0$$

so that the only possible roots are 1 and $-\frac{1}{2}$. Now 1 is a root since $\sqrt{3 - 2} = 3 - \sqrt{2 + 2} = 1$, and if we substitute $-\frac{1}{2}$ for x in our given equation, we have

$$\sqrt{3 - 2(-\tfrac{1}{2})} = 3 - \sqrt{2(-\tfrac{1}{2}) + 2}$$

or

$$\sqrt{3 + 1} = 3 - \sqrt{-1 + 2} = 2$$

Therefore both 1 and $-\frac{1}{2}$ are solutions, and the solution set is $\{1, -\frac{1}{2}\}$.

Exercise 6.8

Find the solution sets of the following equations.

1. $\sqrt{4x+1} = x - 5$
2. $\sqrt{10-x} = 8 + 2x$
3. $\sqrt{2x+7} = 4 - x$
4. $x - \sqrt{2x+1} = 1$
5. $4x - \sqrt{3x+10} = -10$
6. $\sqrt{5x-11} - 1 = \sqrt{3x-8}$
7. $\sqrt{2x+3} - \sqrt{x+1} = 1$
8. $\sqrt{x-1} + \sqrt{5x-1} = 4$
9. $\sqrt{3x+4} - \sqrt{x+5} = 1$
10. $\sqrt{11-x} + \sqrt{x-2} = 3$
11. $\sqrt{2x-5} = 2 + \sqrt{x-2}$
12. $\sqrt{5x+1} = \sqrt{2x+11} + 1$
13. $\sqrt{3x+3} - \sqrt{2x-3} = \sqrt{3x-2}$
14. $\sqrt{2x+5} + \sqrt{3x+7} = \sqrt{x+6}$
15. $\sqrt{2x+5} - \sqrt{x-1} = \sqrt{x+2}$
16. $\sqrt{8x+5} - \sqrt{2x+3} = 1$
17. $2\sqrt{2x+1} - \sqrt{x-3} = \sqrt{6x+1}$
18. $\sqrt{x+3} + \sqrt{2x-3} = 6$
19. $\sqrt{3x+5} + 1 = \sqrt{6x+11}$
20. $\sqrt{3x+4} - \sqrt{x-3} - \sqrt{2x+1} = 0$

6.9 EQUATIONS OF QUADRATIC TYPE

The subject of the solution of polynomial equations is a vast one and our concern here will be with only a small portion of the topic. Those students who go on in mathematics will certainly do further work on this subject. Our work here will be mainly with such equations as are commonly met in courses in analytic geometry, calculus, physics, engineering, etc.

The first type of equations we will discuss is "essentially" quadratic. For example, the equation

$$x^4 - 5x^2 + 4 = 0$$

is actually a quartic, but, by the substitution $y = x^2$, $y^2 = x^4$ it becomes

$$y^2 - 5y + 4 = 0$$

Then

$$y^2 - 5y + 4 = (y-4)(y-1) = 0$$

so that $y = 4$ or 1. But $x^2 = y$, so $x^2 = 4$ or 1; $x = 2, -2, 1,$ or -1. The solution set is $\{2, -2, 1, -1\}$.

Similarly, the equation

$$2x + \sqrt{2x - 5} - 7 = 0$$

may be rewritten as

$$2x - 5 + \sqrt{2x - 5} - 2 = 0$$

suggesting the substitution $y^2 = 2x - 5$, $y = \pm \sqrt{2x - 5}$. Then we have to solve the equation

$$y^2 + y - 2 = 0$$

which gives $y = 1$ or -2. But $y^2 = 2x - 5$ so that

$$x = \frac{y^2 + 5}{2}$$

and hence

$$x = \frac{1 + 5}{2} = 3 \quad \text{or} \quad \frac{4 + 5}{2} = \frac{9}{2}$$

Checking our candidates for solutions (an essential step — recall Section 6.8) we first try $x = 3$ and have

$$2 \cdot 3 + \sqrt{2 \cdot 3 - 5} - 7 = 6 + \sqrt{1} - 7 = 7 - 7 = 0$$

For $\frac{9}{2}$ we have

$$2 \cdot \tfrac{9}{2} + \sqrt{2 \cdot \tfrac{9}{2} - 5} - 7 = 9 + \sqrt{9 - 5} - 7 = 9 + 2 - 7 = 4 \neq 0$$

so that 3 is a root and $\frac{9}{2}$ is not. The solution set is $\{3\}$.

Exercise 6.9

Find the solution sets of the following equations by using a substitution to transform each equation into a quadratic equation.

1. $x^4 - 2x^2 + 1 = 0$ 2. $x^4 + 3x^2 - 28 = 0$

3. $x^4 + x^2 - 30 = 0$ 4. $x^6 - 2x^3 - 15 = 0$

5. $6x^6 + x^3 - 2 = 0$ 6. $6x^6 - 17x^3 + 5 = 0$

7. $3x + \sqrt{3x - 5} - 7 = 0$ 8. $\sqrt{x - 1} + x - 7 = 0$

9. $\sqrt{5x - 4} + 5x = 10$ 10. $\dfrac{1}{x - 1} + \dfrac{1}{\sqrt{x - 1}} - 6 = 0$

11. $\dfrac{1}{3x+2} - \dfrac{2}{\sqrt{3x+2}} - 15 = 0$ **12.** $\dfrac{1}{x-2} - \dfrac{5}{\sqrt{x-2}} + 6 = 0$

13. $\dfrac{2}{(4x-5)^2} + \dfrac{3}{4x-5} + 1 = 0$ **14.** $\dfrac{3}{(2x-1)^2} - \dfrac{16}{2x-1} + 5 = 0$

15. $\dfrac{10}{(3x-2)^2} + \dfrac{13}{3x-2} - 3 = 0$ **16.** $x^2 + x + \dfrac{12}{x^2+x} = 8$

*17. $\dfrac{x^2+2}{x^2-2} + \dfrac{x^2-2}{x^2+2} - \dfrac{170}{77} = 0$ *18. $\dfrac{x^2+3}{x+1} - 3\,\dfrac{x+1}{x^2+3} = 2$

6.10 RATIONAL ROOTS

Skip

As mentioned before, the Babylonians as long ago as 2000 B.C., solved quadratic equations (although they did not recognize negative roots). It was only natural that they, and those who followed, should try next to solve cubic and quartic equations. It was not until the sixteenth century in Italy, however, that this important advance was made. That is, formulas, similar to those for quadratic equations, were developed for the cubic and quartic. However, they are too complicated to be of much practical use, and we will confine our attention here to a method for obtaining solutions to cubic and the higher degree equations which will solve such equations as commonly appear in courses in calculus, physics, and the like.

In such situations, we usually find that our equation has a rational root. This is not true in all practical cases, of course, but for textbook purposes it commonly is. Methods are given in college algebra and calculus for obtaining irrational roots to any desired degree of approximation. The method of obtaining all rational roots is based on the following theorem.

Theorem 6.2. If a rational number $\dfrac{c}{d}$ (in lowest terms) is a root of the polynomial equation $a_n x^n + a_{n-1} x^{n-1} + \cdots + a_0 = 0$, and if all of the coefficients $a_n, a_{n-1}, \ldots, a_0$ are integers, with $a_n \neq 0$, then c is a divisor of a_0 and d is a divisor of a_n.

For example, *if* the equation $6x^3 - 5x^2 - 4x - 3 = 0$ has a rational root $\dfrac{c}{d}$, then c is a divisor of 3 and d is a divisor of 6. Thus $c = \pm 1$ or

± 3 and $d = \pm 1, \pm 2, \pm 3,$ or ± 6. Hence the possible rational roots are

$$\tfrac{1}{1}, -\tfrac{1}{1}, \tfrac{1}{2}, -\tfrac{1}{2}, \tfrac{1}{3}, -\tfrac{1}{3}, \tfrac{1}{6}, -\tfrac{1}{6}, \tfrac{3}{1}, -\tfrac{3}{1}, \tfrac{3}{2}, -\tfrac{3}{2}, \tfrac{3}{3}, -\tfrac{3}{3}, \tfrac{3}{6}, -\tfrac{3}{6}$$

but this list reduces to (because, for example, $\tfrac{1}{2} = \tfrac{3}{6}$)

$$1, -1, \tfrac{1}{2}, -\tfrac{1}{2}, \tfrac{1}{3}, -\tfrac{1}{3}, \tfrac{1}{6}, -\tfrac{1}{6}, 3, -3, \tfrac{3}{2}, -\tfrac{3}{2}$$

Not all these numbers are solutions; in fact, none may satisfy the equation. The theorem merely states that *if* there is a rational root it must be one of these listed here. The easiest way to check these numbers to see whether any are actually roots is by synthetic division combined with the following theorem.

Theorem 6.3 (The Factor Theorem). Let $P(x) = 0$ be a polynomial equation. If $x - a$ is a factor of $P(x)$, then a is a root of $P(x) = 0$.

PROOF: If $x - a$ is a factor of $P(x)$, then $P(x) = (x - a)Q(x)$. But then $P(a) = (a - a)Q(a) = 0$ so that a is a root of $P(x) = 0$.

Now if $x - a$ is a factor of $P(x)$, then $P(x) = (x - a)Q(x)$, and so if $P(x)$ is divided by $x - a$, the remainder is 0. Recall that a polynomial can easily be divided by $x - a$ using synthetic division (see Section 3.3) and that the remainder is the last number in the last row. For example, to see whether $\tfrac{1}{2}$ is a root of $6x^2 - 5x^2 - 4x - 3 = 0$, we divide $6x^3 - 5x^2 - 4x - 3$ by $x - \tfrac{1}{2}$ as indicated below:

$$
\begin{array}{r|rrrr}
\tfrac{1}{2} & 6 & -5 & -4 & -3 \\
  &   & 3 & -1 & -\tfrac{5}{2} \\
\hline
  & 6 & -2 & -5 & \!\!\!\!\| -\tfrac{11}{2}
\end{array}
$$

Since the remainder is $-\tfrac{11}{2} \neq 0$, $\tfrac{1}{2}$ is not a root. Let us try $\tfrac{3}{2}$.

$$
\begin{array}{r|rrrr}
\tfrac{3}{2} & 6 & -5 & -4 & -3 \\
  &   & 9 & 6 & 3 \\
\hline
  & 6 & 4 & 2 & \!\!\!\!\| \; 0
\end{array}
$$

Since the remainder is 0, $x - \tfrac{3}{2}$ is a factor of $6x^3 - 5x^2 - 4x - 3$ and $\tfrac{3}{2}$ is a root of the equation $6x^3 - 5x^2 - 4x - 3 = 0$.

When each of the numbers listed is checked as to whether or not it is a solution the student will find that $x = \tfrac{3}{2}$ is the *only* rational root of the equation

$$6x^3 - 5x^2 - 4x - 3 = 0$$

Example. The only possible rational roots of the equation

$$5x^3 - 2x^2 - 5x - 10 = 0$$

are ±10, ±5, ±2, ±1, $\pm\frac{1}{5}$, $\pm\frac{2}{5}$, but since, as the student should check, *none* of these numbers satisfies the equation, the equation has no rational roots.

Exercise 6.10

Find all of the rational roots of the following equations.

1. $x^4 - x^3 - 7x^2 + x + 6 = 0$ 2. $x^3 + x^2 - 5x - 2 = 0$
3. $x^3 - 2x^2 + x - 2 = 0$
4. $x^4 - 2x^3 - 13x^2 + 14x + 24 = 0$
5. $2x^3 + x^2 - 4x + 6 = 0$ 6. $x^4 - 3x^3 - x + 2 = 0$
7. $2x^3 - x^2 - 4x + 2 = 0$ 8. $3x^3 - 2x^2 - 9x + 6 = 0$
9. $2x^3 - 5x^2 - 10x + 25 = 0$ 10. $2x^3 - 5x^2 - 12x + 30 = 0$
11. $2x^3 - x^2 - 6x - 10 = 0$ 12. $3x^3 + 2x^2 - 3x - 2 = 0$

6.11 FRACTIONAL EQUATIONS

Fractional equations are equations that involve rational algebraic expressions. Thus

$$\frac{4x - 7}{x - 2} = 3 + \frac{1}{x - 2} \quad \text{and} \quad \frac{x(x + 1)}{2} + \frac{x + 1}{x} = \frac{(x + 1)^2}{x}$$

are fractional equations. Such equations, like equations involving radicals, are solved by finding a polynomial equation whose roots are the only possible roots of the given fractional equation. Again, we find that the method used will sometimes produce a root for the final equation that is not a root of the original equation. It is possible, however, to anticipate such an occurrence, as we will show.

Consider first

Example 1. Find the solutions of the equation $\dfrac{4x - 7}{x - 2} = 3 + \dfrac{1}{x - 2}.$

SOLUTION: Assume that there is a number x such that

$$\frac{4x - 7}{x - 2} = 3 + \frac{1}{x - 2}$$

We now multiply both sides of the equation by $x - 2$ to obtain

$$4x - 7 = \left(3 + \frac{1}{x - 2}\right)(x - 2)$$

Thus

$$4x - 7 = 3(x - 2) + 1 = 3x - 5$$

and hence $x = 2$.

We have shown that *if* x is a solution of the equation

$$\frac{4x - 7}{x - 2} = 3 + \frac{1}{x - 2},$$

then $x = 2$. But if we substitute $x = 2$ in the right-hand member of the equation we obtain

$$\frac{4 \cdot 2 - 7}{2 - 2} = \frac{1}{0}$$

and we know that we cannot divide by zero. Hence 2 is *not* a solution.

Before we analyze the process which led to the conclusion that 2 was a possible solution to our equation, let us see exactly why our equation has no solution. To do this, we note that

$$3 + \frac{1}{x - 2} = 3 \cdot \frac{x - 2}{x - 2} + \frac{1}{x - 2} = \frac{3(x - 2) + 1}{x - 2} = \frac{3x - 5}{x - 2}$$

and hence that the original equation is equivalent to

$$\frac{4x - 7}{x - 2} = \frac{3x - 5}{x - 2} \tag{1}$$

Now in Section 1.9 (Definition 1.8) we stated that two fractions, $\frac{a}{b}$ and $\frac{c}{d}$, are equal if and only if $ad = bc$. Thus (1) holds, providing that $x \neq 2$, if and only if

$$(x - 2)(4x - 7) = (x - 2)(3x - 5) \tag{2}$$

holds. But, since $x \neq 2$, $x - 2 \neq 0$, and we can divide both sides of (2) by $x - 2$ (Axiom 2, Section 5.2) and have

$$4x - 7 = 3x - 5$$

which gives $x = 2$, a contradiction. In other words, the only possible solution is a number which we knew in advance could not be a solution, and hence there are no solutions to our given equation.

Generally, what we do to solve a fractional equation is to multiply both sides of the equality by the least common multiple of the denominators.[4] Thus in solving

$$\frac{1}{x-2} + \frac{4}{x+2} = \frac{3}{x-1} \tag{3}$$

we would multiply through by $(x-1)(x-2)(x+2)$. Since we are multiplying through by a common multiple of the denominators, what we obtain will certainly not involve fractions. On the other hand, if we use a common multiple other than the least common denominator, we may introduce additional roots. Thus, for example, if we were to multiply both sides of (3) by $(x-4)(x-1)(x-2)(x+2)$ we would have

$$(x-4)[(x-1)(x+2) + 4(x-1)(x-2)] = (x-4)[3(x-2)(x+2)]$$

Clearly, this last equation has the root $x = 4$, whereas 4 is not a root of (3).

Now, multiplying both sides of an equation by the same number is, by the Axiom 2 just referred to, an operation which yields an equivalent equation *provided* that the multiplier is not zero. Hence, when we are multiplying through by a quantity such as $x - a$ we are assuming that $x \neq a$ and any root of the form $x = a$ subsequently obtained must be rejected. Thus our multiplication above by $(x-4)(x-1)(x-2)$ $(x+2)$ assumed that $x \neq 4$, and we must reject $x = 4$ as a root of our original equation. Similarly, in Example 1, we multiplied through by $x - 2$, and hence cannot possibly consider $x = 2$.

Summarizing this discussion, we are led to the following method. To solve a fractional equation, make a list of all numbers which

---

[4] Assuming that all the fractions involved are "simple" (i.e., not complex) fractions. Thus the equation

$$\frac{1}{1 + \frac{1}{x}} = \frac{1}{x}$$

does not fall under this category. Since complex fractions may always be reduced to simple fractions (Section 3.7), this is no real restriction on our method.

make the denominators of the fractions in the equation equal to zero. Now assume that the unknown quantity for which we are solving is not a number on this list and multiply each side of the equation by the least common denominator of all the fractions in the equation. Then every solution of the resulting polynomial equation which is not on the list is a solution of the original equation.

The following two additional examples illustrate the application of this rule.

Example 2. Solve the equation $\dfrac{4}{x+1} + \dfrac{3}{x} = 2.$

SOLUTION: Our list consists of -1 and 0. We now multiply both sides of our equation by $x(x+1)$ to obtain

$$4x + 3(x + 1) = 2x(x + 1)$$

and then

$$4x + 3x + 3 = 2x^2 + 2x$$

We thus have to solve the quadratic equation

$$2x^2 - 5x - 3 = 0$$

Since $2x^2 - 5x - 3 = (2x + 1)(x - 3)$ our possible roots are $x = -\frac{1}{2}$ and $x = 3$. Since neither of these numbers is on our list, the solutions of the given equation are $-\frac{1}{2}$ and 3. (Assuming, of course, no mistake in computation.) The fact that the numbers obtained are not on the list does not show that all our work is correct, and a check is, as always, desirable. Thus, for example, we check that $-\frac{1}{2}$ is a solution by observing that

$$\frac{4}{-\frac{1}{2}+1} + \frac{3}{-\frac{1}{2}} = \frac{4}{\frac{1}{2}} - \frac{3}{\frac{1}{2}} = 8 - 6 = 2$$

Example 3. Solve for x in the equation

$$\frac{1}{a} + \frac{1}{b} + \frac{1}{x} = \frac{1}{x + a + b} \qquad (a \neq 0, b \neq 0, a + b \neq 0)$$

SOLUTION: Our list consists of 0 and $-(a + b)$. Multiplying both sides of the equation by $abx(x + a + b)$ we have

$$bx(x + a + b) + ax(x + a + b) + ab(x + a + b) = abx$$

and hence

$$bx^2 + abx + b^2x + ax^2 + a^2x + abx + abx + a^2b + ab^2 = abx$$

Collecting together the various powers of x we have

$$(a + b)x^2 + (a^2 + 2ab + b^2)x + a^2b + ab^2 = 0$$

or

$$(a + b)x^2 + (a + b)^2x + ab(a + b) = 0$$

Since our assumption is that $a + b \neq 0$, we may divide each term by $a + b$ to obtain

$$x^2 + (a + b)x + ab = 0$$

Since $x^2 + (a + b)x + ab = (x + a)(x + b)$, we have, as possible solutions, $x = -a$ and $x = -b$. Since these are not on our list they are roots, unless we have made a computational error. That this is not the case is shown by the fact that

$$\frac{1}{a} + \frac{1}{b} + \frac{1}{-a} = \frac{1}{b} = \frac{1}{-a + a + b}$$

and

$$\frac{1}{a} + \frac{1}{b} + \frac{1}{-b} = \frac{1}{a} = \frac{1}{-b + a + b}$$

Exercise 6.11

In problems 1–39, solve the given equations.

1. $\dfrac{x}{5} - \dfrac{1}{4} = \dfrac{2x}{15} + \dfrac{1}{12}$

2. $\dfrac{3x - 2}{4} + \dfrac{2x}{3} = \dfrac{x}{2}$

3. $\dfrac{3x}{8} - \dfrac{2x}{9} - \dfrac{11}{12} = 0$

4. $\dfrac{2x - 5}{2} + \dfrac{3x}{4} = \dfrac{4x}{3}$

5. $\dfrac{4x + 1}{6} + \dfrac{3x}{4} = \dfrac{2x - 4}{3}$

6. $\dfrac{x + 5}{9} - \dfrac{2x + 2}{3} = x + 3$

7. $\dfrac{1}{x} + \dfrac{2}{x + 4} = \dfrac{5}{3x}$

8. $\dfrac{3}{4x} = \dfrac{1}{12x} + \dfrac{2}{5x + 2}$

9. $\dfrac{x}{x + 3} + \dfrac{1}{x + 2} = 1$

10. $\dfrac{1}{x + 1} + \dfrac{1}{x + 4} = \dfrac{2}{x + 2}$

11. $\dfrac{x}{x - 4} - \dfrac{4}{2x - 1} = 1$

12. $\dfrac{1}{x + 2} = \dfrac{2}{x - 1} - \dfrac{1}{x - 2}$

13. $\dfrac{3x-8}{x-2}+\dfrac{4x+3}{2x-1}=5$

14. $\dfrac{2x+1}{x-2}-\dfrac{x-1}{x+2}=\dfrac{x^2+x+7}{x^2-4}$

15. $\dfrac{4x+3}{3x-2}-\dfrac{3x-2}{2x+5}+\dfrac{x^2-10x+3}{6x^2+11x-10}=0$

16. $\dfrac{3x+1}{2x^2-13x+20}+\dfrac{x+2}{x+4}-\dfrac{2x-6}{2x-5}=0$

17. $\dfrac{x}{x-1}+\dfrac{x-1}{x-2}-\dfrac{2x-4}{x-3}=0$ 18. $\dfrac{x+6}{x+5}-\dfrac{2x-1}{x-4}+\dfrac{x+4}{x-2}=0$

19. $\dfrac{2x-3}{2(x-3)}+\dfrac{x+3}{x+2}-\dfrac{4x+9}{2(x+1)}=0$

20. $\dfrac{2x-4}{x-3}-\dfrac{x+3}{x-2}-\dfrac{x-1}{x+2}=0$

21. $\dfrac{2x+8}{3x+2}=\dfrac{1+x}{7}$

22. $\dfrac{3}{x}-\dfrac{2}{1-x}=2$

23. $\dfrac{2}{x-1}+1=\dfrac{4-2x}{x-1}$

24. $\dfrac{1}{x-2}+\dfrac{1}{x-3}=\dfrac{1}{x^2-5x+6}$

25. $\dfrac{2x}{2x-5}-\dfrac{15}{2x+3}=2$

26. $\dfrac{6x}{3x+8}+2x=12$

27. $\dfrac{8}{x}-\dfrac{5}{x+3}=3$

28. $\dfrac{x-1}{x+2}+1=\dfrac{x-1}{x+1}+\dfrac{x^2+1}{x^2+3x+2}$

29. $\dfrac{3x}{x+2}+\dfrac{2}{x}=\dfrac{5}{2}$

30. $\dfrac{x+5}{x-3}-\dfrac{x+6}{2x}=\dfrac{1}{6}$

31. $\dfrac{3(x^2+4)}{x-3}=3+2x+\dfrac{13x}{x-3}$ 32. $\dfrac{x}{x-2}+\dfrac{7}{3}=\dfrac{35}{x-4}$

33. $\dfrac{x-1}{x-2}+\dfrac{x-3}{x+1}=\dfrac{x-1}{x+1}+\dfrac{x^2-x+1}{x^2-x-2}$

34. $\dfrac{7}{1+2x}-\dfrac{5x}{2x^2+3x+1}=\dfrac{1}{3}$ 35. $\dfrac{5(5-4x)}{2x+5}=5(1-x)+\dfrac{12x^2}{2x+5}$

36. $\sqrt{\dfrac{x+41}{x-7}}=\dfrac{x+6}{x-6}$

37. $\sqrt{\dfrac{x+3}{x-1}}-6\sqrt{\dfrac{x-1}{x+3}}=1$

38. $\sqrt{\dfrac{7x+1}{4}}=\dfrac{5x-7}{6}$

39. $2+\sqrt{\dfrac{x}{x-10}}=\dfrac{x-4}{x-14}$

Solve the systems of equations in problems 40–45 by first reducing any fractional equation to an equivalent linear equation.

40. $\dfrac{x+y}{3}+\dfrac{y}{2}=4$

$3x-y=2$

41. $\dfrac{2x-y}{3}+\dfrac{x+y}{4}=2$

$3x+2y=24$

42. $\dfrac{x - 2y + 1}{2x - 5y + 2} = 2$

$x - 2y = 1$

43. $\dfrac{2x - y + 3}{3x - y + 2} = \dfrac{1}{2}$

$7x - 2y = 2$

44. $\dfrac{2x - 2y}{2} + \dfrac{x - y}{3} = -\dfrac{4}{3}$

$\dfrac{x + 2y}{4} + \dfrac{3x + 2y}{3} = 6$

45. $\dfrac{3x - 4y}{5} - \dfrac{2x - 3y}{3} = \dfrac{1}{3}$

$\dfrac{2x + 5y}{3} - \dfrac{2x - y}{6} = \dfrac{7}{6}$

46. Work problem 7 of Exercise 4.3.

47. Work problem 8 of Exercise 4.3.

48. Work problem 11 of Exercise 4.3.

49. Work problem 12 of Exercise 4.3.

50. Work problem 19 of Exercise 4.3.

51. Work problem 20 of Exercise 4.3.

52. Work problem 21 of Exercise 4.3.

53. Work problem 22 of Exercise 4.3.

54. Work problem 24 of Exercise 4.3.

55. Work problem 28 of Exercise 4.3.

56. Work problem 32 of Exercise 4.3.

57. Work problem 33 of Exercise 4.3.

58. Work problem 34 of Exercise 4.3.

59. Work problem 35 of Exercise 4.3.

hardest

60. Twice a number less one half of it is 18. Find the number.

61. The difference between two numbers is 24 and their quotient is $\frac{3}{2}$. Find the numbers.

62. A swimming pool has two intake pipes. One will fill the pool in 10 hours and the other in 15 hours. If both pipes are open, how long will it take the pool to fill?

63. Joe polished one third of his car in 2 hours. He was then joined by Bob and they finished the car in 2 more hours. How long would it have taken Bob to polish the car alone?

64. Three girls were on a decoration committee for a club supper. Each of them could do the decorations alone in 4 hours. Jean started the work at 4:00 P.M., Jane joined her at 4:30, and Janet came at 5:00. At what time were the decorations finished?

65. A pilot flew from his home field to another at the rate of 180 mph and returned at the rate of 150 mph. If the outward trip required 1 hour less time than the return trip, find the distance between the fields.

66. A passenger train left town A for town B 140 miles away at the same time that a freight left B for A and the trains traveled on parallel tracks. If the average speed of the passenger train was 64 mph and that of the freight was 48 mph, how far were the trains from A when they met?

67. How many quarts of distilled water must be added to 5 quarts of a solution that is 30 per cent nitric acid in order to obtain a 20 per cent solution?

68. A 6-gallon radiator contains a solution that is 25 per cent antifreeze. How much must be drained off and replaced by pure antifreeze in order to obtain a solution that is $33\frac{1}{3}$ per cent antifreeze?

69. After 20 pounds of milk containing 3 per cent butterfat was mixed with 30 pounds containing 4 per cent, a sufficient amount of cream that was 30 per cent butterfat was added to produce cereal cream that was 15 per cent butterfat. How many pounds of cream were added?

70. A chemist mixed 20 cubic centimeters of a 20 per cent nitric acid solution with 30 cubic centimeters of a 15 per cent solution. He used a portion of the mixture and replaced it with distilled water. The new solution tested 13.6 per cent nitric acid. How much of the original mixture was used?

71. Two blocks of alloy containing 20 and 40 per cent silver, respectively, were melted together in order to produce 30 pounds of an alloy that was 35 per cent silver. How much did each block weigh?

72. A lot containing 7,500 square feet sold for $30 per front foot. Another lot two thirds as deep containing 6,000 square feet sold for $25 per front foot. Find the dimensions of each lot if the two together sold for $3,000.

The following problems arise in analysis of functions in the calculus. In problems 73–76, solve for x.

73. $-\dfrac{1}{x^2} + 8x = 0$

74. $\dfrac{-4x^2}{x^2+1} + x^2 + 1 = 0$

75. $\dfrac{6x(5x+2) - 5(3x^2+1)}{(5x+2)^2} = 0$

76. $\dfrac{x^2}{2\sqrt{1+2x}} + 2x\sqrt{1+2x} = 0$

In problems 77–80, solve for y' in terms of x and y.

77. $2yy' = \dfrac{4-x}{x^2}$

78. $\dfrac{y - xy'}{y^2} + \dfrac{xy' - y}{x^2} = 0$

79. $2yy' + \dfrac{y - xy'}{y^2} - 3(xy' + y) = 0$

80. $3x^2 + \dfrac{1}{2\sqrt{x}} + \dfrac{5}{2}y\sqrt{y}\,y' + 3y^2y' = 0$

FUNCTIONS, GRAPHS, AND INEQUALITIES

7

7.1 INTRODUCTION

As an aid in discussing the real number system in Chapter 1, we pictured a straight line on which the real numbers were located; i.e., to every real number there corresponded a point on the line, and to every point on the line there corresponded a real number. This one-dimensional geometry was extended to a two-dimensional geometry by the introduction of Cartesian coordinates in Chapter 5, where we used a reference system composed of two perpendicular lines: the horizontal line was called the x-axis and the vertical line, the y-axis. To every ordered pair of real numbers there corresponded a point in the Cartesian plane, and to every point in the plane there corresponded an ordered pair of real numbers. We used this system to picture (linear) relations between two variables x and y such as

$$y = x + 3, \quad 2x + 3y = 10, \quad \text{and} \quad y = 5 \; (= 5 + 0 \cdot x)$$

It is the aim of this chapter to generalize the idea of a relationship between two variables and, likewise, to generalize the idea of the picture or graph of this relationship.

7.2 FUNCTIONS

In graphing linear equations in Chapter 5 we found that the easiest method of determining pairs of numbers to be plotted was to solve the equation for one of the variables in terms of the other. Let us say that

180

Read

we decide to solve for y in terms of x. Then we choose arbitrarily values for x and obtain corresponding values of y. Thus, for example, from the equation $2x + y - 5 = 0$, we get $y = 5 - 2x$. If $x = 0$, $y = 5$; if $x = 1$, $y = 3$; etc.

Analyzing more closely the relationship between x and y in our example, we see that we might say that the equation actually provides a *rule* by which, given any number x, we can find a corresponding number, y. When a number, y, is determined in this fashion by a choice of a number, x, we say that y is a **function of x**. Similarly, if $y = \sqrt{x}$, y is a function of x since, given x, y is determined. Thus, if $x = 4$, $y = 2$; if $x = 9$, $y = 3$; etc. Here, of course, not all values of x are permissible since, for example, $x = -1$ does not give a real number, to which we are restricted in this chapter. Likewise, if $y = \dfrac{1}{x}$, y is a function of x even though the value of $x = 0$ is not permissible here. The question of permissible values will be considered further in Section 7.6.

In all of these examples, y has been related to x by being equal to some algebraic expression in x. This need not be the case. For example, if y is equal to 0 if x is a rational number, and equal to 1 if x is an irrational number, y is then a function of x since y is determined as soon as x is given. Thus, if $x = 1$, $y = 0$; if $x = \frac{5}{3}$, $y = 0$; if $x = \sqrt{2}$, $y = 1$; if $x = \pi$, $y = 1$.

If we want to indicate that y is some function of x without indicating specifically by what rule we determine y from a given x, we write $y = f(x)$, which is read "y equals f of x" and does *not* mean f times x. (A more accurate reading of $y = f(x)$ would be, of course, "y is a *function f of x*," but the former phraseology is the one commonly used.)

Bringing these ideas together, we have

Definition 7.1. A **function** f is a rule which associates with every permissible choice of a number, x, a unique number, y. Then we say y is a function of x and write

$$y = f(x)$$

The values which x is permitted to take on depend upon the particular function being analyzed; the set of these values is known as the **domain of definition** of the function. In the functions that we shall study, the values that x is permitted to take on are those real numbers for which the value of $f(x)$ is also real. The corresponding set of values which y assumes comprise the **range** of the function. The notation $y = f(x)$ is a general statement that x and y are related by some rule.

Example 1. If $y = f(x) = 2x + 3$, then y is a function of x with domain and range both the set of real numbers.

Example 2. If $y = f(x) = \sqrt{x - 4}$, then y is a function of x with domain the set of all real numbers greater than or equal to 4 and range the set of all non-negative real numbers.

Example 3. If $y = f(x)$ when f is the rule which says that y is equal to 0 if x is a rational number, and equal to 1 if x is an irrational number, then y is a function of x with domain the set of all real numbers and range $\{0, 1\}$.

Although $f(x)$ is a popular notation for the functional relationship, other letters are often used. Thus $y = F(x)$, $y = G(x)$, $y = \phi(x)$ (read "y equals phi of x"), etc., are often used. Also, of course, we do not always use the letters x and y. In physics, for example, the equation for the distance traveled by a freely falling body starting from rest is customarily given by

$$s = 16t^2$$

where s is the distance in feet and t is the time in seconds. Here we would say that the distance traveled, s, is a function of the time, t, or

$$s = f(t)$$

Again, more than two quantities might be involved, as in the case of the relationship between the flow of current through an electric circuit with a certain voltage and a certain resistance,

$$I = \frac{E}{R}$$

where I is the current in amperes, E the voltage in volts, and R the resistance in ohms. Here we would say that the current I is a function of the voltage E and the resistance R, or

$$I = f(E, R)$$

read "I equals f of E and R."

Also a relationship between two variables, x and y, can be written in the form

$$f(x, y) = 0$$

by putting all of the terms on the left of the equals sign, leaving zero on the right as in $x + y - 5 = 0$. In this case, x and y are still related by some rule, even though the rule is not explicitly stated. Such a rule is termed an **implicit function** for it is *implied* that y is a function of x. Thus, in our example, $y = 5 - x$ is the *explicit* form of the function whose implicit form is $x + y - 5 = 0$.

Exercise 7.2

In problems 1–10, write an algebraic expression for $f(x)$ or $f(x, y)$.

1. Let f be the rule which associates with every number another number which is 5 greater than the given number.

2. Let f be the rule which associates with every number the square of that number.

3. Let f be the rule which associates with every number greater than 5 the square root of 5 less than the number.

4. Let f be the rule which associates with every nonzero number its reciprocal.

5. Let f be the rule which associates with every number different from 1 or -1 the reciprocal of one less than the square of the number.

6. Let f be the rule which associates with every pair of numbers the sum of the squares of the numbers.

7. Let f be the rule which associates with every pair of numbers the square of the sum of the numbers.

8. Let f be the rule which associates with every positive number the number 5.

9. Let f be the rule which associates with every number the absolute value of the number.

10. Let f be the rule which associates with every number the absolute value of one less than the number.

In problems 11–20, state the rule, f, in words.

11. $f(x) = x + 3$

12. $f(x) = \sqrt{x}$

13. $f(x) = (x + 1)^2$

14. $f(x) = |x| + 1$

15. $f(x) = |x + 1|$

16. $f(x, y) = 2(x + y)$

17. $f(x, y) = x^2 - y^2$

18. $f(x) = 1$ for $x \geqslant 0$, $f(x) = -1$ for $x < 0$
19. $f(x) = x$ for $x > 0$, $f(x) = 1$ for $x = 0$, $f(x) = -1$ for $x < 0$
20. $f(x) = |x|$ for $x < 0$, $f(x) = |x| + x$ if $x \geqslant 0$

In problems 21–30 give the domain and range of f.

21. $f(x) = x$

22. $f(x) = \sqrt{x - 3}$

23. $f(x) = \sqrt{x + 3}$

24. $f(x) = \dfrac{1}{x^2}$

25. $f(x) = 1$

26. $f(x) = 1$ for $x > 0$

27. $f(x) = |x|$

28. $f(x) = \dfrac{1}{x^2 - 4}$

29. $f(x) = \dfrac{1}{\sqrt{x^2 - 4}}$

30. $f(x) = \sqrt{1 - x^2}$

domain = all real numbers ≥ 3

7.3 NOTATION

Let us return to our functional notation, $y = f(x)$. Remembering that this defines a rule which associates with every choice of a number, x in the domain of the function, some unique number, y, let us choose $x = a$; then we associate with the particular value a the particular value for y which is found according to the rule. This value of y obtained when $x = a$ is denoted by the symbol $f(a)$. Thus, if $y = f(x)$, then when $x = a$ we have $y = f(a)$; when $x = 2$, $y = f(2)$; when $x = t^2 - 1$, $y = f(t^2 - 1)$; etc.

Example. Let $y = f(x) = x^2 - 3x + 5$. Then, when

$$
\begin{aligned}
x &= 0, & y &= f(0) = 0^2 - 3(0) + 5 = 5 \\
x &= 2, & y &= f(2) = 2^2 - 3 \cdot 2 + 5 = 3 \\
x &= -1, & y &= f(-1) = (-1)^2 - 3(-1) + 5 = 9 \\
x &= a, & y &= f(a) = a^2 - 3a + 5 \\
x &= t - 1, & y &= f(t - 1) = (t - 1)^2 - 3(t - 1) + 5 \\
& & &= t^2 - 5t + 9
\end{aligned}
$$

Functional notation occurs in many combinations such as $f(a) + f(b)$, $f(x + h) - f(x)$, $\dfrac{f(x)}{g(x)}$, etc. For example, let us find $f(2) - f(-1)$ where $y = f(x)$ is defined in the example above. We have already computed $f(2) = 3$ and $f(-1) = 9$. Thus $f(2) - f(-1) = 3 - 9 = -4$.

To find $f(x + h) - f(x)$ for this function, let us first find $f(x + h) = (x + h)^2 - 3(x + h) + 5 = x^2 + 2xh + h^2 - 3x - 3h + 5$. Then $f(x + h) - f(x) = x^2 + 2xh + h^2 - 3x - 3h + 5 - (x^2 - 3x + 5) = 2xh + h^2 - 3h$.

Exercise 7.3

1. If $f(x) = x + 3$, find $f(0), f(-3), f(-1), f(1)$.
2. If $f(x) = 3x - 2$, find $f(0), f(1), f(2), f(-1), f(-2)$.
3. If $g(x) = x^2$, find $g(0), g(1), g(-1), g(\frac{1}{2}), g(4)$.
4. If $h(x) = 2x^2 - 4$, find $h(0), h(1), h(2), h(\frac{1}{2}), h(-\frac{1}{2})$.
5. If $f(x) = 2x^2 - 3x + 5$, find $f(0), f(1), f(-1), f(a), f(b)$.
6. If $f(x) = x^x$, find $f(1), f(2), f(3), f(4)$.
7. If $h(x) = \dfrac{x^2 - 1}{x + 1}$, find $h(1), h(0), h(x - 1), h(a), h(x^2)$.
8. If $F(x) = x^2$, find $F(y), F(0), F(x + 1), F(x + h)$.
9. If $f(x) = 2x^2 - 3x$, find $\dfrac{f(1)}{f(-1)}, \dfrac{f(a + 1)}{f(a)}, \dfrac{f(x + 1)}{f(x - 1)}$.
10. If $f(x) = 2x^2 - 4x$ and $g(x) = x^2 - 4$, find $\dfrac{f(x)}{g(x)}, \dfrac{f(1)}{g(1)}, \dfrac{f(1)}{g(-1)}, \dfrac{f(-1)}{g(1)}$, $\dfrac{f(0)}{g(0)}$.

7.4 GRAPHS OF FUNCTIONS

Although there are many ways of defining a function, the one most frequently used in elementary mathematics and its applications is by an equation. An equation, $y = f(x)$, gives us a rule for determining the value for y in the range of f for each choice of a value for x in the domain of f. Let us choose a value for x, say x_0, and find the corresponding value for y, $y_0 = f(x_0)$. We could list these two values as an ordered pair of numbers such that $y_0 = f(x_0)$. Thus the rule of correspondence which is defined by the equation $y = f(x)$ gives us a set of ordered pairs of numbers. In fact, we may take the point of view that a function, f, is a set of ordered pairs [1] and write, for example,

---

[1] A function can be defined as a set f of ordered pairs, (x, y), such that no two ordered pairs have the same first component. We cannot have (x_0, y_1) in f and (x_0, y_2) in f with $y_1 \neq y_2$, since Definition 7.1 requires that a *unique* y be associated with x_0.

$$f = \{(x, y) \mid y = f(x)\}.$$

Now we can also consider the set $\{(x, y) \mid y = f(x)\}$ as the solution set of the equation $y = f(x)$. (See Section 5.4.) Solution sets of linear equations were graphed in Chapter 5 by locating all the points in the Cartesian plane that correspond to ordered pairs satisfying the equation. It was proved in Section 5.7 (Theorem 5.3) that the only equations that have a straight line as their graph are those of the form $y = ax + b$. Therefore, in all other cases, we must plot more than two points in order to draw an accurate picture. How many points and which ones depend upon the function, and will become clearer with experience. It is advantageous to tabulate the ordered pairs of numbers to be plotted, as in the following examples.

Example 1. Graph the function f defined by $f(x) = x^2 - 3x + 5$.

x	$f(x)$
-1	9
0	5
1	3
2	3
3	5
4	9

Figure 7.1 shows a smooth curve joining the points whose coordinates are given in the table. We would perhaps have had a more accurate graph had we taken some intermediate points between those that are tabulated, but it would be impossible to take all points because even in a finite interval (say from 0 to 1) there are infinitely many points. On the other hand, if we had omitted the points corresponding to (2, 3) and (3, 5) we might have drawn, erroneously, the graph shown as a dotted line in Figure 7.1. By the use of calculus, it is often possible to obtain an accurate graph with a very few points or, in any event, to get some idea of how many points are needed to obtain a certain desired degree of accuracy. Here we can only say that one can never err in having too many points, but that it is easy to err by having too few.

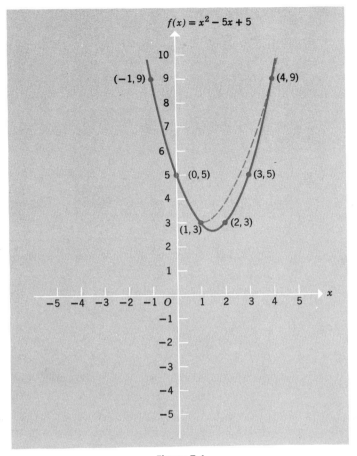

$$f(x) = x^2 - 5x + 5$$

(-1, 9)
(4, 9)
(0, 5)
(3, 5)
(1, 3)
(2, 3)

Figure 7.1

Example 2. Graph $f(x) = x^3 - 5x + 1$ (Figure 7.2).

x	$f(x)$
-3	-11
-2	3
-1	5
0	1
1	-3
2	-1
3	13

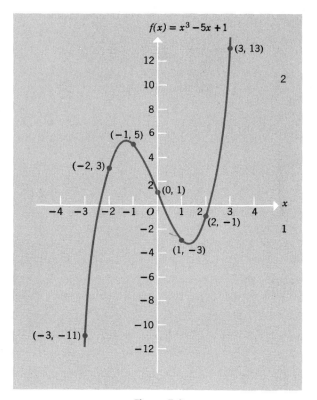

Figure 7.2

Notice that in Figure 7.2 the units on the x-axis are not the same length as the units on the $f(x)$-axis. This is for the purpose of exhibiting the important features of the graph without making it too large. When to use this trick will also become clearer with practice.

Graphs are frequently used to obtain approximate solutions to equations. Thus, for example, if we ask for the solutions to the equation $x^3 - 5x + 1 = 0$ we have only to observe where the graph of $f(x) = x^3 - 5x + 1$ crosses the x-axis since, at these points, $f(x) = x^3 - 5x + 1 = 0$. Thus from Figure 7.2 we see that the roots of $x^3 - 5x + 1 = 0$ are, approximately, -2.3, 0.2, and 2.1.

If the curve does not cross the x-axis, we have no real number solutions of the corresponding equation. This is the case in Figure 7.1, and hence we can conclude that the equation $x^2 - 3x + 5 = 0$ has no real number solutions (as the student may check by the quadratic formula).

Exercise 7.4

1. Locate the following pairs of numbers as points on a graph: (x, x^2) where $x = 0$, (x, x^2) where $x = -1$, (x, x^2) where $x = 1$, (x, x^2) where $x = -2$, (x, x^2) where $x = 2$.

2. Locate the following pairs of numbers as points on a graph: $(x, x^2 - 2x + 1)$ where $x = 0$, $(x, x^2 - 2x + 1)$ where $x = 1$, $(x, x^2 - 2x + 1)$ where $x = 2$, $(x, x^2 - 2x + 1)$ where $x = 3$, $(x, x^2 - 2x + 1)$ where $x = -1$, $(x, x^2 - 2x + 1)$ where $x = -2$.

3. Locate six pairs of numbers as points on a graph by choosing six values for x in $(x, x^2 + 3)$.

4. Locate at least ten pairs of numbers as points on a graph by choosing values of x between -3 and 3 in $(x, x^2 - x)$.

5. Locate at least ten pairs of numbers as points on a graph by letting x take on values between -3 and 3 in $(x, \frac{1}{2}x^2)$.

6. Locate at least nine pairs of numbers as points on a graph by choosing values of x between -5 and 0 in $(x, 2x^2 + 5x)$.

In problems 7–16, graph the given functional relations as (smooth) curves after first making a table, as in the examples of the text, with at least ten entries.

7. $f(x) = 2x^2 - 5$ 　　　　　　**8.** $f(x) = \frac{1}{2}x^2$

9. $f(x) = x^2 + 2x - 3$ 　　　**10.** $f(x) = -x^2$

11. $f(x) = -x^2 + 2x$ 　　　　**12.** $f(x) = -x^2 + 3$

13. $f(x) = x^3$ 　　　　　　　**14.** $f(x) = x^2 + 3x$

15. $f(x) = x^3 + 3x^2$ 　　　　**16.** $f(x) = -x^3 + 2$

In problems 17–24, find, approximately, the real roots, if any, of the given equations by the graphical method described in the text.

17. $f(x) = x^2 - 3x - 5 = 0$ 　　　**18.** $f(x) = x^2 - 5x + 2 = 0$

19. $f(x) = x^2 - 4x + 2 = 0$ 　　　**20.** $f(x) = 2x^2 + 3x + 5 = 0$

21. $f(x) = 3x^2 + 2x + 2 = 0$ 　　**22.** $f(x) = x^3 + 2x - 5 = 0$

23. $f(x) = x^3 + 3x - 20 = 0$ 　　**24.** $f(x) = x^3 + 2x^2 - 7 = 0$

Problems 25–29 occur in the calculus. Obtain your answers in as simple a form as possible.

25. If $f(x) = x^2 + 3x - 5$, find $f(x + h) - f(x)$.

26. If $f(x) = x - \dfrac{1}{x}$, find $f(3 + h) - f(3)$.

27. If $g(x) = \dfrac{x-2}{x-1}$, find $g(2+h) - g(2)$.

28. If $g(x) = \dfrac{1}{x^2}$, find $g(x+h) - g(x)$.

29. If $F(x) = \dfrac{1-x}{x}$, find $F(x+h) - F(x)$.

30. If $y = u^2$, where $u = x + 3$, find y as a function of x.

31. If $y = u^2 + 3$ and $u = x - 1$, find y as a function of x.

32. If $y = \dfrac{u+1}{u-1}$ and $u = \dfrac{1}{x}$, find y as a function of x.

33. If $y = \dfrac{u-1}{u}$ and $u = \dfrac{x-1}{x}$, find y as a function of x.

34. If $y = \dfrac{3u+4}{2u-6}$ and $u = \dfrac{3x+4}{x-3}$, find y as a function of x.

7.5 INEQUALITIES

Many of the principles thus far applied to equations can be extended to inequalities. Recall first the geometrical interpretation of real numbers as points on the real line. We agreed to associate points to the right of 0 with positive numbers and points to the left of 0 with negative numbers. To indicate that a real number, x, is positive (greater than zero) we write $x > 0$, which is read "x is greater than zero." If x is negative (less than zero) we write $x < 0$, "x is less than zero." The order relation between any two real numbers is defined as follows.

Definition 7.2. The real number, a, is **less** than the real number, b, if there exists a positive number, p, such that $b - a = p$. We write $a < b$ which is read, "a is less than b." We write $b > a$ ("b is greater than a") if and only if $a < b$.

Geometrically, $a < b$ if the point corresponding to a on the real number line is to the left of the point corresponding to b. Similarly, $c > d$ if the point corresponding to c is to the right of the point corresponding to d.

Sometimes the equality is included with the inequality. The statement $a \leqslant b$ is read, "a is less than or equal to b" and means a is less than b or a is equal to b. Similarly $c \geqslant d$ is read, "c is greater than or equal to d."

The following axioms characterize the order relation:

Axiom 7.1. (The trichotomy axiom.) If a and b are real numbers, one and only one of the following holds:

$$a < b, \qquad a = b, \qquad a > b$$

Axiom 7.2. If $a > 0$ and $b > 0$, then

$$1.\ a + b > 0$$

and

$$2.\ ab > 0$$

Other properties of the order relation, which can be proved from Definition 7.2 and the above axioms, are given below where a, b, and c are real numbers.

Theorem 7.1. If $a > b$ and $b > c$, then $a > c$.

Theorem 7.2. If $a > b$, then $a + c > b + c$ for all real numbers c.

Theorem 7.3. If $a > b$ and $c > 0$, then $ac > bc$.

Theorem 7.4. If $a > b$ and $c < 0$, then $ac < bc$.

Let us prove Theorem 7.1. $a > b$ implies $a - b = p$ where p is a positive number. $b > c$ implies $b - c = q$ where q is a positive number. Thus

$$(a - b) + (b - c) = p + q$$

or

$$a - c = p + q$$

Since p and q are positive numbers, $p + q$ is a positive number r. (Why?) Thus, $a - c = r$ where r is a positive number and this means that $a > c$.

Proofs of the other theorems are left as exercises. The student should also prove the following corollaries to Theorems 7.1–7.4.

Corollary 7.1. If $a < b$ and $b < c$, then $a < c$.

Corollary 7.2. If $a < b$, then $a + c < b + c$ for all real numbers c.

Corollary 7.3. If $a < b$ and $c > 0$, then $ac < bc$.

Corollary 7.4. If $a < b$ and $c < 0$, then $ac > bc$.

Another useful consequence of our axioms is the following theorem.

Theorem 7.5. If a is any real number, then $a^2 \geq 0$.

PROOF: By Axiom 7.1 we know that $a > 0$, $a = 0$, or $a < 0$. Now if $a = 0$, $a^2 = 0$ and so $a^2 \geq 0$ in this case. If $a > 0$ we can use Theorem 7.3 with $b = 0$ and $c = a$ to obtain $a \cdot a > 0 \cdot a = 0$ so that $a^2 > 0$ in this case. Finally, if $a < 0$ we can use Corollary 7.4 with $b = 0$ and $c = a$ to obtain, again, $a \cdot a > 0 \cdot a = 0$. Thus $a^2 \geq 0$ in every case.

One final theorem is another useful one.

Theorem 7.6. If $a > 0$, then $\dfrac{1}{a} > 0$ and if $a < 0$, then $\dfrac{1}{a} < 0$.

PROOF (by *reductio ad absurdum*): Assume $a > 0$ and $\dfrac{1}{a} < 0$. Then, multiplying both sides of $\dfrac{1}{a} < 0$ by a, we have, by Corollary 7.3,

$$a\,\frac{1}{a} < 0$$

or

$$1 < 0$$

which is false. The second part of the theorem can be proved similarly and is left as an exercise for the student.

Any statement involving one of the symbols $>, <, \geq, \leq$ is called an **inequality**. Let us say that two inequalities have the same sense if their inequality signs "point" in the same direction. Then Theorem 7.2 and Corollary 7.2 say that we may add the same number to both sides of an inequality and the resulting inequality will have the same sense as the original inequality. Theorem 7.3 and Corollary 7.3 say that we may multiply both sides of an inequality by the same positive number and again preserve the sense of the inequality. But Theorem 7.4 and Corollary 7.4 say that if we multiply both sides of an inequality by the same *negative* number, then the *sense of the inequality is reversed.*

We also see from Theorem 7.6 in conjunction with Theorem 7.3 and Corollary 7.3 that we may divide both sides of an inequality by the same positive number (i.e., multiply by its reciprocal) without changing the sense of the inequality. But if we divide both sides of an inequality by the same negative number, we see from Theorem 7.6 in conjunction with Theorem 7.4 and Corollary 7.4 that the sense of the inequality is reversed.

Inequalities, like equalities, need not be true statements. For example, $5 > 3$ is true, while $-2 > 3$ is false. A statement such as $x > 3$ is true only if x is replaced by a real number that is greater than 3. On the other hand, $a^2 \geqslant 0$ is true for any real number a. Thus, like equations, an inequality involving variables can be classed as either an **absolute inequality,** which is true for all permissible substitutions for the variables involved, or as a **conditional inequality,** which is true for only certain substitutions. (That is, a conditional inequality is false for at least one permissible substitution for the variables involved.) Thus an absolute inequality is similar to an identical equation (Section 5.1).

Absolute inequalities can be established by using the properties of inequalities to reduce the inequality to an obviously true statement.

Example 1. Prove $a^2 + b^2 \geqslant 2ab$ for all real numbers a and b.

PROOF: We subtract $2ab$ from both sides (i.e., add $-2ab$ to both sides using Theorem 7.2) to obtain

$$a^2 - 2ab + b^2 \geqslant 0$$

Now we factor the left-hand side obtaining

$$(a - b)^2 \geqslant 0,$$

a statement which is true for all real numbers a and b by Theorem 7.5.

The student should be warned, however, that when we start with the statement to be proved and reduce it to a true statement, all steps in the proof must be reversible in order for the proof to be valid.

Thus in Example 1 we must check to see that it is possible to *begin* with $(a - b)^2 \geq 0$ and *end* at $a^2 + b^2 \geq 2ab$. Since $(a - b)^2 = a^2 - 2ab + b^2$ we see that $(a - b)^2 \geq 0$ implies that $a^2 - 2ab + b^2 \geq 0$. Now we add $2ab$ to both sides of this last inequality to obtain, as desired, $a^2 + b^2 \geq 2ab$ for all real numbers a and b.

To see the need for the requirement of reversibility, observe that we could, if it were not for this requirement, "prove" that $a > 0$ for all real numbers a! All we would need to do would be to write

$$a > 0$$

and then use Theorem 7.3 with $b = 0$ and $c = a$ to arrive at

$$a^2 > 0$$

which, by Theorem 7.5, is a true statement for all real numbers a. We cannot, however, by applying any of our axioms or theorems, go from $a^2 > 0$ to $a > 0$, i.e., the steps are not reversible.

Example 2. Prove that if $a > b > 0$, then

$$\frac{1}{a} < \frac{1}{b}$$

PROOF: Since a and b are both positive (given), ab is positive (why?) and we may divide both sides of $a > b$ by $ab > 0$ to obtain

$$\frac{a}{ab} > \frac{b}{ab}$$

Therefore,

$$\frac{1}{b} > \frac{1}{a}$$

To complete the proof, the student should check that the steps are reversible.

Exercise 7.5

In problems 1–7, prove the given inequalities where a and b are real numbers.

1. If $a > 0$, then $a + \dfrac{1}{a} \geq 2$.

2. If $a > b > 0$, then $a^2 > b^2$.

3. If $a > 0$ and $b > 0$, then $a^2 \geqslant b^2$ implies $a \geqslant b$.

4. If $a < b < 0$, then $a^2 > b^2$.

5. If $a < 0$ and $b < 0$, then $a^2 > b^2$ implies $a < b$.

6. If $a < b$, then $-a > -b$.

7. If $a \geqslant b \geqslant 0$, then $\sqrt{a} \geqslant \sqrt{b}$.

8. Prove that $|x| + |y| \geqslant |x + y|$ for all real numbers x and y by considering the four cases: $x \geqslant 0, y \geqslant 0$; $x \geqslant 0, y < 0$; $x < 0, y \geqslant 0$; $x < 0, y < 0$.

9. Prove that $|x| < a$ implies $-a < x < a$ by considering the two cases $x \geqslant 0$ and $x < 0$.

10. Prove that $a < b$ implies $a < \dfrac{a+b}{2} < b$.

11. Prove that $ab > 0$ if and only if $a > 0$ and $b > 0$ or $a < 0$ and $b < 0$.

12. Prove that $ab < 0$ if and only if $a > 0$ and $b < 0$ or $a < 0$ and $b > 0$.

13. Prove Theorem 7.2.

14. Prove Theorem 7.3.

15. Prove Theorem 7.4.

16. Prove Corollary 7.1.

17. Prove Corollary 7.2.

18. Prove Corollary 7.3.

19. Prove Corollary 7.4.

20. If $0 < x < 1$, what can be said of x^2; of x^3?

21. If $0 < x < \frac{1}{2}$, what can be said of x^2; of x^3?

22. If $-1 < x < 0$, what can be said of x^2; of x^3?

23. If $-\frac{1}{2} < x < \frac{1}{2}$, what can be said of x^2; of x^3?

24. If $-2 < x < 2$, what can be said of x^2; of x^3?

25. If $x < 10$, what can be said of x^2; of x^3?

26. If $-x > 1$, what can be said of x^2; of x^3?

7.6 CONDITIONAL INEQUALITIES

The **solution set** of a conditional inequality in one variable is the set of all its solutions. Thus a conditional inequality defines a subset of the set of real numbers. For example, the solution set of $x > 3$ can be written as $X = \{x | x > 3\}$. The **graph** of a conditional inequality can be shown by drawing a dark line over the part of the real line that represents the solution set. The graph of $x > 3$ is the half line shown in Figure 7.3. If the equality is included with the inequality, this can be indicated graphically by a large dot at the end point of the graph. For example, the graph of $x \leqslant \frac{3}{2}$ is shown in Figure 7.4.

Two inequalities are said to be **equivalent** if they have the same solution sets. To solve an inequality means to represent its solution set in as simple a form as possible by an equivalent inequality. Linear inequalities can be solved by following a procedure similar to that used for solving equations. We apply one or more of Theorems 7.1–7.6 and Corollaries 7.1–7.4 of the preceding section.

Example 1. Solve the inequality $2x + 5 > 9$.

SOLUTION:

$2x + 5 + (-5) > 9 + (-5)$	Adding -5 to both sides.
$2x > 4$	Simplifying.
$\frac{1}{2}(2x) > \frac{1}{2} \cdot 4$	Multiplying both sides by $\frac{1}{2}$.
$x > 2$	

The solution set is

$$X = \{x \mid 2x + 5 > 9\}$$
$$= \{x \mid x > 2\}$$

Example 2. Solve the inequality $-5(x - 1) \geqslant 3(x - 3)$.

SOLUTION:

$-5x + 5 \geqslant 3x - 9$	Removing parentheses.
$-5x + 5 - 3x - 5 \geqslant 3x - 9 - 3x - 5$	Adding -5 and $-3x$ to both sides.
$-8x \geqslant -14$	Simplifying.
$(-\frac{1}{8})(-8x) \leqslant (-14)(-\frac{1}{8})$	Multiplying both sides by $-\frac{1}{8}$. (Notice the reversal of ">" to "≤").
$x \leqslant \frac{7}{4}$	

The solution set is

$$X = \{x \mid -5(x - 1) \geqslant 3(x - 3)\}$$
$$= \{x \mid x \leqslant \frac{7}{4}\}$$

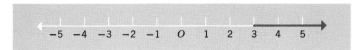

Figure 7.3

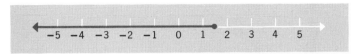

Figure 7.4

Students should be very careful to reverse the sense of the inequality symbol when multiplying or dividing both sides of an inequality by a negative number. In particular, if it is necessary to multiply both sides of an inequality by an expression containing the variable, and it is not known whether the expression is positive or negative, we must consider two cases as illustrated in the following example.

Example 3. Solve the inequality

$$\frac{1}{x-1} > \frac{1}{3}$$

CASE 1. If $x - 1 > 0$, then when we multiply both sides of the given inequality by $3(x - 1)$ to clear of fractions, we have

$$3 > x - 1$$

Now we have the double restrictions

$$x - 1 > 0 \quad and \quad 3 > x - 1$$

and the solution set for Case 1 is the *intersection* of the solution sets of these two inequalities. Solving each of the two inequalities, we find that

$$x > 1 \quad and \quad x < 4$$

Thus the solution set for Case 1 is

$$X_1 = \{x|x > 1\} \cap \{x|x < 4\}$$
$$= \{x|x > 1 \text{ and } x < 4\}$$
$$= \{x|1 < x < 4\}$$

CASE 2. If $x - 1 < 0$, then when we multiply both sides of $\frac{1}{x-1} > \frac{1}{3}$

by $3(x-1)$ we must reverse the inequality symbol to obtain

$$3 < x - 1$$

Thus for Case 2 we have the double restrictions

$$x - 1 < 0 \quad \text{and} \quad 3 < x - 1$$

Solving, we find that

$$x < 1 \quad \text{and} \quad x > 4$$

Thus the solution set for Case 2 is

$$\begin{aligned} X_2 &= \{x \mid x < 1\} \cap \{x \mid x > 4\} \\ &= \{x \mid x < 1 \text{ and } x > 4\} \\ &= \varnothing \end{aligned}$$

Now since either $x-1 > 0$ (Case 1) or $x-1 < 0$ (Case 2), $(x-1 \neq 0$ for if $x-1 = 0, \dfrac{1}{x-1}$ is undefined) the solution set, X, of the given inequality is the *union* of the solution sets for the two cases.

$$\begin{aligned} X &= X_1 \cup X_2 \\ &= \{x \mid 1 < x < 4\} \cup \varnothing \\ &= \{x \mid 1 < x < 4\} \end{aligned}$$

If the absolute value symbol occurs with an inequality, there are again two cases to be considered.

Example 4. Find the solution set of

$$|2x + 5| \leqslant x + 3$$

CASE 1. If $2x + 5 \geqslant 0$, then $|2x + 5| = 2x + 5$ and the inequality becomes

$$2x + 5 \leqslant x + 3$$
$$x \leqslant -2$$

For Case 1 we have the simultaneous restrictions

$$2x + 5 \geqslant 0 \quad \text{and} \quad x \leqslant -2$$

or

$$x \geqslant -\tfrac{5}{2} \quad \text{and} \quad x \leqslant -2$$

The solution set for Case 1 is

$$X_1 = \{x | x \geq -\tfrac{5}{2} \text{ and } x \leq -2\}$$
$$= \{x | -\tfrac{5}{2} \leq x \leq -2\}$$

CASE 2. If $2x + 5 < 0$, then $|2x + 5| = -(2x + 5)$ and the inequality becomes

$$-(2x + 5) \leq x + 3$$
$$2x + 5 \geq -(x + 3) = -x - 3$$
$$3x \geq -8$$
$$x \geq -\tfrac{8}{3}$$

For Case 2 we have

$$2x + 5 < 0 \qquad \text{and} \qquad x \geq -\tfrac{8}{3}$$

or

$$x < -\tfrac{5}{2} \qquad \text{and} \qquad x \geq -\tfrac{8}{3}$$

The solution set is

$$X_2 = \{x | x < -\tfrac{5}{2} \text{ and } x \geq -\tfrac{8}{3}\}$$
$$= \{x | -\tfrac{8}{3} \leq x < -\tfrac{5}{2}\}$$

Finally, the solution set, X, of the given inequality is the union of X_1 and X_2.

$$X = X_1 \cup X_2$$
$$= \{x | -\tfrac{8}{3} \leq x < -\tfrac{5}{2}\} \cup \{x | -\tfrac{5}{2} \leq x \leq -2\}$$
$$= \{x | -\tfrac{8}{3} \leq x \leq -2\}$$

Exercise 7.6

In problems 1–16, solve the given linear inequality.

1. $x - 5 < 3$ 2. $2 \geq x + 2$
3. $2x - 3 \leq x - 5$ 4. $2 - 3x > 5x - 1$
5. $x + 2 \geq 0$ 6. $2 - x < 4$
7. $2 - x > 4$ 8. $3x + 1 \geq 0$
9. $1 - 2x \geq 5$ 10. $x + 2 > -4$
11. $2x - 5 < -3$ 12. $4 - 2x \geq 4$
13. $2 - x \leq 2$ 14. $4 - 3x \geq 3$
15. $3x - 3 > 5$ 16. $15 - 3x > 5$

In problems 17–25, define the given set using a single set-builder and sketch its graph.

17. $\{x|x < 5\} \cap \{x|x > 2\}$ 18. $\{x|x < 5\} \cup \{x|x > 2\}$

19. $\{x|x \leqslant 3\} \cap \{x|x \leqslant -1\}$ 20. $\{x|x \leqslant 3\} \cup \{x|x < -1\}$

21. $\{x|x - 1 \leqslant 0\} \cap \{x|x - 4 \geqslant 0\}$

22. $\{x|x - 1 \leqslant 0\} \cup \{x|x - 4 \geqslant 0\}$

23. $\{x|x \geqslant -3\} \cap \{x|x \leqslant -3\}$

24. $\{x|0 \leqslant x \leqslant 3\} \cup \{x|x < -2\}$

25. $\{x|x > 0\} \cup \{x|x = 0\} \cup \{x|x < 0\}$

In problems 26–31, find the solution set of the given inequality.

26. $\dfrac{2x - 3}{x + 2} \leqslant 1$ 27. $\dfrac{3 - 5x}{x^2} > 2$

28. $\dfrac{4}{2 - x} < 1$ 29. $|2x - 4| > 3$

30. $\left|\dfrac{2x + 3}{-2}\right| < 1$ 31. $|2x - 5| \leqslant 9$

32. $\dfrac{1}{x} + \dfrac{1}{x + 3} > 0$

7.7 INEQUALITIES OF HIGHER DEGREE

We now take up higher degree inequalities and will consider two approaches, one graphical and the other algebraic. First we give some examples of the graphical method.

Example 1. To solve the inequality $x^2 - 5x + 4 \leqslant 0$ let

$$y = x^2 - 5x + 4$$

and consider the graph of this equation as shown in Figure 7.5.

Now we see from the graph that $y \leqslant 0$ precisely when $1 \leqslant x \leqslant 4$, but since $y = x^2 - 5x + 4$ we conclude that $x^2 - 5x + 4 \leqslant 0$ precisely when $1 \leqslant x \leqslant 4$. Hence the solution set of the inequality $x^2 - 5x + 4 \leqslant 0$ is $X = \{x|1 \leqslant x \leqslant 4\}$.

Example 2. To solve the inequality $x^3 > 2x - x^2$ we first add $-2x + x^2$ to both sides of the inequality obtaining

$$x^3 + x^2 - 2x > 0$$

Then we graph $y = x^3 - 2x + x^2$ as shown in Figure 7.6. From the graph we see that $y > 0$ when $x > 1$ or $-2 < x < 0$. Hence these are the numbers which satisfy the inequality $x^3 - 2x + x^2 > 0$. But this inequality is equivalent to the original one, and so the solution to our problem is $x > 1$ or $-2 < x < 0$; that is, the solution set is $X = \{x | x > 1 \text{ or } -2 < x < 0\}$.

The algebraic method of solution of a quadratic inequality in x consists of factoring (if possible) the polynomial after obtaining zero on one side of the inequality. Then we can apply the following rule. (See problems 11 and 12 of Exercise 7.5.)

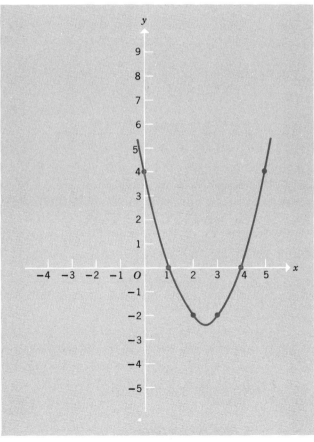

Figure 7.5

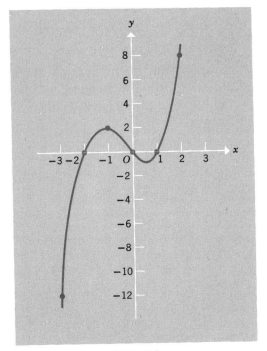

Figure 7.6

Rule. 1. The product of two numbers is positive if and only if the numbers are both positive or both negative.

2. The product of two numbers is negative if and only if one number is positive and the other is negative.

Example 1. To solve $x^2 - 5x + 4 \leqslant 0$ we factor $x^2 - 5x + 4$ and have

$$(x - 1)(x - 4) \leqslant 0$$

Then, according to the rule, either

$$\begin{matrix} x - 1 \leqslant 0 & & x - 1 \geqslant 0 \\ and & or & and \\ x - 4 \geqslant 0 & & x - 4 \leqslant 0 \end{matrix}$$

Solving for x in each inequality, we obtain

$$\begin{matrix} x \leqslant 1 & & x \geqslant 1 \\ and & or & and \\ x \geqslant 4 & & x \leqslant 4 \end{matrix}$$

Now we see that the first set of inequalities is impossible since x cannot be less than or equal to 1 *and* greater than or equal to 4 at the same time. The second set, however, can be rewritten as $1 \leqslant x \leqslant 4$, and hence the solution set of $x^2 - 5x + 4 \leqslant 0$ is $X = \{x | 1 \leqslant x \leqslant 4\}$.

The rule just given can be generalized to a product of more than two factors.

Rule. A product of several numbers is positive if and only if an even number (possibly zero) of them are negative. A product is negative if and only if an odd number of its factors are negative.

Example 2. To solve $x^3 + x^2 - 2x > 0$, we factor $x^3 + x^2 - 2x$ and have

$$x(x + 2)(x - 1) > 0$$

Now we have four possibilities: (I) all factors positive; (II) the last two negative; (III) the first and third negative; and (IV) the first two negative. In tabular form this gives:

I	II	III	IV
$x > 0$	$x > 0$	$x < 0$	$x < 0$
and	and	and	and
$x + 2 > 0$ or	$x + 2 < 0$ or	$x + 2 > 0$ or	$x + 2 < 0$
and	and	and	and
$x - 1 > 0$	$x - 1 < 0$	$x - 1 < 0$	$x - 1 > 0$

Solving for x in each inequality we obtain:

I	II	III	IV
$x > 0$	$x > 0$	$x < 0$	$x < 0$
and	and	and	and
$x > -2$ or	$x < -2$ or	$x > -2$ or	$x < -2$
and	and	and	and
$x > 1$	$x < 1$	$x < 1$	$x > 1$

However, (I) simplifies to $x > 1$, for if $x > 1$, it certainly follows that $x > 0$ and $x > -2$ also. On the other hand, since we cannot have $x > 0$ *and* $x < -2$ it follows that (II) is impossible. Similarly, (IV) is also impossible since we cannot have $x < 0$ *and* $x > 1$. Finally in (III), we note that $x < 0$ certainly implies $x < 1$ so that we are left with $x < 0$ *and* $x > -2$. This can be written as $-2 < x < 0$. Hence,

$x^3 + x^2 - 2x > 0$ if and only if

$$x > 1 \quad or \quad -2 < x < 0$$

That is, the solution set of $x^3 + x^2 - 2x > 0$ is

$$X = \{x|x > 1\} \cup \{x|-2 < x < 0\}$$
$$= \{x|x > 1 \text{ or } -2 < x < 0\}$$

Exercise 7.7

In problems 1–10, find the solution set of the given inequality by the graphical method.

1. $2x - 5 > 0$
2. $4 - 5x \leqslant 2$
3. $3 - 2x \leqslant 4$
4. $8 - 2x^2 \geqslant 0$
5. $-4x^2 > 3x - 1$
6. $x^2 + 2x < 3$
7. $x^2 \geqslant 3x$
8. $x^3 - 5x > -x$
9. $x^3 - 2x^2 > 3x$
10. $7x - 6x^2 \leqslant x^3$

In problems 11–26, find the solution set of the given inequality by the algebraic method.

11. $x^2 \geqslant -3x$
12. $2 - 2x^2 \geqslant 3x$
13. $3x^2 - x < 2$
14. $x^2 + 1 \geqslant 2x$
15. $1 < -4x(x + 1)$
16. $x^2 < 6x - 9$
17. $x^2 \geqslant -10x - 25$
18. $x^2 < 2x + 3$
19. $-x^3 + 4x^2 \leqslant 3x$
20. $x^2(x - 1) > 2x$
21. $(x - 3)(x - 2)(x + 1) > 0$
22. $(x - 4)(x + 2)(x - 1) > 0$
23. $(2x - 3)(5x - 4)(3x - 1) < 0$
24. $(3x + 1)(2x - 1)(5x + 4) < 0$
*25. $(2x - 1)(5x + 2)(3x - 4)(x - 5) \geqslant 0$
*26. $(3x - 1)(2x + 3)(x - 4)(3x - 5) < 0$

7.8   RESTRICTIONS ON THE DOMAIN OF A FUNCTION

We have already observed in Section 7.2 that, if y is a function of x, some values of x may not be permissible. We pointed out that, if $y = \dfrac{1}{x}$, the value $x = 0$ must be excluded, since division by zero is impossible. Again, if $y = \dfrac{x^2 + 1}{x - 1}$, $x = 1$ is not permissible, for this value makes the denominator zero. In general, the domain of any function may not include any number which would give rise to a division by zero.

The other restriction on the domain of a function arises because we are restricting ourselves to real numbers [2] so that we can utilize our rectangular coordinate system to picture these functions. Now, of course, if $y = f(x)$, there is no problem in choosing x to be real, but sometimes a choice of a particular real number for x will not lead us to a real number for y, as we pointed out in Section 7.2 by the example $y = \sqrt{x - 4}$. In fact, any time that we have an *even* root involving x we must be careful. We can, however, easily determine the restrictions in such cases by simply insisting that the expression under each radical sign must never be negative. That is, we insist that such expressions be greater than or equal to zero and solve the resulting inequalities by the methods of the previous sections.

Example 1. In the function defined by $y = \sqrt{x + 1}$, $x + 1$ must be greater than or equal to zero. That is, $x + 1 \geqslant 0$ and hence $x \geqslant -1$. Thus the domain of the function consists of all real numbers greater than or equal to -1.

Example 2. Determine the restrictions on x if y is given implicitly in terms of x by the equation $6x^2 + 9y^2 + x - 6y = 0$.

SOLUTION: By referring to Example 3 of Section 6.6, we see that

$$y = \frac{1 \pm \sqrt{1 - 6x^2 - x}}{3}$$

and, if the expression under the radical sign is to be non-negative, we must have

$$1 - 6x^2 - x \geqslant 0$$

or, multiplying both sides of our inequality by -1,

$$6x^2 + x - 1 \leqslant 0$$

---

[2] There is a branch of mathematics known as the theory of functions of a complex variable in which this restriction does not exist—but no simple picture can be drawn of such functions.

Since $6x^2 + x - 1 = (3x - 1)(2x + 1)$ we have, by the method described in the preceding section,

I		II
$3x - 1 \geqslant 0$		$3x - 1 \leqslant 0$
and	or	and
$2x + 1 \leqslant 0$		$2x + 1 \geqslant 0$

From (I) we obtain $x \geqslant \frac{1}{3}$ and $x \leqslant -\frac{1}{2}$ which is impossible, whereas from (II) we obtain $x \leqslant \frac{1}{3}$ and $x \geqslant -\frac{1}{2}$. Therefore a number x is in the domain of this implicit function if and only if $-\frac{1}{2} \leqslant x \leqslant \frac{1}{3}$.

We should note that the equation $6x^2 + 9y^2 + x - 6y = 0$ actually defines *two* functions, one by

$$y = \frac{1 + \sqrt{1 - 6x^2 - x}}{3}$$

and the other by

$$y = \frac{1 - \sqrt{1 - 6x^2 - x}}{3}$$

Exercise 7.8

Determine, in each problem, the restrictions on x if $f(x)$ or y is to be a real number.

1. $f(x) = \dfrac{1}{x^2}$

2. $f(x) = \dfrac{x}{x + 1}$

3. $f(x) = \dfrac{1}{x^2 - 1}$

4. $f(x) = \dfrac{x - 5}{x^2 - 3x + 2}$

5. $x^2 + y^2 = 25$

6. $x^2 - y^2 = 25$

7. $y^2 = 4x$

8. $y^2 = -x$

9. $y^2 + 2y + 8x - 15 = 0$

10. $y^2 - 4y + 4x = 0$

11. $x^2 + y^2 + 4x - y + 4 = 0$

12. $x^2 + y^2 = 2x$

13. $x^2 + y^2 = 3y$

14. $x^2 - y^2 + 2x = 0$

15. $\dfrac{(x - 1)^2}{9} - \dfrac{(y + 3)^2}{16} = 1$

16. $4x^2 - 5y^2 - 16x + 10y - 9 = 0$

17. $x^2 + y^2 + 4x - 2y = 3$

18. $(x - 1)^2 + (y + 2)^2 = 25$

19. $2x^2 + 4xy + 2y^2 = 9$

20. $3xy - 2x + y - 6 = 0$

7.9 QUADRATIC EQUATIONS IN TWO VARIABLES

The **general quadratic equation in two variables** can be written as follows:

$$ax^2 + by^2 + cxy + dx + ey + f = 0 \tag{1}$$

Notice that there is no term of higher degree than the second even though linear terms occur as well as a constant term.

If $b = 0$, this equation is linear in y and it can be solved for y in terms of x to obtain a function of x, which can then be graphed. If $b \neq 0$, the equation can be regarded as a quadratic equation in y, and it is still possible to solve for y by using the quadratic formula if necessary. The graph of any quadratic equation in two variables can be classified as [3] one of the varieties shown in Figure 7.7.

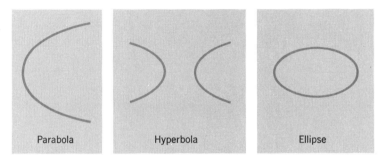

Parabola   Hyperbola   Ellipse

Figure 7.7

These graphs are known as **conic sections** since they can all be obtained by taking a cross section of a cone by a plane in various positions. In analytic geometry, quadratic equations in two variables and their corresponding graphs are thoroughly analyzed, but it is not our aim to do so here. We can, however, consider two elementary cases where the graphing does not require a complicated analysis.

First, we shall consider the case when $b = c = 0$ but $e \neq 0$ in (1). This always gives us a **parabola.** Since $b = c = 0$ and $e \neq 0$, we have a linear equation in y, and it is a simple matter to solve for y in terms of x to obtain a quadratic function in x:

$$y = f(x) = a_2 x^2 + a_1 x + a_0 \tag{2}$$

---

[3] With the exception of the so-called "degenerate" cases of a point, a straight line, or two straight lines.

This function can be graphed as discussed in Section 7.4, by assigning a certain number of values to x and computing the corresponding functional values.

Example 1. Graph $y = 2x^2 - 5$.

SOLUTION: From the table

x	-2	-1	0	1	2
y	3	-3	-5	-3	3

we obtain the graph shown in Figure 7.8.

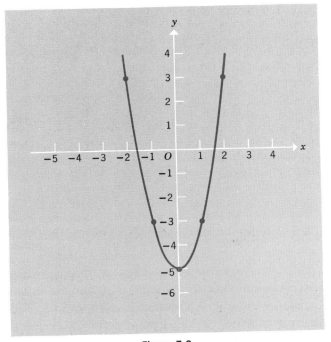

Figure 7.8

Example 2. Graph $x^2 + y - 2x = 0$.

SOLUTION: First we solve for y to obtain $y = -x^2 + 2x$. Then we construct the table

x	-2	-1	0	1	2	3	4
y	-8	-3	0	1	0	-3	-8

from which we obtain Figure 7.9.

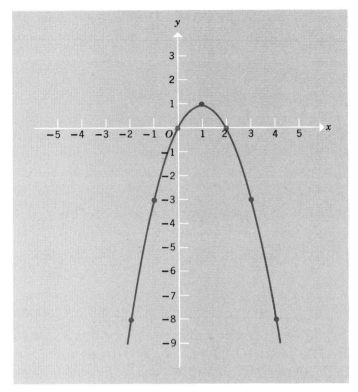

Figure 7.9

Now we may notice several interesting facts about the quadratic function (2):

1. The parabola opens either up or down; this can be determined from the sign of a_2 since when x becomes large in absolute value it is the quadratic term that will dominate the function. Thus, if a_2 is positive, the parabola will open up, but if a_2 is negative, the parabola will open down.

2. If there is no linear term ($a_1 = 0$), the parabola will be "symmetric" with respect to the y-axis. That is, the **axis** of the parabola, which is the vertical line passing through its **vertex** (the highest or lowest point), coincides with the y-axis.

3. If there is no constant term ($a_0 = 0$), the graph will pass through the origin, for, in this case, when $x = 0$, $f(x)$ is also zero.

Now let us consider another special case of equation (1) where $a = b \neq 0$ and $c = d = e = 0$:

$$ax^2 + ay^2 + f = 0 \tag{3}$$

We can then subtract f from both sides of this equation, divide both sides by a, and, finally, replace the constant, $-\dfrac{f}{a}$, by g. Therefore, we are really considering equations of the form

$$x^2 + y^2 = g \tag{4}$$

Then $y^2 = g - x^2$ and hence

$$y = \sqrt{g - x^2} \qquad \text{or} \qquad y = -\sqrt{g - x^2}$$

Rather than actually plotting $y = \sqrt{g - x^2}$ and $y = -\sqrt{g - x^2}$, let us return to equation (4) and determine what the expression $x^2 + y^2$ means geometrically.

In the first place, the student has probably noticed that we have used $f(x)$ for y and *vice versa*. This is because when we solve for y in terms of x we mean y is a function of x, or

$$y = f(x)$$

Thus let us label the vertical axis of our graph the y-axis and note that, if we join the point with coordinates (x, y) to the origin by a straight line, we obtain a right triangle, the horizontal and vertical sides of which are of lengths x and y respectively. But then, from the theorem of Pythagoras, we see that the distance from the origin to the point with coordinates (x, y) is the hypotenuse of a right triangle, and hence is equal to $\sqrt{x^2 + y^2}$. That is, the expression $x^2 + y^2$ is the square of the distance of the point (x, y) from the origin.

Returning now to equation (4), we see that any pair of numbers (x, y) which satisfies the condition $x^2 + y^2 = g$ with $g > 0$ must correspond to a point in the Cartesian plane which is at a distance of $\sqrt{g}$ units from the origin and that the collection of all such points is obviously a circle of radius $\sqrt{g}$ with its center at the origin.

Example 3. Graph the equation $2x^2 + 2y^2 - 13 = 0$.

SOLUTION: To see that this is the equation of a circle, we put the equation in the form (4):

$$2x^2 + 2y^2 = 13$$
$$x^2 + y^2 = \tfrac{13}{2}$$

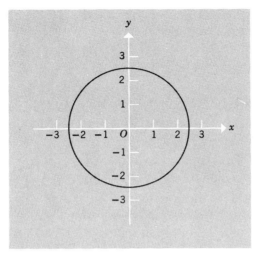

Figure 7.10

Hence, by the analysis just made, this equation is the equation of a circle (see Figure 7.10) with a radius of length

$$\sqrt{\frac{13}{2}} = \sqrt{\frac{13}{2} \cdot \frac{2}{2}} = \frac{\sqrt{26}}{2}$$

Notice that here the constant term f in (3) must be different in sign from a since when we subtract f from both sides of the equation and divide both sides by a, we must have a positive number because $x^2 + y^2 > 0$ except when $x = y = 0$.

Exercise 7.9

In problems 1–10, graph the quadratic equation (1) when a, b, c, d, e, and f have the values indicated.

1. $a = 1, e = 1, f = -4, b = c = d = 0$
2. $a = -1, e = 1, f = 4, b = c = d = 0$
3. $a = -1, e = 2, d = 2, b = c = f = 0$
4. $a = 1, e = 2, d = -2, b = c = f = 0$
5. $a = 2, d = 1, f = -3, b = c = e = 0$
6. $a = 2, d = -2, f = 5, b = c = e = 0$
7. $a = b = 1, f = -9, c = d = e = 0$
8. $a = b = -2, f = 8, c = d = e = 0$

9. $a = 1, d = 2, e = 1, f = -5, b = c = 0$

10. $a = -2, d = 3, e = 2, f = -1, b = c = 0$

If all but one of the constants in (1) are fixed, we have a (one-parameter) "family of curves"; that is, a set (infinite in number) of curves which resemble each other in some way. In problems 11–18, graph the "members" of each family indicated in parts (*a*), (*b*), (*c*), and (*d*) and determine the influence that the "parameter" has on the curves.

11. $y = 2x - p$ (*p* is the parameter). (*a*) $p = 0$, (*b*) $p = 1$, (*c*) $p = 2$, (*d*) $p = -1$, (*e*) how does *p* influence the curve?

12. $y = sx + 2$ (*s* is the parameter). (*a*) $s = 0$, (*b*) $s = 1$, (*c*) $s = \frac{1}{2}$, (*d*) $s = -1$, (*e*) how does *s* influence the curve?

13. $x^2 + y - q = 0$ (*q* is the parameter). (*a*) $q = 0$, (*b*) $q = 1$, (*c*) $q = -1$, (*d*) $q = 2$, (*e*) how does *q* influence the curve?

14. $y = tx^2$ (*t* is the parameter). (*a*) $t = 1$, (*b*) $t = 2$, (*c*) $t = 3$, (*d*) $t = 0$, (*e*) how does *t* influence the curve?

15. $-x^2 + px + y = 0$ (*p* is the parameter). (*a*) $p = 0$, (*b*) $p = 1$, (*c*) $p = 2$, (*d*) $p = -1$, (*e*) how does *p* influence the curve?

16. $x^2 - 2x + ky = 1$ (*k* is the parameter). (*a*) $k = 1$, (*b*) $k = 2$, (*c*) $k = 3$, (*d*) $k = \frac{1}{2}$, (*e*) how does *k* influence the curve?

17. $x^2 + y^2 - u = 0$ (*u* is the parameter). (*a*) $u = 1$, (*b*) $u = 2$, (*c*) $u = 4$, (*d*) $u = 0$, (*e*) how does *u* influence the curve?

18. $x^2 + y^2 + rx = 0$ (*r* is the parameter). (*a*) $r = 1$, (*b*) $r = 2$, (*c*) $r = 0$, (*d*) $r = -1$, (*e*) how does *r* influence the curve?

7.10 SYSTEMS OF QUADRATIC EQUATIONS

In Chapter 5, we saw that the solution of a pair of linear equations in two variables corresponded to the point of intersection of their graphs, both of which were straight lines.

Now when we consider the graphs of quadratic equations in two variables, we would expect a straight line to intersect any of these conic sections in, at most, two points. Of course, a line might be tangent to the conic section giving only one solution, or it might miss the conic section completely, resulting in no (real) solution. Thus a system consisting of a quadratic equation and a linear equation will have at most two solutions. The algebraic method of solution involves solving the linear equation for one of the variables in terms of the other and substituting the result in the quadratic equation of the system to obtain a quadratic equation in only one variable.

Example 1. Solve the system

$$x^2 + y^2 = 10$$
$$x + 2y = 1$$

SOLUTION: We solve the linear equation for x in terms of y to obtain $x = 1 - 2y$ and substitute the result, $1 - 2y$, for x in the quadratic equation to obtain

$$(1 - 2y)^2 + y^2 = 10$$

Then we have

$$1 - 4y + 4y^2 + y^2 = 10$$

so that

$$5y^2 - 4y - 9 = 0$$
$$(5y - 9)(y + 1) = 0$$

Therefore

$$y = \tfrac{9}{5} \quad \text{or} \quad y = -1$$

Substituting these values in turn in the linear equation, we find the corresponding values for x:

$$x + 2(\tfrac{9}{5}) = 1, x = 1 - \tfrac{18}{5} = -\tfrac{13}{5}$$

or

$$x + 2(-1) = 1, x = 3$$

The solutions of the system are therefore

$$(x = -\tfrac{13}{5}, y = \tfrac{9}{5}) \quad \text{and} \quad (x = 3, y = -1).$$

To consider the corresponding graphs of this system, we notice that $x^2 + y^2 = 10$ represents a circle with radius $\sqrt{10}$ and $x + 2y = 1$ is the line passing through the points $(5, -2)$ and $(1, 0)$ (Figure 7.11).

Two conic sections may intersect in as many as four points; therefore, when both equations are quadratic we may have anywhere from no solution to four solutions. There are several methods for solving simultaneous quadratic systems, some of which will be illustrated by the following examples, but there are other systems which are im-

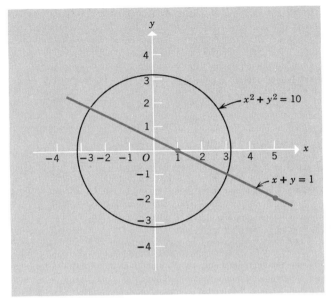

Figure 7.11

possible to solve without methods for solving a general cubic or quartic equation in one variable.

Our next example does involve a quartic, but one which is easy to solve by elementary methods. (Compare Section 6.9.)

Example 2. Solve

$$xy = 3$$
$$x^2 + y^2 = 10$$

SOLUTION: We solve the first equation for y to obtain $y = \dfrac{3}{x}$ and substitute in the second equation to obtain

$$x^2 + \left(\frac{3}{x}\right)^2 = 10$$

Multiplying both sides of this equation by x^2 we have

$$x^4 + 9 = 10x^2$$

or

$$x^4 - 10x^2 + 9 = 0$$

This quartic factors into $(x^2 - 1)(x^2 - 9)$. Therefore $x^2 = 1$ or $x^2 = 9$. Hence

$$x = 1, -1, 3 \text{ or } -3$$

Substituting these values in turn in the first equation, we obtain the corresponding values for y:

$$
\begin{array}{cccc}
x = 1 & x = -1 & x = 3 & x = -3 \\
y = 3 & y = -3 & y = 1 & y = -1
\end{array}
$$

To consider this system graphically, we notice that the second equation is the equation of a circle with radius $\sqrt{10}$, whereas the graph of the first equation is a **hyperbola** (see Figure 7.12) obtained from the following table.

x	−5	−4	−3	−2	−1	1	2	3	4	5
y	$-\frac{3}{5}$	$-\frac{3}{4}$	-1	$-\frac{3}{2}$	-3	3	$\frac{3}{2}$	1	$\frac{3}{4}$	$\frac{3}{5}$

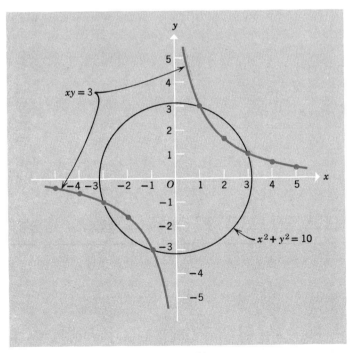

Figure 7.12

Sometimes one of the quadratic terms can be eliminated by adding or subtracting, leaving a quadratic function in the other unknown, as in the following example.

Example 3. Solve

$$2x^2 - y + 11 = 0$$
$$x^2 - y^2 - 5 = 0$$

SOLUTION: We multiply both sides of the second equation by 2 to obtain $2x^2 - 2y^2 - 10 = 0$; we subtract this equation from the first to get $2y^2 - y + 21 = 0$. Now, using the quadratic formula, we find that

$$y = \frac{1 \pm \sqrt{1 - 4 \cdot 2 \cdot 21}}{2 \cdot 2} = \frac{1 \pm \sqrt{-167}}{4}$$

But the square root of a negative number is not a real number and therefore there are no real solutions to this system. Furthermore, this means that the graphs of the two equations do not intersect, as is shown in Figure 7.13.

In some cases we can, by adding or subtracting, obtain a third equation which is solvable with one of the two given equations by the methods thus far illustrated.

Example 4. Solve

$$x^2 + y^2 = 5$$
$$x^2 - xy + y^2 = 7$$

SOLUTION: Subtracting the second equation from the first, we have

$$xy = -2$$

Thus let us consider the system

$$x^2 + y^2 = 5$$
$$xy = -2$$

Solving the second equation for y and substituting the result in the first, we obtain

$$x^2 + \left(-\frac{2}{x}\right)^2 = 5$$

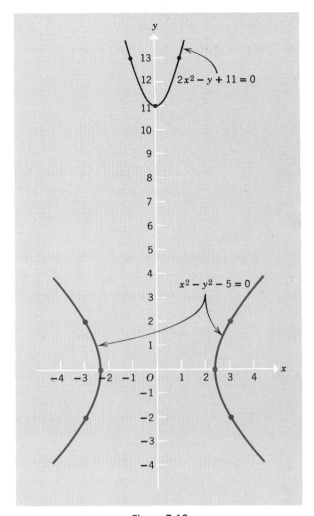

Figure 7.13

Then we multiply both sides by x^2 to get

$$x^4 + 4 = 5x^2$$

or

$$x^4 - 5x^2 + 4 = 0$$

Hence $(x^2 - 1)(x^2 - 4) = 0$, and, therefore, $x^2 = 1$ or $x^2 = 4$. Thus $x = \pm 1$

or $x = \pm 2$. Substituting these values in turn in the equation $xy = -2$ we obtain the solutions

$$
\begin{array}{cc}
x = 1 & x = -1 \\
y = -2 & y = 2 \\[2mm]
x = 2 & x = -2 \\
y = -1 & y = 1
\end{array}
$$

The graph of this system is shown in Figure 7.14. It is left to the student to check these number pairs in the original system.

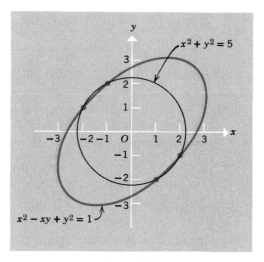

Figure 7.14

Exercise 7.10

In problems 1–10, find the points of intersection of the graphs of the given equations by algebraic methods and illustrate graphically.

1. $x^2 + y^2 = 16$
   $x + y = 4$

2. $x^2 + y^2 = 20$
   $x - y = 2$

3. $x^2 + y^2 = 16$
   $x + y = 10$

4. $y = x^2 - 4x$
   $2x - y - 5 = 0$

5. $x^2 + y = 3$
   $5x + y = 7$

6. $x + 2y = 7$
   $y^2 + 3x = 12$

7. $x^2 + y^2 = 25$
$x^2 + 4x + y^2 - 21 = 0$

8. $x^2 + y^2 = 4$
$y^2 - 2x = 5$

9. $x^2 + y^2 = 9$
$2x - y^2 + 1 = 0$

10. $x^2 + y^2 = 1$
$x^2 + y^2 - 2x + 3 = 0$

In problems 11–20, find the real solutions of the given system of equations.

11. $xy = 15$
$x^2 - y^2 = 16$

12. $2x^2 + xy + 10 = 0$
$xy = 2$

13. $4x - y^2 = 0$
$xy - 2 = 0$

14. $xy = 6$
$x^2 = xy - 2$

15. $x^2 - xy + y^2 = 3$
$x^2 + y^2 = 5$

16. $x^2 + y^2 - 10y + 24 = 0$
$x^2 + 2xy + y^2 = 25$

17. $x^2 + 2y^2 = 3$
$x^2 + y^2 = 2x$

18. $x^2 + y^2 = 25$
$x^2 - y^2 = 24$

19. $2x^2 + 3y^2 = 5$
$2x + y^2 = 3$

20. $3x^2 - 4y^2 - 73 = 0$
$x^2 = 5y^2 - 16$

EXPONENTS AND LOGARITHMS

8

8.1 POSITIVE AND NEGATIVE INTEGRAL EXPONENTS

The symbol a^n was defined in Chapter 2 for n a natural number:

$$a^n = a \cdot a \cdot \cdots \cdot a \ (n \text{ factors})$$

The rules for operating with positive integral exponents follow easily from this definition. For m and n positive integers these rules are:

1. $a^n a^m = a^{n+m}$

2. $\left(\dfrac{a}{b}\right)^n = \dfrac{a^n}{b^n} \qquad (b \neq 0)$

3. $(a^m)^n = a^{mn}$

4. $(ab)^n = a^n b^n$

5. $(a) \ \dfrac{a^m}{a^n} = a^{m-n} \text{ if } m > n \qquad (a \neq 0)$

    $(b) \ \dfrac{a^m}{a^n} = \dfrac{1}{a^{n-m}} \text{ if } m < n \qquad (a \neq 0)$

Rules (1), (3), and (4) were established in Section 2.3. The other rules follow from the definition of a^n and the definition $\dfrac{a}{b} \cdot \dfrac{c}{d} = \dfrac{ac}{bd}$. Thus we have

$$\left(\frac{a}{b}\right)^n = \frac{a}{b} \cdot \frac{a}{b} \cdot \cdots \cdot \frac{a}{b} \left(n \text{ factors } \frac{a}{b}\right)$$

$$= \frac{a \cdot a \cdot \cdots \cdot a \ (n \text{ factors } a)}{b \cdot b \cdot \cdots \cdot b \ (n \text{ factors } b)}$$

$$= \frac{a^n}{b^n}$$

Similarly we have, for $m > n$

$$\frac{a^m}{a^n} = \frac{a \cdot a \cdot \ldots \cdot a \ (m \text{ factors } a)}{a \cdot a \cdot \ldots \cdot a \ (n \text{ factors } a)}$$

$$= \frac{a \cdot a \cdot \ldots \cdot a \ (n \text{ factors } a)}{a \cdot a \cdot \ldots \cdot a \ (n \text{ factors } a)} \cdot \frac{a \cdot a \cdot \ldots \cdot a \ (m - n \text{ factors } a)}{1}$$

$$= 1 \cdot \frac{a^{m-n}}{1} = a^{m-n}$$

The proof of $5b$ is left as an exercise for the student.

In this chapter we shall attempt to extend the definition of a^m for m a positive integer to a^x for x any real number. Our first step in this extension is to define the zero exponent.

Definition 8.1. $a^0 = 1$ if $a \neq 0$.

This means that any nonzero number raised to the zero power equals 1: $2^0 = 1$, $(-3)^0 = 1$, $\pi^0 = 1$, etc. The reason for making such a definition is that our rules remain valid under this agreement. For example, Rule 1 says that $a^n a^m = a^{n+m}$ for n and m positive integers. But what happens if $m = 0$? We have

$$a^n a^0 = a^{n+0} = a^n$$

and since we have just defined $a^0 = 1$ we also have

$$a^n a^0 = a^n \cdot 1 = a^n$$

Thus Rule 1 continues to hold even if m or n is now zero. The student should check the other rules to see that they, too, hold for zero exponents if we agree that $a^0 = 1$.

Similarly, we make the following definition.

Definition 8.2. $a^{-n} = \dfrac{1}{a^n}$ if n is an integer and $a \neq 0$.

For example,

$$3^{-2} = \frac{1}{3^2} = \frac{1}{9}, \quad a^{-3} = \frac{1}{a^3}$$

Again we are led to make this definition by our desire to have all rules hold even when m and n are negative integers. Thus if Rule 1 is to continue to hold, we must have

$$a^n a^{-n} = a^{n+(-n)} = a^0$$

Since we have already agreed that $a^0 = 1$ if $a \neq 0$, it follows that

$$a^n a^{-n} = 1, \quad a^{-n} = \frac{1}{a^n} \text{ for } a \neq 0$$

We may now dispense with Rule 5b as 5a is meaningful even when $m > n$ because

$$\frac{a^n}{a^m} = \frac{1}{a^{m-n}} = a^{-(m-n)} = a^{n-m}$$

Thus, for example,

$$\frac{a^2}{a^5} = \frac{1}{a^3} = a^{-3} = a^{2-5}$$

It is frequently desirable to write expressions without the use of negative exponents. This can always be done by using our Fundamental Principle (Section 1.9) as the following three examples show.

Example 1. $4x^2 y^{-2} z^{-3} = (4x^2 y^{-2} z^{-3}) \dfrac{y^2 z^3}{y^2 z^3} = \dfrac{4x^2}{y^2 z^3}.$

Example 2. $\dfrac{2x^{-3} y^{-2} z^4}{xy^4 z^{-2}} = \dfrac{2x^{-3} y^{-2} z^4}{xy^4 z^{-2}} \cdot \dfrac{x^3 y^2 z^2}{x^3 y^2 z^2} = \dfrac{2z^6}{x^4 y^6}.$

Example 3. $\dfrac{a^{-2} b^{-2}}{a^{-1} - b^{-1}} = \dfrac{a^{-2} b^{-2}}{a^{-1} - b^{-1}} \cdot \dfrac{a^2 b^2}{a^2 b^2} = \dfrac{1}{ab^2 - a^2 b} = \dfrac{1}{ab(b - a)}.$

Before going on to a consideration of fractional exponents it may be well to emphasize that we have not *proved* that $a^0 = 1$ and $a^{-n} = \dfrac{1}{a^n}$, but, rather, we have *defined* them as such. Thus while

$$\frac{a^n}{a^n} = 1$$

it does not necessarily follow that

$$\frac{a^n}{a^n} = a^{n-n} = a^0$$

since the law of exponents on which the second statement is based,

$$\frac{a^n}{a^m} = a^{n-m},$$

was proved only for $n > m$. Hence if we want to have the law hold when $n = m$, we must have $a^0 = 1$. But, on the other hand, if we were not willing to have $a^0 = 1$, we could simply continue to insist that $n > m$ in the formula $\dfrac{a^n}{a^m} = a^{n-m}$. This insistence, of course, would be inconvenient, and it is for this reason that we decide to say that $a^0 = 1$.

We should also notice that, with the above definition of the negative exponent, even Rule 5a can be eliminated since it becomes a special case of Rule 1:

$$\frac{a^n}{a^m} = a^n \cdot \frac{1}{a^m} = a^n(a^{-m}) = a^{n-m}$$

Exercise 8.1

In problems 1–50, perform the indicated operations by using the rules for exponents and Definitions 8.1 and 8.2. Express all answers without negative exponents.

1. $2^3 \cdot 2^2$
2. $3^3 \cdot 3^2$
3. $(-5)^0$
4. $(2^3)^2$
5. $(3^3)^2$
6. $4^2 \cdot 4^0$
7. $a^3 \cdot a^5$
8. $x \cdot x^7$
9. $(y^2)^3$
10. $(-x^3)^3$
11. $(ab)^3$
12. $(2 \cdot 5)^2$
13. $(3 \cdot 4)^3$
14. $(abc)^2$
15. $2^3 \cdot 2^k$
16. $a^m \cdot a^k$
17. $(4a^3)^2$
18. $\left(\dfrac{x}{y}\right)^2$
19. $\left(\dfrac{a}{b}\right)^3$
20. $a^2 \left(\dfrac{a}{b}\right)^3$
21. $(a^2)^{3k}$
22. $\dfrac{1}{b^2}\left(\dfrac{b}{a}\right)^2$
23. $\left(\dfrac{1}{3}\right)^3$
24. $\left(\dfrac{1}{5}\right)^a$
25. $(-2)^5$
26. $(-2)^{-3}$
27. $a^{-1}b$
28. ab^{-1}
29. $(ab)^{-2}$
30. $(-3)^{-3}(-3)^2$
31. $\dfrac{x^2 y^5}{x^4 y^4}$
32. $\left(\dfrac{2x}{3y^2}\right)^3$
33. $\left(\dfrac{2x}{3y}\right)^2 \left(\dfrac{3y}{4x}\right)^3$
34. $(a^{-2})^3 \cdot (b^2)^{-4}$
35. $(4a^2 b^{-3})^2$
36. $\left(\dfrac{a^x b^y}{z^n}\right)^k$
37. $\left(\dfrac{a^x k^z}{b^2}\right)^m$
38. $\left(\dfrac{c}{2d}\right)^{-3} \left(\dfrac{2d}{3a}\right)^3$
39. $(-5a^{-1})^{-2}$
40. $\dfrac{a^{n+1}}{a}$
41. $a^n a^{-2n}$
42. $\left(\dfrac{a^{n+1}}{a^n}\right)^{-2}$

43. $(-2xy^2)^3$

44. $x^{-5} \cdot x^{a+3}$

45. $\dfrac{a^{3k}x^{k+1}}{a^k x^2}$

46. $a^{4n}b^m a^{-n}b^{-2}$

47. $(-a^{-2}x^3)^3$

48. $(x^{-2}y^{-3})^{-2}$

49. $\dfrac{a^{-2}b^{-1}}{a^{-1}b^{-2}}$

50. $\dfrac{(x^{-3}y^{-5})^{-2}}{(x^{-1}y)^{-2}}$

In problems 51–75, simplify the given expression. Express all answers without negative exponents.

51. $\dfrac{2^0 p^3 q^2 r}{2^3 p q^0 r^2}$

52. $\dfrac{(x^2 y^3)^2 (xy^2)^2}{(x^2 y)^3}$

53. $\left(\dfrac{-3a^2 b}{2c^3 d^2}\right)^3 \left(\dfrac{a^2 b^3}{c^4 d^2}\right)^{-2}$

54. $\dfrac{6^2}{2^3}$

55. $\dfrac{(x^2)^3 y}{x^6}$

56. $\dfrac{x^{n+2} y^{2n+1}}{x^{n-2} y^{1-2n}}$

57. $1^{-1} + 2^{-2}$

58. $(1^{-1} + 2^{-2})^0$

59. $(a^0 + b^0)^2$

60. $(x^0 + y^0)^0$

61. $3a^{-n} + \dfrac{2}{a^n}$

62. $x^{-1} + y^{-1}$

63. $(x + y)^{-1}$

64. $ab^{-1} + a^{-1}b$

65. $\dfrac{a^{-2}b^2}{a^{-1} + b^{-1}}$

66. $\dfrac{ab}{a^{-1} + b^{-1}}$

67. $(2x^{-2} - 3)(x^{-2} + 2)$

68. $\dfrac{1 + 3x^{-1}}{x^{-2}}$

69. $\dfrac{x + y^{-1}}{x^{-1} + y}$

70. $\dfrac{x^{-1} + y^{-2}}{x^{-1}}$

71. $\dfrac{a^{-1} + b^{-1}}{a^{-2} - b^{-2}}$

72. $(a^{-1} + b^{-1})(a^{-2} - a^{-1}b^{-1} + b^{-2})$

73. $\dfrac{(a + b)^{-1}}{a^{-1} + b^{-1}}$

74. $\dfrac{1 + (3x)^{-1}}{x}$

75. $\dfrac{(x^{-2} - y^{-2})^{-1}}{(x^{-1} - y^{-1})^{-1}}$

8.2   THE PRINCIPAL nth ROOT OF A NUMBER

In the next section we will base our extension of exponents to cover fractional exponents on the idea of roots of numbers. In Chapter 6 we defined $\sqrt{a}$ for $a > 0$ as that positive number x such that $x^2 = a$. We generalize this concept by the definition that follows.

Definition 8.3. (1) If n is an odd positive integer, then $\sqrt[n]{a}$ is that real number x such that $x^n = a$. (2) If n is an even positive integer and $a \geq 0$, then $\sqrt[n]{a}$ is that nonnegative real number x such that $x^n = a$. The number x just described is called the **principal *n*th root** of a. The symbol $\sqrt[n]{a}$ is called a **radical**; n is the **index** of the radical, and a is the **radicand**.

Example 1. $\sqrt[3]{8} = 2$ since $2^3 = 8$.

Example 2. $\sqrt[3]{-216} = -6$ since $(-6)^3 = -216$.

Example 3. $\sqrt[2]{9} = \sqrt{9} = 3$ since $3^2 = 9$ and $3 > 0$.

Example 4. $\sqrt[n]{0} = 0$ for all positive integers n.

Several comments are in order regarding this definition. First of all, we have not proved that $\sqrt[n]{a}$ is unique. That is, for example, we have not shown that there is no number x other than -6 such that $x^3 = -216$ nor that there is no positive number x other than 3 such that $x^2 = 9$. Actually, there are n distinct nth roots of any nonzero number but, in the case where n is odd, only one of the roots is a real number. On the other hand, when n is even, there are only two real roots and, since we agree to choose the positive one, we again have uniqueness if we restrict ourselves to real numbers.

Second, we note the restriction that, when n is even, a be nonnegative. This is essential since $x^n \geq 0$ if n is even. Again, however, if a is negative none of the roots is a real number. (See examples in Section 11.4.)

Finally, we should observe that the roots of some numbers are irrational; $\sqrt{2}$, $\sqrt[3]{5}$, $\sqrt[4]{3}$ are irrational numbers.

In what follows, it should always be assumed that when we write $\sqrt[n]{a}$ we are assuming that $a \geq 0$ when n is even.

Our basic theorems concerning roots are

$$\sqrt[n]{ab} = \sqrt[n]{a}\sqrt[n]{b} \tag{1}$$

$$\sqrt[n]{\frac{a}{b}} = \frac{\sqrt[n]{a}}{\sqrt[n]{b}} \text{ if } b \neq 0 \tag{2}$$

To prove (1) we simply observe that, by Definition 8.3, $(\sqrt[n]{ab})^n = ab$. But similarly, $(\sqrt[n]{a}\sqrt[n]{b})^n = (\sqrt[n]{a})^n(\sqrt[n]{b})^n$ by Rule 4 for the positive integral exponent n. Since $(\sqrt[n]{a})^n = a$ and $(\sqrt[n]{b})^n = b$ by Definition 8.3,

$\sqrt[n]{a}\sqrt[n]{b}$ is also an nth root of ab. By the uniqueness of principal roots it follows that (1) holds. The proof of (2) is left as an exercise for the student.

These laws and Definition 8.3 can be used to simplify **radical expressions** (expressions involving roots) as is illustrated in the following examples.

Example 5. $\sqrt{8} = \sqrt{4 \cdot 2} = \sqrt{4}\,\sqrt{2} = 2\,\sqrt{2}$.

Example 6. $\sqrt{x^2} = |x|$.

Before continuing with additional examples let us comment on the absolute value sign in the answer to Example 6. To see the necessity for it suppose that $x = -1$. Then

$$\sqrt{x^2} = \sqrt{(-1)^2} = \sqrt{1} = 1 = |-1| \neq -1$$

In general, since the principal even roots of numbers are, by Definition 8.3, always positive, we must always use an absolute value sign when we are working with a principal even root of an algebraic expression unless we are certain that the algebraic expression is positive for all possible values of the letters involved (for example, $\sqrt[4]{x^4} = \sqrt{x^2} = |x|$). This precaution is not necessary, however, when we work with odd roots since, for example, $\sqrt[3]{a^3} = a$ whether $a = 1$ or $a = -1$ and, in fact, $\sqrt[3]{a} \neq |a|$ when a is negative. (Cf. Section 6.3.)

Example 7.

$$\sqrt[4]{16a^5x^5y^2} = \sqrt[4]{(16a^4x^4)(axy^2)} = \sqrt[4]{16a^4x^4}\,\sqrt[4]{axy^2} = |2ax|\,\sqrt[4]{axy^2}$$

Actually, in this case the absolute value sign, while correct, is not necessary since if a and x are both positive or both negative, $ax = |ax|$. On the other hand, we cannot have only one of a or x negative, for then $16a^5x^5y^2$ would be negative contrary to our agreement that the radicand for an even root should be nonnegative.

Example 8. $\sqrt[3]{4a^2b}\,\sqrt[3]{16a^2b^4} = \sqrt[3]{64a^4b^5} = \sqrt[3]{(64a^3b^3)(ab^2)}$
$$= \sqrt[3]{64a^3b^3}\,\sqrt[3]{ab^2} = 4ab\sqrt[3]{ab^2}$$

Example 9. $\sqrt{\dfrac{1}{3}} = \sqrt{\dfrac{1}{3}\cdot\dfrac{3}{3}} = \sqrt{\dfrac{3}{9}} = \dfrac{\sqrt{3}}{3}$

Example 10.
$$\frac{\sqrt{2}}{\sqrt{3}} \cdot \frac{\sqrt{3}}{\sqrt{3}} = \frac{\sqrt{6}}{3}$$

Example 11.
$$\frac{\sqrt[5]{a^2}}{\sqrt[5]{b^3}} = \sqrt[5]{\frac{a^2}{b^3}} = \sqrt[5]{\frac{a^2b^2}{b^5}} = \frac{\sqrt[5]{a^2b^2}}{\sqrt[5]{b^5}} = \frac{\sqrt[5]{a^2b^2}}{b}$$

Example 12.

$$\frac{x^2 - y^2}{2ax\sqrt{x+y}} = \frac{(x^2-y^2)\sqrt{x+y}}{2ax\sqrt{x+y}\sqrt{x+y}} = \frac{(x^2-y^2)\sqrt{x+y}}{2ax(x+y)}$$
$$= \frac{(x+y)(x-y)\sqrt{x+y}}{2ax(x+y)} = \frac{(x-y)\sqrt{x+y}}{2ax}$$

When the denominator is the sum or difference of two terms where at least one of the terms has a square root factor, we proceed as follows:

Example 13.

$$\frac{\sqrt{3}+\sqrt{2}}{\sqrt{3}-\sqrt{2}} = \frac{(\sqrt{3}+\sqrt{2})(\sqrt{3}+\sqrt{2})}{(\sqrt{3}-\sqrt{2})(\sqrt{3}+\sqrt{2})} = \frac{(\sqrt{3}+\sqrt{2})^2}{(\sqrt{3})^2 - (\sqrt{2})^2}$$
$$= \frac{(\sqrt{3}+\sqrt{2})^2}{3-2} = 3 + 2\sqrt{6} + 2 = 5 + 2\sqrt{6}$$

Example 14.

$$\frac{1}{\sqrt{2}+1} = \frac{1 \cdot (\sqrt{2}-1)}{(\sqrt{2}+1)(\sqrt{2}-1)} = \frac{\sqrt{2}-1}{(\sqrt{2})^2 - 1^2} = \frac{\sqrt{2}-1}{2-1} = \sqrt{2} - 1$$

In Examples 9–14 the process of simplification used is known as **rationalizing the denominator.** In numerical problems it is clearly advantageous to rationalize the denominator. For example, to find the decimal value, to 4 places, of the number given in Example 13 we would have to work with

$$\frac{1.7321 + 1.4142}{1.7321 - 1.4142} = \frac{3.1463}{0.3179}$$

if we approach the problem directly, and with

$$5 + 2 \times 2.4495$$

if we first rationalize the denominator.

On the other hand, there are cases (in, for example, the calculus) where it is desirable to **rationalize the numerator.**

Example 15.

$$\frac{\sqrt{a+h}-\sqrt{a}}{h} = \frac{\sqrt{a+h}-\sqrt{a}}{h} \cdot \frac{\sqrt{a+h}+\sqrt{a}}{\sqrt{a+h}+\sqrt{a}}$$

$$= \frac{(\sqrt{a+h})^2-(\sqrt{a})^2}{h(\sqrt{a+h}+\sqrt{a})} = \frac{a+h-a}{h(\sqrt{a+h}+\sqrt{a})}$$

$$= \frac{1}{\sqrt{a+h}+\sqrt{a}}$$

Our last two examples illustrate the simplification of sums and differences of radical expressions.

Example 16.

$$\sqrt{18}+5\sqrt{2}-\sqrt{8}+\sqrt[3]{16}-\sqrt{3} = 3\sqrt{2}+5\sqrt{2}-2\sqrt{2}+2\sqrt[3]{2}-\sqrt{3}$$
$$= (3+5-2)\sqrt{2}+2\sqrt[3]{2}-\sqrt{3} = 6\sqrt{2}+2\sqrt[3]{2}-\sqrt{3}$$

Example 17.

$$\sqrt{a^5 b^2}+3b\sqrt{a^3}-5a\sqrt{b^4 a} = a^2|b|\sqrt{a}+3ab\sqrt{a}-5ab^2\sqrt{a}$$
$$= (a|b|+3b-5b^2)a\sqrt{a}$$

Notice again the use of the absolute value sign in $a^2|b|\sqrt{a}$. This is because b may be negative whereas $\sqrt{a^5 b^2}$ is nonnegative. On the other hand, we do not need to write $|a|$ since if a were negative the three radicands, $a^5 b^2$, a^3, $b^4 a$ would be negative contrary to our agreement that radicands of even roots must be nonnegative. Finally, in $5ab^2\sqrt{a}$ we do not, of course, need $|b^2|$ since b^2 itself is always nonnegative.

Exercise 8.2

In problems 1–20, simplify the given radical utilizing the procedures given in Examples 1–11.

1. $\sqrt{75}$

2. $\sqrt{27}$

3. $\sqrt[3]{16}$

4. $\sqrt[3]{4^3}$

5. $\sqrt{4^3}$

6. $\sqrt{8a^3}$

7. $\sqrt[3]{27a^6 b^3}$

8. $\sqrt[4]{16a^4 b^0}$

9. $\sqrt[5]{-1}$

10. $\sqrt[3]{-x^3 y^6}$

11. $\sqrt{8a^7b^2}$

12. $\sqrt{50a^2x^4}$

13. $\sqrt[3]{x^{3k}}$

14. $\sqrt[4]{16x^{8n}}$

15. $\sqrt{\dfrac{2x^6}{9y^4}}$

16. $\sqrt{\dfrac{4x^5}{y^4}}$

17. $\sqrt[3]{-\dfrac{16}{x^3y^9}}$

18. $\sqrt[3]{\dfrac{-1}{64a^3b^6}}$

19. $\sqrt{3 + \dfrac{6x}{y^2} + \dfrac{3x^2}{y^4}}$

20. $\sqrt{\dfrac{x}{b^2} + \dfrac{2x}{ab} + \dfrac{x}{a^2}}$

In problems 21–30, simplify the given radicals and combine radicals by addition or subtraction whenever possible.

21. $3\sqrt{2} + 2\sqrt{2}$

22. $5\sqrt{8} - 3\sqrt{18}$

23. $2\sqrt{x} - \sqrt{4x} + \sqrt{8x}$

24. $\sqrt{2a} + \sqrt{18a} - \sqrt{50a}$

25. $\sqrt[3]{3} - \sqrt[3]{81} + \sqrt[3]{24}$

26. $\sqrt[3]{2} - 6\sqrt[3]{2} + 4\sqrt[3]{8}$

27. $4a\sqrt{x^3} - 2\sqrt{a^2x^3} + 3x\sqrt{a^2x}$

28. $x\sqrt{2x^3} - 8\sqrt{8x^5} + x^2\sqrt{98x}$

29. $3\sqrt{b^3} + 4\sqrt{a^2bc^4} + \sqrt{4b^5c^2}$

30. $a\sqrt{8a^2b} - \sqrt{18a^4b} + a^2\sqrt{32b}$

In problems 31–55, rationalize the denominators and simplify.

31. $\sqrt{\tfrac{2}{3}}$

32. $\sqrt{\tfrac{5}{12}}$

33. $\sqrt{\tfrac{4}{5}}$

34. $\dfrac{\sqrt{8}}{\sqrt{6}}$

35. $\sqrt{\dfrac{2}{xy}}$

36. $\dfrac{\sqrt[3]{3}}{\sqrt[3]{9}}$

37. $\sqrt{\dfrac{16}{3}}$

38. $\dfrac{\sqrt[3]{6}}{\sqrt[3]{5}}$

39. $2\sqrt{\dfrac{1}{2}}$

40. $\sqrt{\dfrac{x^4}{z}}$

41. $\sqrt{\dfrac{2a^3}{5bc^5}}$

42. $\sqrt{\dfrac{2}{7x^3}}$

43. $\sqrt{\dfrac{4a}{x+y}}$

44. $\sqrt{\dfrac{x-y}{x+y}}$

45. $\dfrac{\sqrt{4b}}{\sqrt{b(x-y)}}$

46. $\sqrt{\dfrac{a+b}{a-b}}$

47. $\sqrt{\dfrac{a}{3} + \dfrac{5}{b}}$

48. $\sqrt{\dfrac{1}{a} - \dfrac{9}{5b^2}}$

49. $\sqrt{16 + \dfrac{3}{7x^3}}$

50. $\sqrt{\dfrac{ab^2}{2}} + \sqrt{\dfrac{16a^2b}{3}} - \sqrt{\dfrac{9ab^2}{2}} - \sqrt{\dfrac{a^2b}{3}}$

51. $\dfrac{1}{\sqrt{8x}} - \dfrac{5\sqrt{2x}}{x} + \dfrac{3}{\sqrt{18x}}$

52. $\sqrt{\dfrac{3a}{b}} - \sqrt{\dfrac{b}{3a}} - \dfrac{5}{3}\sqrt{\dfrac{3b}{a}}$

53. $\sqrt{3a} + \sqrt{12a} - \dfrac{6a}{\sqrt{3a}}$

54. $\dfrac{3}{2}\sqrt{\dfrac{3a}{x}} + 5\sqrt{\dfrac{3x}{4a}} - \sqrt{\dfrac{3a}{x}} - \dfrac{1}{2}\sqrt{\dfrac{3x}{a}}$

55. $2\sqrt{\dfrac{2a}{b}} - 4\sqrt{\dfrac{b}{2a^3}} + 6\sqrt{\dfrac{a^3b}{8}}$

In problems 56–75, multiply and express in as simple a form as possible.

56. $\sqrt{2}\sqrt{5}$

57. $\sqrt{6}\sqrt{12}$

58. $\sqrt{3}\sqrt{15}$

59. $\sqrt[3]{5}\sqrt[3]{25}$

60. $2\sqrt{5}\sqrt{10}$

61. $\sqrt[3]{x}\sqrt[3]{x^4}$

62. $6\sqrt{m} \cdot 5\sqrt{m^3}$

63. $\sqrt{3}\sqrt{x}\sqrt{6x^2}$

64. $(2\sqrt{2x})^2$

65. $\sqrt[3]{-4}\sqrt[3]{18}$

66. $\sqrt{a}\sqrt{2a}\sqrt{6a^3}$

67. $(\sqrt{3}-\sqrt{2})(\sqrt{3}+\sqrt{2})$

68. $(2\sqrt{3}-\sqrt{5})(2\sqrt{3}+\sqrt{5})$

69. $(5\sqrt{2}+3\sqrt{3})(5\sqrt{2}-3\sqrt{3})$

70. $(\sqrt{x}-2\sqrt{y})(\sqrt{x}+2\sqrt{y})$

71. $(\sqrt{x}+3\sqrt{y})^2$

72. $(\sqrt{a}+3\sqrt{b})^2$

73. $(3\sqrt{x}+2\sqrt{y})(\sqrt{x}-3\sqrt{y})$

74. $(2\sqrt[3]{x}-1)(4\sqrt[3]{x}+2)$

75. $(a\sqrt{x}-b\sqrt{y})^3$

In problems 76–93, rationalize the denominator.

76. $\dfrac{2-\sqrt{3}}{3-\sqrt{3}}$

77. $\dfrac{1}{2+5\sqrt{2}}$

78. $\dfrac{1+\sqrt{2}}{1-\sqrt{2}}$

79. $\dfrac{1}{\sqrt{2}-1}$

80. $\dfrac{\sqrt{2}-\sqrt{3}}{2\sqrt{2}-\sqrt{3}}$

81. $\dfrac{1}{3-2\sqrt{2}}$

82. $\dfrac{\sqrt{a}-\sqrt{b}}{\sqrt{a}+\sqrt{b}}$

83. $\dfrac{a-2\sqrt{x}}{a+2\sqrt{x}}$

84. $\dfrac{\sqrt{3a}-\sqrt{b}}{\sqrt{a}-\sqrt{b}}$

85. $\dfrac{\sqrt{xy}}{\sqrt{x}-\sqrt{y}}$

86. $\dfrac{a+b-\sqrt{a-b}}{a+b+\sqrt{a-b}}$

87. $\dfrac{\sqrt{a+b}+\sqrt{a-b}}{\sqrt{a+b}-\sqrt{a-b}}$

88. $\dfrac{1}{x-\sqrt{x^2-1}}$

*89. $\dfrac{\sqrt{2}}{\sqrt{3}-2\sqrt{2}+1}$

*90. $\dfrac{2}{\sqrt{3}-\sqrt{5}+2}$

*91. $\dfrac{a}{\sqrt{a}+\sqrt{b}-1}$

*92. $\dfrac{1}{\sqrt[3]{3}-\sqrt[3]{2}}$

*93. $\dfrac{2}{\sqrt[3]{5}+\sqrt[3]{2}}$

In problems 94–99, rationalize the numerator.

94. $\dfrac{\sqrt{5}-\sqrt{2}}{3}$

95. $\dfrac{\sqrt{3}+\sqrt{2}}{5}$

96. $\dfrac{1 - \sqrt{2}}{3}$ **97.** $\dfrac{1 + \sqrt{3}}{6}$

98. $\dfrac{\sqrt{a} - \sqrt{b}}{5}$ **99.** $\dfrac{\sqrt{x + h} - \sqrt{x}}{h}$

8.3   FRACTIONAL EXPONENTS

If we assume that Rule 3 will hold for the (as yet undefined) fractional exponent $\dfrac{1}{n}$ (n a positive integer), we must have

$$(a^{1/n})^n = a^{n/n} = a^1 = a$$

But since $(\sqrt[n]{a})^n = a$, the uniqueness of our principal nth roots gives us the next definition.

Definition 8.4. $a^{1/n} = \sqrt[n]{a}$ for all positive integers n.

Again, as was the case for a^0 and a^{-n}, we have not *proved* that $a^{1/n} = \sqrt[n]{a}$. Rather, we are saying that, if we want Rule 3 to hold for the exponent $1/n$ as well as for the integer m, we must *define* $a^{1/n} = \sqrt[n]{a}$. For other rational number exponents we have

Definition 8.5. If p and q are integers with $q > 0$ and if p and q are not both even when a is negative, then

$$a^{p/q} = (a^{1/q})^p = (\sqrt[q]{a})^p$$

or

$$a^{p/q} = (a^p)^{1/q} = \sqrt[q]{a^p}$$

Thus $a^{p/q}$ means the qth root of the number a raised to the pth power. If q is odd or a is nonnegative, it makes no difference whether we just take the qth root and raise this result to the pth power, or whether we raise the number to the pth power before taking the qth root. The result is the same. The case when q is even and a negative leads, as pointed out before, to the complex numbers to be considered in Chapter 11.

Example 1.

$$8^{2/3} = (8^{1/3})^2 = (\sqrt[3]{8})^2 = 2^2 = 4 \qquad \text{or} \qquad 8^{2/3} = (8^2)^{1/3} = (64)^{1/3} = \sqrt[3]{64} = 4$$

Example 2. $(-32)^{2/5} = (\sqrt[5]{-32})^2 = (-2)^2 = 4.$

Here the use of the alternate definition involves obtaining $\sqrt[5]{(-32)^2}$ and evaluating this is certainly far more work. In general, $(\sqrt[q]{a})^p$ is the best form to use in evaluating $a^{p/q}$.

The reason for insisting in Definition 8.5 that not both p and q be even when a is negative can be illustrated by an example. Consider the expression $(-8)^{2/6}$. If we use $a^{p/q} = \sqrt[q]{a^p}$, then $(-8)^{2/6} = \sqrt[6]{(-8)^2} = \sqrt[6]{64} = 2.$ But if we use $a^{p/q} = (\sqrt[q]{a})^p$, then $(-8)^{2/6} = (\sqrt[6]{-8})^2$ which is not defined. On the other hand, $\frac{2}{6} = \frac{1}{3}$, and $(-8)^{2/6} = (-8)^{1/3} = \sqrt[3]{-8} = -2.$ Thus if m and n are both even, we can write $\dfrac{m}{n} = \dfrac{p}{q}$ where p and q are not both even and take $a^{m/n} = a^{p/q}$.

Before using Definition 8.5, we should note that the rules for exponents do not always apply when $a < 0$. For example,

$$[(-2)^2]^{3/2} = 4^{3/2} = (\sqrt{4})^3 = 2^3 = 8$$

But, if we try to use Rule 3, we have

$$[(-2)^2]^{3/2} = (-2)^{2 \cdot \frac{3}{2}} = (-2)^3 = -8$$

To avoid such contradictory results let us consider only expressions $a^{p/q} \left(\text{for } \dfrac{p}{q} \text{ not an integer}\right)$ when $a > 0$.

We are now in a position to prove the following theorem.

Theorem 8.1. Let a and b be positive real numbers and x and y be rational numbers. Then

(1) $a^x a^y = a^{x+y}$,   (2) $(a^x)^y = a^{xy}$,   (3) $(ab)^x = a^x b^x$.

We will consider only the proof of (1) here leaving the proofs of (2) and (3) as exercises for the student.

PROOF: Let $x = \dfrac{m}{n}$, $y = \dfrac{r}{s}$, m, n, r, s integers, $n > 0, s > 0$. Then

$a^{m/n} a^{r/s} = a^{ms/ns} a^{rn/sn} = \sqrt[ns]{a^{ms}} \sqrt[ns]{a^{rn}}$	Definition 8.5
$= \sqrt[ns]{a^{ms} a^{rn}}$	Property (1), Section 8.2
$= \sqrt[ns]{a^{ms+rn}}$	Rule 1 for integral exponents
$= a^{(ms+rn)/ns}$	Definition 8.5
$= a^{m/n+r/s}$	Definition of addition of rational numbers

The student may note that we have stated only three of the five rules originally given. The other two, however, follow from these three and the definition of a negative exponent:

$$a^{-x} = \frac{1}{a^x} \quad (a \neq 0)$$

Thus, for example,

$$\left(\frac{a}{b}\right)^x = (ab^{-1})^x = a^x b^{-x} = \frac{a^x}{b^x} \quad (b \neq 0)$$

by our definition and (3) of our theorem.

The use of fractional exponents often facilitates the simplification of radical expressions. We illustrate by the following examples.

Example 3. $\sqrt[3]{\sqrt{5}} = \sqrt[3]{5^{1/2}} = (5^{1/2})^{1/3} = 5^{1/2 \cdot 1/3} = 5^{1/6} = \sqrt[6]{5}$

Example 4. $\dfrac{\sqrt[3]{a^2}}{\sqrt[5]{a^3}} = \dfrac{a^{2/3}}{a^{3/5}} = a^{2/3 - 3/5} = a^{1/15} = \sqrt[15]{a}$

Example 5. $(\sqrt[3]{x^2})^6 = (x^{2/3})^6 = x^4$

Example 6.

$$\left(\frac{2\sqrt{x^3}}{\sqrt[3]{y^4}}\right)^6 = \left(\frac{2x^{3/2}}{y^{4/3}}\right)^6 = \frac{(2x^{3/2})^6}{(y^{4/3})^6} = \frac{2^6(x^{3/2})^6}{(y^{4/3})^6} = \frac{64x^9}{y^8}$$

Exercise 8.3

In problems 1–10, rewrite the given expression in radical form and simplify if possible.

1. $4^{1/2}$ 2. $8^{1/3}$ 3. $16^{3/4}$

4. $27^{2/3}$ 5. $(a^5)^{1/3}$ 6. $(4x^2)^{1/2}$

7. $16^{-1/2}$ 8. $x^{5/4}x^{-1/4}$ 9. $8^{1/2}2^{-1/2}$

10. $\left(\dfrac{x^3}{27}\right)^{-1/3}$

In problems 11–20, reduce the given radical to a radical of lower index by the use of fractional exponents and then simplify further if possible.

11. $\sqrt[4]{a^2}$ 12. $\sqrt[9]{a^3}$ 13. $\sqrt[4]{9}$

14. $\sqrt[4]{25}$

15. $\sqrt[6]{8a^3}$

16. $\sqrt[4]{16a^8y^4}$

17. $\sqrt[4]{4a^2x^6}$

18. $\sqrt[10]{32a^5b^{15}}$

19. $\sqrt[9]{\dfrac{a^3x^6}{27}}$

20. $\sqrt[16]{16x^{-8}}$

In problems 21–30, apply the laws of exponents and then change to radical form and simplify if possible.

21. $x^{1/4}x^{3/4}$

22. $x^{-2}x^{2/3}$

23. $a^{1/2}a^{-1}$

24. $(32)^{-3/5}$

25. $x^{1/2}x^{1/3}x^{1/4}$

26. $(x^{1/2}x^2x^{1/3})^6$

27. $(x^{-1/2}x^{1/4})^{-1}$

28. $\left(\dfrac{a^{1/2}}{a^{1/3}}\right)^2$

29. $\left(\dfrac{16}{a^2b^4}\right)^{1/2}$

30. $\left(\dfrac{-27}{a^6b^6}\right)^{-1/3}$

In problems 31–50, simplify the given expression by the use of rational exponents and then change back to radical form.

31. $\sqrt{2}\,\sqrt[3]{4}$

32. $\sqrt{3}\,\sqrt[3]{3}$

33. $\sqrt[3]{x}\,\sqrt{x}$

34. $\sqrt[3]{a^2}\,\sqrt[4]{a^3}$

35. $\sqrt[3]{\sqrt{x}}$

36. $\sqrt{\sqrt{3}}$

37. $\sqrt[4]{\sqrt[4]{5}}$

38. $\sqrt{a}\,\sqrt[4]{a}$

39. $\sqrt[3]{a^2}\,\sqrt[4]{a^2}$

40. $(\sqrt[3]{ax^2})^5$

41. $(2\sqrt[3]{a})^4$

42. $\sqrt{\sqrt{2}}$

43. $(a\,\sqrt[5]{4})^3$

44. $\dfrac{\sqrt[3]{a}}{\sqrt{a}}$

45. $\dfrac{\sqrt[4]{8}}{\sqrt{2}}$

46. $\dfrac{\sqrt[4]{3}}{\sqrt{81}}$

47. $\dfrac{3\sqrt{y}}{\sqrt[3]{x^2y^4}}$

48. $\sqrt[4]{\dfrac{27a^2y^6}{18b^4}}$

49. $\sqrt{\sqrt{9a^4b^6}\,\sqrt{a^8b^2}}$

50. $\sqrt[5]{\dfrac{243x^5y^9}{32xy^4}}$

* **51.** Prove that $(a^x)^y = a^{xy}$ when x and y are rational numbers and a is a positive real number.

* **52.** Prove that $(ab)^x = a^xb^x$ when x is a rational number and a and b are positive real numbers.

In problems 53–58, the given expression arises in the calculus. Simplify into polynomials or simple fractions and express in terms of radicals with denominators free of radicals.

53. $\dfrac{2x\sqrt[5]{x} - (x^2 - 5)\frac{1}{5}x^{-4/5}}{(\sqrt[5]{x})^2}$

54. $\dfrac{(x^2 - 1)^{3/2} - x \cdot \frac{3}{2}(x^2 - 1)^{1/2}2x}{[(x^2 - 1)^{3/2}]^2}$

55. $(x - 1)^{2/3} \frac{1}{3}(x + 2)^{-2/3} + (x + 2)^{1/3} \frac{2}{3}(x - 1)^{-1/3}$

56. $\frac{3}{2}(a^{2/3} - x^{2/3})^{1/2}(-\frac{2}{3}x^{-1/3})$

57. $\dfrac{2}{3}\left(\dfrac{1-x}{1+x}\right)^{-1/3}\left[\dfrac{-(1+x)-(1-x)}{(1+x)^2}\right]$

58. $\dfrac{\frac{1}{3}x^{-2/3} + \left(-\dfrac{x^{1/3}}{y^{1/3}}\right)^2\left(\dfrac{1}{3}\right)(y^{-2/3})}{-\dfrac{1}{3}x^{-2/3}}$

8.4 EXPONENTIAL FUNCTIONS

If b is a positive real number $\neq 1$, the rule which associates with each real number x the positive real number b^x is called an **exponential function.** Actually, we have only shown so far that b^x is defined if x is any rational number. However, for x any real number, b^x can be defined so that the laws of exponents hold, that is, so that $b^x b^y = b^{x+y}$, etc., for all real numbers x and y.

For example, if $b = 2$ the expression 2^x is defined for $x = \frac{2}{3}$; that is, $2^{2/3} = \sqrt[3]{4}$, while if x is an irrational number such as $\sqrt{2}$ or π, then 2^x (and, in general, b^x) can be defined so that, just as $2^3 \cdot 2^5 = 2^{3+5} = 2^8$, $2^{\sqrt{2}} \cdot 2^\pi = 2^{\sqrt{2}+\pi}$. Thus we can consider such numbers as $2^{\sqrt{2}}$ and 2^π and even approximate them by rational numbers. For example, $2^{\sqrt{2}}$ is approximately equal to $2^{1.4} = 2^{14/10} = 2^{7/5} = \sqrt[5]{2^7} = 2\sqrt[5]{4}$.

The basic properties of the exponential functions are summarized in the following theorem.

Theorem 8.2. If a and b are positive real numbers, then

(1) $a^x a^y = a^{x+y}$; $(a^x)^y = a^{xy}$; and $(ab)^x = a^x b^x$ for all real numbers x and y.

(2) If x and y are any real numbers and $b \neq 1$, then $b^x = b^y$ if and only if $x = y$.

(3) If r and b are given positive real numbers, then the equation $b^x = r$ (with x the variable) always has a solution. (For example, $2^x = 8$

has the solution $x = 3$; $27^x = 3$ has the solution $x = \frac{1}{3}$; $5^x = 7$ has a solution even though x is not a rational number.)

We shall not attempt a proof of this theorem here. The significance and usefulness of the properties (2) and (3) will become clearer when we discuss the logarithmic function.

8.5   THE GRAPHING OF EXPONENTIAL FUNCTIONS

Let $f(x) = 3^x$. Then $f(0) = 1$, $f(\frac{1}{2}) = 3^{1/2} = \sqrt{3} = 1.732$ (approximately), $f(1) = 3$, $f(2) = 9$, $f(-1) = 3^{-1} = \frac{1}{3}$. Plotting the points with coordinates $(0, 1)$, $(\frac{1}{2}, 1.732)$, $(1, 3)$, $(2, 9)$, and $(-1, \frac{1}{3})$ and drawing a smooth curve through these points, we obtain the graph of $y = 3^x$ in Figure 8.1.

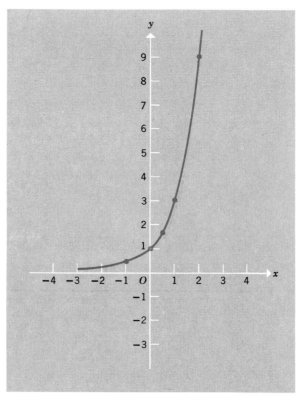

Figure 8.1

In general, the graph of $y = b^x$ where b is a real number greater than 1 is similar to the graph shown in Figure 8.1 in that the curve always intersects the y-axis at $(0, 1)$, gets closer and closer to the x-axis as x takes on the values $-1, -2, -3, \ldots$, and rises as x becomes larger. The larger the value of b, the steeper will be this rise.

Similarly, if b is a positive real number less than 1, the graph of $y = b^x$ will be of the same general shape as the graph of $y = \left(\dfrac{1}{2}\right)^x = \dfrac{1}{2^x} = 2^{-x}$ which is shown in Figure 8.2.

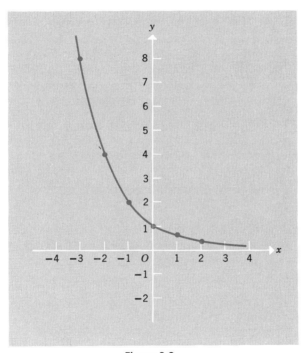

Figure 8.2

Exercise 8.5

1. Graph $f(x) = 2^x$ 2. Graph $f(x) = (\tfrac{3}{2})^x$
3. Graph $f(x) = (\tfrac{1}{3})^x$ 4. Graph $f(x) = (\tfrac{2}{3})^x$
*5. Prove that, if $f(x) = a^x$, $f(x + y) = f(x)f(y)$

8.6   DEFINITION OF A LOGARITHM

If x and b are positive real numbers and $b \neq 1$, then the **logarithm** of x to the base b, $\log_b x$, is defined to be the number y such that b^y is equal to x. For example, $\log_{10} 100 = 2$ because $10^2 = 100$; $\log_3 27 = 3$ because $3^3 = 27$; $\log_4 \dfrac{1}{16} = -2$ because $4^{-2} = \dfrac{1}{4^2} = \dfrac{1}{16}$; $\log_3 1 = 0$ because $3^0 = 1$, etc. The following definition is a somewhat more formal statement.

Definition 8.6. If x and b are positive real numbers and $b \neq 1$, then $\log_b x = y$ if and only if $b^y = x$.

Thus the theory of logarithms and the theory of exponents are just two different ways of looking at the same thing. Every statement about logarithms can be translated into a statement about exponents and vice versa.

Example 1. $27^{2/3} = 9$ and therefore $\log_{27} 9 = \frac{2}{3}$.

Example 2. $\log_{32} \frac{1}{4} = -\frac{2}{5}$ and therefore $32^{-2/5} = \frac{1}{4}$.

Example 3. Find $\log_{1/2} 4$.

SOLUTION: If $\log_{1/2} 4 = y$, then $\left(\dfrac{1}{2}\right)^y = 4$. Hence $\dfrac{1^y}{2^y} = 4$, $1 = 4 \cdot 2^y$, $2^y = \frac{1}{4}$, and thus $y = -2$.

Example 4. If $\log_9 x = \frac{1}{2}$, find x.

SOLUTION: If $\log_9 x = \frac{1}{2}$, then $9^{1/2} = x = 3$.

Example 5. If $\log_b 3 = -1$, find b.

SOLUTION: If $\log_b 3 = -1$, then $b^{-1} = 3$ and $b = \frac{1}{3}$.

Example 6. Find $\log_{10} 7$.

SOLUTION: If $\log_{10} 7 = x$, then $10^x = 7$. Here no rational number gives $\log_{10} 7$. In general, the logarithms of numbers are irrational. Their approximate value can be found from tables, as discussed in Section 8.10, while the method of obtaining such tables is described in calculus texts.

Exercise 8.6

In problems 1–16, express the given equality in logarithmic form.

1. $4 = 2^2$ **2.** $27 = 3^3$

3. $8 = 2^3$ **4.** $\frac{1}{5} = 5^{-1}$

5. $4 = 64^{1/3}$ **6.** $27 = 9^{3/2}$

7. $\frac{1}{8} = 2^{-3}$ **8.** $8 = (\frac{1}{2})^{-3}$

9. $125 = 5^3$ **10.** $1 = 7^0$

11. $18 = 3^p$ **12.** $A = 10^{1.25}$

13. $1 = \pi^0$ **14.** $B = (1.718)^{15}$

15. $a = b^{\log_b a}$ **16.** $17 = 10^{\log_{10} 17}$

In problems 17–32, express the given equality in exponential form.

17. $\log_2 16 = 4$ **18.** $\log_3 27 = 3$

19. $\log_5 125 = 3$ **20.** $\log_{1/2} 4 = -2$

21. $\log_{1/2} \frac{1}{8} = 3$ **22.** $\log_7 1 = 0$

23. $\log_2 \frac{1}{16} = -4$ **24.** $\log_{9/16} \frac{3}{4} = \frac{1}{2}$

25. $\log_{1/125} 5 = -\frac{1}{3}$ **26.** $\log_{10} 1{,}000 = 3$

27. $\log_{10} \frac{1}{10} = -1$ **28.** $\log_a 32 = 5$

29. $\log_{10} 1 = 0$ **30.** $\log_{a^2} 81 = 2$

31. $\log_7 49 = p$ **32.** $\log_b a = x$

In problems 33–46, find the given logarithm by changing to exponential form.

33. $\log_2 8$ **34.** $\log_9 81$

35. $\log_3 27$ **36.** $\log_{36} 6$

37. $\log_2 2$ **38.** $\log_2 \frac{1}{32}$

39. $\log_{10} 10{,}000$ **40.** $\log_{10} 0.001$

41. $\log_{b^{1/2}} b^6$ **42.** $\log_b b^{1.4}$

43. $\log_{a^2} a^{10}$ **44.** $\log_{1/2} 8$

45. $\log_{1/5} 125$ **46.** $\log_{1/b} b^7$

In problems 47–63, find the value of $x, A, B, \ldots$, in the given equation by changing to exponential form.

47. $\log_2 x = 5$ 48. $\log_3 A = 4$

49. $\log_{1/2} B = -3$ 50. $\log_{1/5} x = 3$

51. $\log_{36} N = -\frac{3}{2}$ 52. $\log_{81} P = -\frac{3}{4}$

53. $\log_7 p = -3$ 54. $\log_x 32 = 5$

55. $\log_b 5 = \frac{1}{2}$ 56. $\log_a 4 = \frac{1}{2}$

57. $\log_a 7 = \frac{1}{3}$ 58. $\log_b 2 = \frac{1}{6}$

59. $\log_x 8 = \frac{3}{5}$ 60. $\log_{10} 10^2 = L$

61. $\log_{10} 10^{1.25} = L$ 62. $\log_{10} \left(\frac{1}{10}\right)^5 = x$

63. $\log_{10} \left(\frac{1}{100}\right)^{1/2} = x$

8.7 THE LOGARITHMIC FUNCTION

If b is a positive real number and $b \neq 1$, the rule which associates with each positive real number x, the real number $\log_b x$, is a function, called the **logarithmic function**, whose domain is the set of all positive real numbers.

Example. If $f(x) = \log_2 x$, then $f(8) = 3, f(4) = 2, f(2) = 1, f(1) = 0, f(\sqrt{2}) = \frac{1}{2}, f(\frac{1}{2}) = -1, f(\frac{1}{4}) = -2$. If we graph this function, plotting the points whose coordinates are $(8, 3), (4, 2), (2, 1), (1, 0), (\sqrt{2}, \frac{1}{2}), (\frac{1}{2}, -1), (\frac{1}{4}, -2)$ and drawing a smooth curve through these points, we obtain the graph in Figure 8.3.

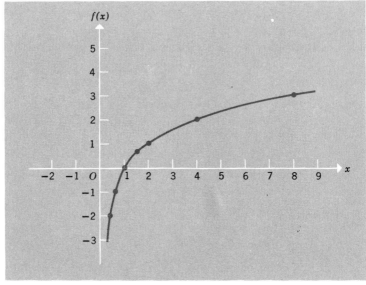

Figure 8.3

In general, if b is a real number greater than 1, the graph of $y = \log_b x$ is similar to that shown in Figure 8.3. The student should observe how the graphs of $y = b^x$ and $y = \log_b x$ are related and consider the graph of $y = \log_b x$ when $0 < b < 1$.

Exercise 8.7

1. Graph the function f defined by $f(x) = \log_b x$ for $b = \frac{1}{2}$.

2. Do as in problem 1 for $b = 3$.

3. Draw the graphs of $y = 2^x$ and $y = \log_2 x$ on the same coordinate system. Also draw the line $y = x$. (*Note:* We can think of one of these graphs as the "mirror image" of the other in a mirror placed on the line $y = x$.)

4. Draw the graphs of $y = (\frac{1}{2})^x$ and $y = \log_{1/2} x$ on the same graph. Also draw the line $y = x$. (See note in problem 3.)

8.8 PROPERTIES OF LOG$_b$ X

If b, x, and y are positive real numbers with $b \neq 1$, then the following identities hold:

(1) $b^{\log_b x} = x$;

(2) $\log_b (xy) = \log_b x + \log_b y$;

(3) $\log_b \left(\dfrac{x}{y}\right) = \log_b x - \log_b y$;

(4) $\log_b x^r = r \log_b x$ for any real number r;

(5) $\log_b x = \log_b y$ if and only if $x = y$.

Proof of properties (1) and (2):

(1) By our definition of $\log_b x$, $\log_b x = y$ means $b^y = x$ or $b^{\log_b x} = x$.

(2) $b^{\log_b (xy)} = xy$ by (1). But $b^{\log_b x} = x$ and $b^{\log_b y} = y$ also by (1). Hence

$$b^{\log_b (xy)} = b^{\log_b x} b^{\log_b y} = b^{\log_b x + \log_b y}$$

by the laws of exponents. By (2) of Theorem 8.2 of Section 8.4, $\log_b (xy) = \log_b x + \log_b y$.

The proofs of (3) and (4) are left as exercises for the student. Finally, (5) follows from the fact that, if $\log_b x = s$ and $\log_b y = t$, then $b^s = x$ and $b^t = y$ by definition of a logarithm. But if $s = t$, then $x = b^s = b^t = y$ and conversely, if $x = b^s = b^t = y$, then $s = t$ by (2) of Theorem 8.2.

Exercise 8.8

In problems 1–15, find the given logarithm by using the properties 2–4 of the text given that $\log_2 3 = 1.585$ and $\log_2 5 = 2.322$.

1. $\log_2 2$ 2. $\log_2 6$ 3. $\log_2 10$

4. $\log_2 12$ 5. $\log_2 \frac{1}{2}$ 6. $\log_2 \frac{1}{3}$

7. $\log_2 0.1$ 8. $\log_2 \frac{3}{4}$ 9. $\log_2 \sqrt{2}$

10. $\log_2 \sqrt[3]{2}$ 11. $\log_2 \sqrt{6}$ 12. $\log_2 \sqrt[3]{5}$

13. $\log_2 15$ 14. $\log_2 \frac{5}{2}$ 15. $\log_2 \frac{2}{5}$

In problems 16–21, given that $\log_{10} A = 2.1234$, find the logarithm without computing A.

16. $\log_{10} A^2$ 17. $\log_{10} A^3$ 18. $\log_{10} \sqrt{A}$

19. $\log_{10} \sqrt[3]{A}$ 20. $\log_{10} 1/A$ 21. $\log_{10} 1/A^2$

In problems 22–30, use the properties of logarithms to write the given expression as the logarithm of a single number.

22. $\log_a 7 + \log_a 5$ 23. $\log_a 14 - \log_a 2$

24. $\frac{1}{2} \log_a 16$ 25. $3 \log_a 2\sqrt[3]{3}$

26. $\log_a \frac{1}{3} - \log_a \frac{5}{6} + \log_a \frac{5}{4}$

27. $\log_a x - 2 \log_a y + \log_a xy^2$

28. $\log_a (x^2 - 4y^2) - \log_a xy - \log_a (x + 2y)$

29. $2 \log_a 2 - \log_a 5x^2 - (\log_a 12 - \log_a 25x^3)$

30. $\log_a (y^2 + 2y - 3) - \log_a (y - 5) + \log_a (y^2 - 3y - 10)$
$- \log_a (y^2 + 5y - 6)$

In problems 31–38, use the fact that $\log_{10} 5 = 0.6990$.

31. Find $\log_{10} 25$ **32.** Find $\log_{10} 125$

33. Find $\log_{10} 50$ **34.** Find $\log_{10} 0.25$

35. Find $\log_{10} 4$ **36.** Find $\log_{10} 20$

37. Find $\log_{10} 1.25$ **38.** Find $\log_{10} 12.5$

39. Prove property 3 of the text.

40. Prove property 4 of the text.

The identities in problems 41 and 42 are needed in the calculus.

***41.** Show that $-\log_b (x - \sqrt{x^2 - 1}) = \log_b (x + \sqrt{x^2 - 1})$ for $x \geqslant 1$.

***42.** Show that $\dfrac{\log_b (x + h) - \log_b x}{h} = \log_b \left(1 + \dfrac{h}{x}\right)^{1/h}$ for $x > 0$ and

$h > 0$.

8.9 SOLUTION OF EQUATIONS

There are two very common techniques for solving a wide variety
of equations involving logarithmic and exponential expressions. They
are the operations of

(1) taking the logarithms of both sides, and

(2) exponentiating both sides.

We will give examples of each technique and note that (1) makes use
of the fact guaranteed by property (5) of the logarithm function that,
if two numbers are equal, their logarithms are equal (that is, if $x = y$
then $\log_b x = \log_b y$), while (2) makes use of the fact that, if $r = s$ then
$b^r = b^s$.

Example 1. Solve the equation $27^{x^2+1} = 243$.

SOLUTION: As usual, we seek all numbers which satisfy the equation. If x is such a number, then

$$27^{x^2+1} = 243$$

Then, taking logarithms to the base 3 of both sides we have

$$\log_3 27^{x^2+1} = \log_3 243$$

Since $\log_b x^r = r \log_b x$, it follows that

$$(x^2 + 1) \log_3 27 = \log_3 243$$

But $\log_3 27 = 3$ and $\log_3 243 = 5$ so that we have

$$3(x^2 + 1) = 5$$
$$x^2 + 1 = \tfrac{5}{3}$$
$$x^2 = \tfrac{2}{3}$$
$$x = \sqrt{\tfrac{2}{3}} \text{ or } x = -\sqrt{\tfrac{2}{3}}$$

It is easily seen by checking that each of these numbers satisfies the equation.

The choice of logarithms to the base 3 was, of course, motivated by the fact that both 27 and 243 are powers of 3. If no such happy choice is available to us we take logarithms to the base 10 as in the next example.

Example 2. Solve the equation $5^x = 12$.

SOLUTION: Taking logarithms to the base 10 of both sides we have

$$\log_{10} 5^x = \log_{10} 12$$

But $\log_{10} 5^x = x \log_{10} 5$ so that

$$x \log_{10} 5 = \log_{10} 12$$

and

$$x = \frac{\log_{10} 12}{\log_{10} 5}$$

In the next section we will show how to use tables to find approximate values of logarithms to the base 10 and hence show how an approximate value of x can be obtained as a solution to equations such as that of Example 2.

Example 3. Solve the equation $\log_3 (x^2 - 8x) = 2$.

SOLUTION: This time the technique consists of exponentiation of both sides to the base 3 to obtain

$$3^{\log_3 (x^2 - 8x)} = 3^2$$

Then since $b^{\log_b x} = x$ by identity (1) of Section 8.8, we have

$$x^2 - 8x = 9$$
$$x^2 - 8x - 9 = 0$$
$$(x - 9)(x + 1) = 0$$

and hence

$$x = 9 \quad \text{or} \quad x = -1$$

Check: Since

$$\log_3 [9^2 - 8(9)] = \log_3 9 = 2$$

and

$$\log_3 [1 - 8(-1)] = \log_3 9 = 2$$

the solution set of the given equation is $\{-1, 9\}$.

Example 4. Solve the equation $\log_{10} (x^2 + 3x) + \log_{10} 5x = 1 + \log_{10} 2x$.

SOLUTION: We first subtract $\log_{10} 2x$ from both sides of our equation so as to have the right-hand side free of logarithmic expressions and obtain

$$\log_{10} (x^2 + 3x) + \log_{10} 5x - \log_{10} 2x = 1$$

Then by (2) and (3) of Section 8.8 we obtain

$$\log_{10} \frac{5x(x^2 + 3x)}{2x} = 1$$

or, since $x \neq 0$ (why?),

$$\log_{10} \tfrac{5}{2}(x^2 + 3x) = 1$$

Now we exponentiate to the base 10 to obtain

$$10^{\log_{10} 5/2(x^2+3x)} = 10^1$$

and, since $b^{\log_b x} = x$,

$$\tfrac{5}{2}(x^2 + 3x) = 10$$

Thus $5x^2 + 15x = 20$ and hence

$$x^2 + 3x - 4 = (x + 4)(x - 1) = 0$$

Therefore

$$x = 1 \quad \text{or} \quad x = -4$$

But $x = -4$ is not a solution (why?) whereas $x = 1$ is a solution because when $x = 1$ we have

$$\log_{10}(x^2 + 3) + \log_{10} 5x = \log_{10} 4 + \log_{10} 5 = \log_{10}(4 \cdot 5) = \log_{10} 20$$

and also, since $\log_{10} 10 = 1$,

$$1 + \log_{10} 2x = \log_{10} 10 + \log_{10} 2 = \log_{10}(2 \cdot 10) = \log_{10} 20$$

Exercise 8.9

In problems 1–10, take the logarithms of both sides of the given equation to the base given and use the properties of logarithms to solve the equation. In the cases where base 10 is used, the answer will be in terms of logarithms of numbers to the base 10 as in Example 2 of the text.

1. $3^x = 9$ (base 3)
2. $2^{x^2} = 16$ (base 2)
3. $5^{2x-x^2} = \frac{1}{125}$ (base 5)
4. $\dfrac{9^{x^2}}{3^{5x}} = \dfrac{1}{9}$ (base 3)
5. $5^{2x} = 28$ (base 10)
6. $2^{3x} = 75(3^{2x})$ (base 10)
7. $18^{x^2} = 15$ (base 10)
8. $(1.5)^{-x} = 0.35$ (base 10)
9. $4^{3x-1} = (\frac{1}{2})^{x-1}$ (base 2)
10. $2^{x^2-4x} = \frac{1}{16}$ (base 2)

In problems 11–18, exponentiate both sides of the given equation and use the properties of logarithms to solve the equation.

11. $\log_{10} x = 3$
12. $\log_3 x = 5$
13. $\log_2 x^2 = -2$
14. $\log_5 x^2 = 0$
15. $\log_2(x^2 - 2x) = 3$
16. $\log_{1/2}(x^2 - 15x) = -4$
17. $\log_3(x + 4) + \log_3(x - 2) = 3$
18. $\log_5(x - 1) + \log_5(4x + 1) = 3$

In problems 19–22, prove that the given equation is an identity.

*19. $(\log_b a)(\log_a c) = \log_b c$ for $c > 0$
*20. $10^{\log_b c} = c^{\log_b 10}$ for $c > 0$
*21. $(ax)^s = a^t$ where $s = 1/\log_a x$ and $t = 1 + 1/\log_a x$ for $x > 0$ and $x \neq 1$
*22. $\log_a \dfrac{1}{x^2} = \left(\log_a \dfrac{1}{b^2}\right)(\log_b x)$ for $x > 0$ and $x \neq 1$

5.0   .6990

log₁₀ 5.14

5.1

8.10 FINDING LOG₁₀ X BY TABLES

To find $\log_{10} x$ for any positive number x we use the table of four-place logarithms in the back of this book, part of which is reproduced here for convenience:

	0	1	2	3	4	5	6	7	8	9
50	6990	6998	7007	7016	7024	7033	7042	7050	7059	7067
51	7076	7084	7093	7101	7110	7118	7126	7135	7143	7152
52	7160	7168	7177	7185	7193	7202	7210	7218	7226	7235
53	7243	7251	7259	7267	7275	7284	7292	7300	7308	7316
54	7324	7332	7340	7348	7356	7364	7372	7380	7388	7396

The entry 50 at the top of the left-hand column is understood to be the number 5.0; the number beneath it, 5.1; 5.2; etc. The other entries are understood to be 0.6990, 0.6998, etc.

Example 1. $\log_{10} 5 = 0.6990$, $\log_{10} 5.01 = 0.6998$, $\log_{10} 5.02 = 0.7007$ while $\log_{10} 5.1 = 0.7076$, $\log_{10} 5.11 = 0.7084$, $\log_{10} 5.12 = 0.7093$. (Although the above results are only approximations, we will write the equality sign for convenience.)

Example 2. $\log_{10} 5.103$ is taken to be $\frac{3}{10}$ of the way between $\log_{10} 5.10 = 0.7076$ and $\log_{10} 5.11 = 0.7084$. Hence $\log_{10} 5.103 = 0.7076 + \frac{3}{10}(0.7084 - 0.7076) = 0.7076 + (0.3)(0.0008) = 0.7076 + 0.00024 = 0.7078$ approximately. This process is known as **linear interpolation** and although it is not strictly accurate it does give a better approximation than we would get by just using $\log_{10} 5.10$ or $\log_{10} 5.11$.

Our tables, then, may be used quite directly to obtain the logarithms to the base 10 of numbers between 1 and 10. To find the logarithm to the base 10 of a number less than 1 or greater than 10 we first write the number in its so-called **scientific form** as a number between 1 and 10 multiplied by 10 raised to an appropriate power. For example,

$34.08 = 3.408 \times 10^1; \quad 340.8 = 3.408 \times 10^2; \quad 0.03408 = 3.408 \times 10^{-2}$. After doing this we can then find the logarithm of this number by using the facts that $\log_{10}(xy) = \log_{10} x + \log_{10} y$ and $\log_{10} 10^m = m \log_{10} 10 = m \cdot 1 = m$.

Example 3. $\log_{10} 54.2 = \log_{10}(5.42 \times 10^1) = \log_{10} 5.42 + \log_{10} 10$
$$= 0.7340 + 1 = 1.7340$$

Example 4. $\log_{10} 531.7 = \log_{10}(5.317 \times 10^2) = \log_{10} 5.317 + \log_{10} 10^2$
$$= 0.7257 + 2 = 2.7257$$

Here we interpolate to find $\log_{10} 5.317$. Since 5.317 is $\frac{7}{10}$ of the way between 5.31 and 5.32, and $\log_{10} 5.31 = 0.7251$ and $\log_{10} 5.32 = 0.7259$, we have

$$\log_{10} 5.317 = 0.7251 + \tfrac{7}{10}(0.7259 - 0.7251)$$
$$= 0.7251 + 0.7(0.0008)$$
$$= 0.7251 + 0.00056 = 0.7257$$

In our last two examples we find the logarithms of numbers less than 1 and also refer to the full table at the back of the book.

Example 5. $\log_{10} 0.314 = \log_{10}(3.14 \times 10^{-1})$
$$= \log_{10} 3.14 + \log_{10} 10^{-1}$$
$$= 0.4969 - 1$$

The student should note here that, while we can, and normally do, rewrite a logarithm such as $0.7257 + 2$ as 2.7257, we do *not* have $0.4969 - 1$ equal to -1.4969 since $-1.4969 = -(0.4969 + 1) = -0.4969 - 1 \neq 0.4969 - 1$.

Example 6. $\log_{10} 0.007146 = \log_{10}(7.146 \times 10^{-3})$
$$= \log_{10} 7.146 + \log_{10} 10^{-3}$$
$$= 0.8541 - 3$$

Here we find $\log_{10} 7.146$ by taking it to be $\frac{6}{10}$ of the way between $\log_{10} 7.14 = 0.8537$ and $\log_{10} 7.15 = 0.8543$. Hence $\log_{10} 7.146 = 0.8537 + \frac{6}{10}(0.8543 - 0.8537) = 0.8537 + 0.6(0.0006) = 0.8537 + 0.00036 = 0.8541$ approximately.

Notice that we always write our logarithms as the sum of a positive decimal less than 1 and an integer: $1.7340 = 1 + 0.7340$, $2.7257 = 2 + 0.7257$, $0.4969 - 1 = 0.4969 + (-1)$, $0.8541 - 3 = 0.8541 + (-3)$. The positive decimal part is called the **mantissa** of the logarithm and the integral part, the **characteristic.** The mantissas in our four examples are, respectively, 0.7340, 0.7257, 0.4969, and 0.8541; the characteristics are, respectively, 1, 2, −1, and −3. In $\log_{10} 5 = 0.6990$, the mantissa is 0.6990 and the characteristic is 0.

Exercise 8.10

In problems 1–15, write the given number in the scientific form.

1. 136	2. 235	3. 27.68
4. 0.234	5. 67.21	6. 78,660
7. 523.2	8. 31,415	9. 0.030103
10. 1.005	11. 0.000375	12. 0.0135
13. 0.265	14. $\frac{1}{100}$	15. $\frac{1}{1000}$

In problems 16–39, use the table given in this section to find the logarithm of the given number to the base 10 after first writing it in scientific form. Use properties (2) to (4) of logarithms (Section 8.8) and interpolation if needed.

16. 5.3	17. 50	18. 0.511
19. 5.07	20. 523	21. 53.9
22. 0.0527	23. 0.00510	24. $\frac{1}{50}$
25. $\frac{1}{52.3}$	26. 0.005	27. 52
28. 520	29. 0.520	30. $\sqrt{53.3}$
31. $\sqrt{523}$	32. 5245	33. 53.64
34. 0.5001	35. 0.05052	36. 5.222
37. 5.346	38. 52.88	39. 529.7

In problems 40–51, use the tables in the back of the book to find the logarithms to the base 10 of the given numbers.

40. 23.7	41. 0.891	42. 1776
43. 1.835	44. 0.05007	45. 7.989
46. 0.1958	47. 0.007963	48. 3.794
49. 0.1999	50. 8.397	51. 60.03

8.11 THE ANTILOGARITHM OF A NUMBER

A number r is given and the question is asked: What number x satisfies the equation $\log_{10} x = r$? We say that x is the **antilogarithm** to the base 10 of r and write $x = \text{antilog}_{10} r$.

If $\log_{10} x = r$, then, by definition of the logarithmic function, $x = 10^r$. That is

$$\text{antilog}_{10} r = 10^r$$

or, in general,

$$\text{antilog}_b r = b^r$$

Thus the antilogarithmic functions are really nothing but the exponential functions.

Example 1. $\text{Antilog}_5 2 = 5^2 = 25$.

Example 2. $\text{Antilog}_{10} 0.7084 = 10^{0.7084}$. Now we look for the number 0.7084 in our logarithm tables and find that $0.7084 = \log_{10} 5.11$. Hence

$$\text{antilog}_{10} 0.7084 = 10^{\log_{10} 5.11} = 5.11$$

since, by (1) of Section 8.8, $b^{\log_b x} = x$.

Example 3. $\text{Antilog}_{10} 3.7168 = 10^{3.7168} = (10^3)(10^{0.7168})$
$$= 10^3 10^{\log_{10} 5.21} = 10^3(5.21) = 5210$$

So far, our given number has been such that the mantissa (the positive decimal part) has always been a number which can be found in the logarithm tables. When it is not, we resort to a process of linear interpolation similar to that described in Section 8.10.

Example 4. $\text{Antilog}_{10} 1.4850 = 10^{1.4850} = (10)(10^{0.4850})$. Now we observe that $\log 3.05 = 0.4843$ while $\log 3.06 = 0.4857$. The difference between 0.4857 and 0.4843 is 0.0014, while the difference between 0.4843 and 0.4850 is 0.0007. Thus 0.4850 lies $\dfrac{0.0007}{0.0014} = \dfrac{1}{2}$ of the way between 0.4843 and 0.4857 and we assume [1], that, likewise, anti-

---

[1] Again, as was the case in Section 8.10, this assumption is not completely justified.

$\log_{10} 0.4850$ lies halfway between 3.05 and 3.06 so that antilog$_{10}$ $0.4850 = 3.055$. Hence

$$\text{antilog}_{10}\ 1.4850 = (10)(3.055) = 30.55$$

Example 5. Antilog $2.3625 = 10^{2.3625} = (10^2)(10^{0.3625})$. Since $\log_{10}$ $2.30 = 0.3617$ and $\log_{10} 2.31 = 0.3636$ and 0.3625 is $\dfrac{0.0008}{0.0019} = \dfrac{8}{19}$ of the way between 0.3617 and 0.3636, we conclude that antilog$_{10}$ 0.3625 is $\frac{8}{19} = 0.4$ (to the nearest tenth) of the way between 2.30 and 2.31. Hence

$$\text{antilog}_{10}\ 2.3625 = (10^2)(2.304) = 230.4$$

There remains to consider the case when our given number is negative. Generally (although not always) such negative numbers will appear in the form of, for example, $0.8762 - 2$, and we will leave for the concluding section of this chapter numbers written as, for example, -3.2731.

Example 6. Antilog$_{10}$ $0.8762 - 2 = 10^{0.8762-2} = (10^{-2})(10^{0.8762})$
$$= (10^{-2})(10^{\log_{10} 7.52})$$
$$= (10^{-2})7.52 = 0.0752$$

Example 7. Antilog$_{10}$ $0.5579 - 1 = 10^{0.5579-1} = (10^{-1})(10^{0.5579})$. Again, we must resort to linear interpolation to complete our solution. Since $\log_{10} 3.61 = 0.5575$, $\log_{10} 3.62 = 0.5587$, and 0.5579 is $\dfrac{0.0004}{0.0012} = \dfrac{1}{3}$ of the way between 0.5575 and 0.5587, we conclude that antilog$_{10}$ 0.5579 is $\frac{1}{3} = 0.3$ (to the nearest tenth) of the way between 3.61 and 3.62. Hence

$$\text{antilog}_{10}\ 0.5579 - 1 = (10^{-1})(3.613) = 0.3613$$

Exercise 8.11

In problems 1–10, evaulate the given antilogarithms without using tables.

1. antilog$_2$ 5
2. antilog$_5$ (-2)
3. antilog$_{10}$ 2
4. antilog$_8$ -1

5. antilog$_7$ 0   **6.** antilog$_a$ x
7. antilog$_4$ $(x^2 + 3)$ **8.** antilog$_2$ (-5)
9. antilog$_{10}$ 1   **10.** antilog$_{10}$ (-5)

In problems 11–38, use the tables in the back of the book to find the given antilogarithms, interpolating when necessary.

11. antilog$_{10}$ 0.3010 $2,0$ **12.** antilog$_{10}$ 1.3010
13. antilog$_{10}$ 2.3010   **14.** antilog$_{10}$ 0.3010 $-$ 2
15. antilog$_{10}$ 1.4440   **16.** antilog$_{10}$ 0.6937
17. antilog$_{10}$ 0.9400 $-$ 2   **18.** antilog$_{10}$ 0.1761 $-$ 1
19. antilog$_{10}$ 0.2718 $-$ 2   **20.** antilog$_{10}$ 4.5809
21. antilog$_{10}$ 0.8312   **22.** antilog$_{10}$ 0.9965 $-$ 2
23. antilog$_{10}$ 0.1035   **24.** antilog$_{10}$ 1.9892
25. antilog$_{10}$ 3.5617   **26.** antilog$_{10}$ 0.8263 $-$ 2
27. antilog$_{10}$ 0.0123   **28.** antilog$_{10}$ 0.4804 $-$ 1
29. antilog$_{10}$ 1.7320   **30.** antilog$_{10}$ 2.6666
31. antilog$_{10}$ 0.3172   **32.** antilog$_{10}$ 0.4545 $-$ 4
33. antilog$_{10}$ 2.1496   **34.** antilog$_{10}$ 1.1066
35. antilog$_{10}$ 0.9326 $-$ 1   **36.** antilog$_{10}$ 0.3741 $-$ 2
37. antilog$_{10}$ 3.1456   **38.** antilog$_{10}$ 6.1234

8.12 COMPUTATIONS BY THE USE OF LOGARITHMS

By use of the tables and the properties of the logarithmic and exponential functions a wide variety of arithmetic computations can be accomplished. We will, in the examples, make extensive use of the following two basic facts about antilogarithms:

   (a) antilog$_{10}$ $x = 10^x$;
   (b) antilog$_{10}$ [log$_{10}$ x] $= 10^{\log_{10} x} = x$;

and the laws of logarithms given in Section 8.8:

   (1) log$_{10}$ $(xy) = $ log$_{10}$ $x + $ log$_{10}$ y;

   (2) log$_{10}$ $\dfrac{x}{y} = $ log$_{10}$ $x - $ log$_{10}$ y;

   (3) log$_{10}$ $x^r = r$ log$_{10}$ x.

Our first example illustrates the use of logarithms for multiplication.

Example 1.

$(5.06)(71.32)$

$= \text{antilog}_{10} \left[\log_{10} (5.06)(10 \times 7.132) \right]$ ⟶ by (b)

$= \text{antilog}_{10} \left[\log_{10} 5.06 + \log_{10} 10 + \log_{10} 7.132 \right]$ ⟶ by (1)

$= \text{antilog}_{10} (0.7042 + 1 + 0.8532)$ ⟶ by tables and interpolation

$= \text{antilog}_{10} 2.5574$ ⟶ by addition

$= 10^{2.5574}$ ⟶ by (a)

$= 10^{2+0.5574}$ ⟶ by addition

$= (10^2)(10^{0.5574})$ ⟶ by laws of exponents

$= (10^2)(10^{\log_{10} 3.609})$ ⟶ by tables and interpolation

$= (10^2)(3.609) = 360.9$ ⟶ by (b)

At this point the student may well say that it would be much easier to do the problem in the ordinary way! Two comments may be made about this fact. First of all, a problem such as $(5.62)(3.97)(4.61)(7.54)$ can be done in the same fashion and, in longer computations, and especially those involving division and root extraction, the value of logarithms is more apparent. Second, many of the steps given above may be omitted by the experienced computer and the whole problem laid out in a more mechanical fashion as follows. (It should be pointed out, however, that the preceding method illustrates the reasoning behind the process of computing with logarithms.)

Example 1'. To multiply 5.06×71.32 we write, as before,

$5.06 \times 71.32 = \text{antilog} \left[\log (5.06)(7.132)(10) \right]$ ⟶ by (b)

$= \text{antilog} \left[\log 5.06 + \log 7.132 + \log 10 \right]$ ⟶ by (1)

Thus the first task is to find the logarithms of the numbers and add them together. This is presented in tabular form below.

$$
\begin{array}{ll}
\log 5.06 & = 0.7042 \\
\log 7.132 & = 0.8532 \\
\underline{\log 10} & = \underline{1.0000} \\
\log 5.06 + \log 7.132 + \log 10 & = 2.5574 = 0.5574 + 2
\end{array}
$$

Thus we see that

$$5.06 \times 71.32 = \text{antilog} \ (0.5574 + 2)$$
$$= 3.609 \times 10^2 = 360.9$$

Several remarks are necessary. First, notice that the base 10 is no longer indicated. Second, notice that the number 71.32 was immediately written in scientific notation as 7.132×10. This is because it must be remembered that the tables only give the logarithms of numbers between 1 and 10. For the same reason the logarithm of the answer, 2.5574, was rewritten as $0.5574 + 2$, which leads to the answer in scientific notation. These conventions will be adhered to in the remaining examples of this chapter.

Our second example deals with division and uses the corresponding rule (2).

Example 2.

$$\frac{50.73}{2.42} = \text{antilog} \left[\log \frac{5.073 \times 10}{2.42} \right] \qquad \text{by } (b)$$
$$= \text{antilog} \ [\log 5.073 + \log 10 - \log 2.42] \qquad \text{by (1) and (2)}$$
$$\log 5.073 = \ \ 0.7053$$
$$\log 10 \ \ \ = \ \ 1.0000$$
$$-\log 2.42 \ \ = -0.3838$$
$$\log 5.073 + \log 10 - \log 2.42 \ \ = \ \ 1.3215 = 0.3215 + 1$$

Thus

$$\frac{50.73}{2.42} = \text{antilog} \ (0.3215 + 1)$$
$$= 2.097 \times 10 = 20.97$$

It is now possible to evaluate the expression $\dfrac{\log_{10} 12}{\log_{10} 5}$ obtained in Example 2 of Section 8.9.

Example 3. To evaluate $\dfrac{\log_{10} 12}{\log_{10} 5}$ we first calculate

$$\log_{10} 12 = \log_{10} (1.2 \times 10)$$
$$= \log_{10} 1.2 + \log_{10} 10 \qquad \text{by (1)}$$
$$= 0.0792 + 1.0000 = 1.7092$$

and

$$\log_{10} 5 = 0.6990$$

Thus

$$\frac{\log_{10} 12}{\log_{10} 5} = \frac{1.0792}{0.6990}$$

which must be rounded off to $\dfrac{1.079}{0.6990}$ since our tables are only four-

place tables. Now we have a problem of division and write

$$\frac{1.079}{0.6990} = \text{antilog} \left[\log \frac{1.079}{6.990 \times 10^{-1}} \right] \qquad \text{by } (b)$$

$$= \text{antilog} \left[\log \frac{1.079 \times 10}{6.990} \right]$$

$$= \text{antilog} [\log 1.079 + \log 10 - \log 0.6990] \qquad \text{by } (1) \text{ and } (2)$$

$$\begin{aligned}
\log 1.079 &= & 0.0330 \\
\log 10 &= & 1.0000 \\
-\log 6.990 &= & -0.8445 \\
\log 1.079 + \log 10 - \log 6.990 &= & 0.1885
\end{aligned}$$

Thus

$$\frac{1.079}{0.6990} = \text{antilog } 0.1885 = 1.544$$

Notice that $\dfrac{\log_{10} 2}{\log_{10} 5} \neq \log_{10} 2 - \log_{10} 5 = \log_{10} \dfrac{2}{5}$.

Our next two examples deal with roots.

Example 4.

$$\begin{aligned}
\sqrt{2} = 2^{1/2} &= \text{antilog } (\log 2^{1/2}) & \text{by } (b) \\
&= \text{antilog } (\tfrac{1}{2} \log 2) & \text{by } (3) \\
&= \text{antilog } \tfrac{1}{2}(0.3010) = \text{antilog } 0.1505 = 1.414
\end{aligned}$$

Example 5.

$$\begin{aligned}
\sqrt[3]{100.4} = (100.4)^{1/3} & \\
&= \text{antilog } [\log (100.4)^{1/3}] & \text{by } (b) \\
&= \text{antilog } [\tfrac{1}{3}(\log 1.004 \times 10^2)] & \text{by } (3) \\
&= \text{antilog } [\tfrac{1}{3}(\log 1.004 + \log 10^2)] & \text{by } (1) \\
&= \text{antilog } [\tfrac{1}{3}(0.0017 + 2)] = \text{antilog } [\tfrac{1}{3}(2.0017)] \\
&= \text{antilog } 0.6672 = 4.647
\end{aligned}$$

Next we illustrate "combination" problems.

Example 6.

$$\sqrt{\frac{72.1 \times 5.74}{1.29 \times 20.6}}$$

$$= \left(\frac{72.1 \times 5.74}{1.29 \times 20.6}\right)^{1/2}$$

$$= \text{antilog}\left[\log\left(\frac{7.21 \times 10 \times 5.74}{1.29 \times 2.06 \times 10}\right)^{1/2}\right] \qquad \text{by } (b)$$

$$= \text{antilog }[\tfrac{1}{2}(\log 7.21 + \log 5.74 - \log 1.29 - \log 2.06)] \qquad \text{by (1), (2),}$$
$$\text{and (3)}$$

$$\begin{array}{rl}
\log 7.21 = & 0.8579 \\
\log 5.74 = & 0.7589 \\
-\log 1.29 = & -0.1106 \\
-\log 2.06 = & -0.3139 \\ \hline
& 1.1923
\end{array}$$

Thus

$$\sqrt{\frac{72.1 \times 5.74}{1.29 \times 20.6}} = \text{antilog }[\tfrac{1}{2}(1.1923)]$$

$$= \text{antilog } 0.5962 = 3.946$$

The student may have observed that, in all of the examples, negative characteristics never appeared, or was it ever necessary to subtract a larger number from a smaller number so as to obtain a negative number. Although it is perfectly possible to do problems involving negative characteristics, certain difficulties are avoided by the device shown in the following three examples.

Example 7. $(0.0054)(0.0071)(0.1732) = \dfrac{5.4 \times 7.1 \times 1.732}{10^7}$. We then compute $5.4 \times 7.1 \times 1.732$ and divide the result by 10^7.

Example 8. $\dfrac{5.3}{5.42} = \dfrac{53}{5.42} \div 10$. Here 5.3 is smaller than 5.42 so that $\log_{10} 5.3 < \log_{10} 5.42$. Hence we compute $\dfrac{53}{5.42}$ and divide our answer by 10.

Example 9. $\sqrt[3]{0.0143} = \sqrt[3]{\dfrac{14.3}{10^3}} = \dfrac{\sqrt[3]{14.3}}{10}$.

Finally, we mention that problems involving negative numbers cannot be done directly by logarithms since, for example, $\log_{10}(-1.2) = x$ means that $10^x = -1.2$, whereas 10^x is always positive. But if, for example, we wish to compute $(-1.314)(5.73)$ we observe that $(-1.314) \times$

$(5.73) = -(1.314 \times 5.73)$, compute 1.314×5.73 by logarithms, and take the negative of this number. Similarly, $\sqrt[3]{-2.74} = -\sqrt[3]{2.74}$.

Exercise 8.12

In problems 1–20, perform the indicated operations by the use of logarithms.

1. 357.1×82.73

2. 10.44×0.7652

3. $(-2.173)(0.03247)$

4. $371.4 \div 3.129$

5. $0.003471 \div (-72.92)$

6. $7.839 \div 342.5$

7. $-1 \div 72.35$

8. $1 \div 0.003791$

9. $1 \div 81.43$

10. $\sqrt{7.942}$

11. $(0.01324)^{-2}$

12. $(-7.235)^{-2}$

13. $\sqrt[3]{13.21}$

14. $\sqrt{\dfrac{1.212 \times 3.714}{0.09321 \times 12.93}}$

15. $\sqrt[3]{\dfrac{10.91 \times 15.23}{27.93}}$

16. $\sqrt[3]{\dfrac{9.146 \times 0.02197}{5.061}}$

17. $\sqrt{\dfrac{9.632 \times 80.31}{0.09361 \times 5.739}}$

18. $\dfrac{\log 5.32}{\log 1.72}$

19. $\dfrac{\log 23.5}{\log 2.17}$

20. $\dfrac{\log 45.7}{\log 31.4}$

In problems 21–26, solve the given exponential equation by taking logarithms to the base 10.

21. $2^x = 25$

22. $5^{x-1} = 4$

23. $3^x = 37.5$

24. $2(5^x) = (1.75)^{x-3}$

25. $3(15^x) = 7^{x-1}$

26. $5^{x-2} = 3^{3-2x}$

27. Radium decays at a rate which satisfies the law

$$Q = P(2.72)^{-0.00385t}$$

where P is the original amount in grams (at time $t = 0$) and Q is the amount left after t seconds. How long will it take 2 grams to disintegrate to one gram? (This length of time is known as the half-life of radium.)

28. If a certain amount of money P (called the principal) is invested in a fund at 6 per cent interest compounded continuously, the amount of money, A, after n years is given by the formula

$$A = P(2.72)^{0.06n}$$

How much will $100 amount to after 100 years?

29. Using the formula given in problem 28, find out how long it will take $1,000 invested at 6 per cent interest to double itself.

30. The formula of problem 28 becomes, when we substitute r for 0.06,

$$A = P(2.72)^{rn}$$

This is known as the law of natural growth (for populations, bacteria cultures, etc.) where A is the population after n years if we start with a population P which is increasing continuously at the rate r. In a certain city of 20,000 it is found that the rate of growth is 0.033. What will the population be after 10 years?

8.13 THE ANTILOGARITHM OF A NEGATIVE NUMBER

In Section 8.11, we considered the antilogarithm of a negative number only when the number was written in such a fashion as $0.8762 - 2$. Suppose now that we wish to find $\text{antilog}_{10}\,(-3.2731)$. What we do is to write

$$-3.2731 = -3 - 0.2731 = -4 + (1 - 0.2731)$$
$$= -4 + 0.7269 = 0.7269 - 4$$

and then apply the techniques of Section 8.11.
Similarly,

$$\text{antilog}_{10}\,(-1.547) = \text{antilog}_{10}\,(-1 - 0.547)$$
$$= \text{antilog}_{10}\,[-2 + (1 - 0.547)]$$
$$= \text{antilog}_{10}\,(-2 + 0.453)$$
$$= \text{antilog}_{10}\,(0.4530 - 2)$$

Exercise 8.13

Evaluate the given antilogarithms.

1. $\text{antilog}_{10}\,(-1.7385)$
2. $\text{antilog}_{10}\,(-2.7373)$
3. $\text{antilog}_{10}\,(-0.1762)$
4. $\text{antilog}_{10}\,(-0.0176)$
5. $\text{antilog}_{10}\,(-1.3000)$
6. $\text{antilog}_{10}\,(-3.4762)$
7. $\text{antilog}_{10}\,(-5.5241)$
8. $\text{antilog}_{10}\,(-0.9259)$
9. $\text{antilog}_{10}\,(-3.5427)$
10. $\text{antilog}_{10}\,(-2.1742)$

FAMILIES OF CURVES

9

9.1 FAMILIES OF CURVES

The equation $y = x^2 + k$ may be regarded as the equation of a **family** of parabolas (Section 7.9) in the sense that each time a real number is substituted for k the graph of the equation obtained is a parabola. If $k = 1$ we have $y = x^2 + 1$, for $k = -2$ we have $y = x^2 - 2$, etc. Some of the parabolas of the family are shown in Figure 9.1.

Example 1. $y = \dfrac{k}{x^2}$ is the equation of a family of curves. Some of the members of this family are shown in Figure 9.2.

Example 2. The graph of the function f defined by $f(x) = kx$ is a straight line through the origin for any given value of k. Some of the members of the family of lines defined by $f(x) = kx$ are shown in Figure 9.3.

Example 3. If k is a nonzero real number, the function f defined by $f(x) = \dfrac{k}{x}$ has for its graph a hyperbola (Section 7.10). Some members of the family of hyperbolas defined by $f(x) = \dfrac{k}{x}$ are shown in Figure 9.4.

Example 4. If k is a positive real number $\neq 1$, the function f defined by

$$f(x) = \log_k x$$

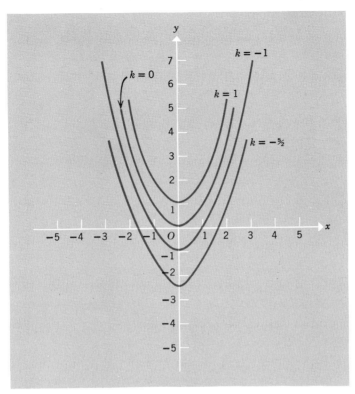

Figure 9.1

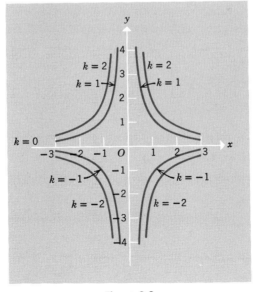

Figure 9.2

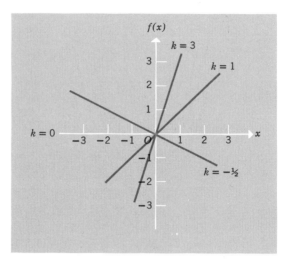

Figure 9.3

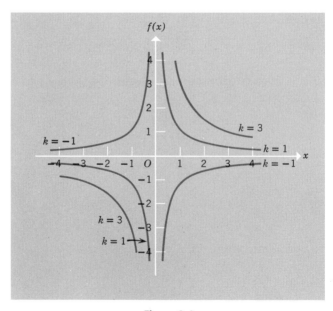

Figure 9.4

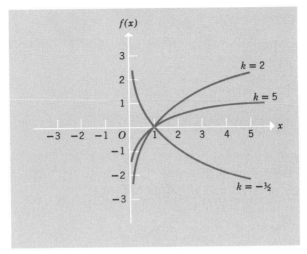

Figure 9.5

has for its graph a logarithmic curve (Section 8.7). Some members of the family of logarithmic curves are shown in Figure 9.5.

Exercise 9.1

Graph the members of the family of curves defined by each of the following equations for $k = 0, 1, -1$, and 2.

1. $y = x + k$
2. $y = kx + 1$
3. $y = 2x + k$
4. $y = kx + 2$
5. $y = 2x^2 + k$
6. $y = x^2 + k$
7. $y = kx^2$
8. $y = x^3 + k$
9. $y = kx^3 + 5$
10. $y = \dfrac{k}{x^2}$
11. $y = \dfrac{k}{x^3}$
12. $y = \dfrac{1}{x^2} + k$
13. $y = -\dfrac{1}{x^2} + k$
14. $y = k\sqrt{x}$
15. $y = \sqrt{x} + k$
16. $f(x) = x^4 + k$
17. $f(x) = kx^4$
18. $f(x) = kx^4 + 1$
19. $f(x) = kx^4 + k$
20. $f(x) = kx^4 - k$
21. $g(x) = kx + k$
22. $g(x) = kx - k$

23. $h(x) = kx^2 - k$ **24.** $t(x) = \dfrac{k}{1-x}$

25. $q(x) = \dfrac{k}{(1-x)^2}$ **26.** $r(x) = \dfrac{k}{1-x^2}$

9.2   DETERMINING THE CONSTANT

In this section, we will see that we can find an equation of a particular curve C if we know two things:

    (1) a family of curves to which C belongs,

and

    (2) a point on C.

Example 1. Find an equation for a curve C that possesses the following two properties:

    (1) C is a member of the family defined by $y = kx^3 + 1$,

and

    (2) the point with coordinates $(2, 17)$ is on C.

SOLUTION: By (1), an equation of the form $y = kx^3 + 1$ is an equation for C. We now utilize (2) to **determine the constant,** k. By (2), the equation $y = kx^3 + 1$ has the solution $(2, 17)$. Hence $17 = k \cdot 2^3 + 1$ so that $k = 2$. Thus $y = 2x^3 + 1$ is an equation for C.

Example 2. Find an equation of a curve C that is a member of the family defined by $f(x) = k\sqrt{x}$ if the point with coordinates $(4, 9)$ is on C.

SOLUTION: We have given that $f(x) = k\sqrt{x}$, and since the point with coordinates $(4, 9)$ is on C, it follows that $f(4) = k\sqrt{4} = 9$. Hence $k = \frac{9}{2}$ and the curve has an equation of the form $f(x) = \frac{9}{2}\sqrt{x}$.

Example 3. Suppose that $f(x) = \dfrac{k}{x^3}$ and $f(1) = 4$. What is $f(2)$?

SOLUTION: $f(1) = \dfrac{k}{1^3} = 4$. Hence $k = 4$ and $f(x) = \dfrac{4}{x^3}$. Then

$$f(2) = \frac{4}{2^3} = \frac{4}{8} = \frac{1}{2}$$

Geometrically, this means that we determine which one of the family of curves defined by $f(x) = \dfrac{k}{x^3}$ passes through the point with coordinates (1, 4) and then find the y-coordinate of the point on this curve that has x-coordinate 2.

Exercise 9.2

In problems 1–18, find an equation for the curve or curves passing through the point whose coordinates are given and belonging to the family defined by the given equation.

1. (3, 0), $y = kx + 3$ **2.** (1, 1), $y = 2x + k$

3. (2, 5), $y = x^2 + k$ **4.** (1, 3), $y = kx^3 + 1$

5. (2, 1), $y = \dfrac{k}{x^2}$ **6.** (4, −1), $y = \dfrac{k}{x^3}$

7. (0, 0), $y = 2x + k$ **8.** (0, 1), $y = 3x^2 + k$

9. (2, 9), $y = k^2(x + 1)$ **10.** (1, 2), $y = k^2(x^2 + 1)$

11. (−1, 0), $y = \dfrac{1}{k}x + k$ **12.** (1, 1), $f(x) = kx + 1$

13. (3, 10), $f(x) = x^2 + k$ **14.** $\left(2, \dfrac{1}{16}\right)$, $f(x) = \dfrac{k}{x^5}$

15. (2, 1), $g(x) = \dfrac{1}{x + k}$ **16.** (2, 64), $h(t) = kt^5$

17. (4, 1), $f(z) = k\sqrt{z}$ **18.** (4, 8), $g(w) = w^{3/2} + k$

19. Given $f(x) = kx^2$ and $f(2) = 1$, what is $f(3)$?

20. Given $g(t) = at^2 + 5$ and $g(1) = 30$, what is $g(2)$?

21. Given $f(s) = cs^2 + 2$ and $f(-\frac{1}{2}) = 5$, what is $f(-\frac{1}{3})$?

22. Given $g(y) = dy^{1/2} + d$ and $g(4) = 3$, what is $g(9)$?

23. Given $f(w) = \dfrac{k}{\sqrt{w} + 1}$ and $f(4) = 5$, what is $f(16)$?

24. Given $f(x) = kx^{3/2} + k$ and $f(\frac{1}{4}) = 1$, what is $f(1)$?

25. What is the y-coordinate of the point whose x-coordinate is 3 on the curve passing through the point with coordinates (1, −13) and belonging to the family of curves defined by the equation $y = k^2x^2 + 6k - 4$?

26. A curve C has an equation of the form $y = 1 + x^2 + kx$ and passes through the point with coordinates $(1, 4)$. Which of the following points whose coordinates are given are on C: $(2, 3)$, $(-1, 1)$, $(0, 1)$, $(4, 3)$, $(\frac{1}{2}, \frac{2}{3})$, $(-\frac{1}{2}, \frac{1}{4})$, $(\frac{1}{3}, \frac{7}{9})$?

*27. Write an equation for the curve belonging to the family defined by $y = kx^{2/3} + 1$ and passing through the point with coordinates $(8, 17)$. Sketch this curve.

*28. What are the equations of the curves in the family defined by $y = k^2x^3 + kx$ which pass through the point with coordinates $(1, 2)$? Sketch each of these curves.

9.3   VARIATION AND PROPORTION

In physics and chemistry phrases such as "y is inversely proportional to x," or "force varies inversely as the square of the distance," etc., occur. Associated with each such phrase is a family of curves. In the following table we "translate" some of the more common phrases.

Statement	Translation
1. "y is proportional to x." "y is directly proportional to x." "y varies directly with x."	1. $y = kx$ (or "We are considering a curve which is a member of the family defined by the equation $y = kx$.")
2. "y is inversely proportional to x." "y varies inversely with x."	2. $y = k/x$ (or "The curve is a member of the family defined by the equation $y = k/x$.")
3. "y is inversely proportional to the square of x." "y varies inversely as the square of x."	3. $y = k/x^2$ (or "The curve is a member of the family defined by the equation $y = k/x^2$.")

Example 1. "Volume is inversely proportional to pressure."

TRANSLATION: $V = k/p$, where V is the volume and p is the pressure.

Example 2. "Force varies inversely as the square of the distance."

TRANSLATION: $F = k/d^2$, where F is the force and d is the distance.

In each of the above examples a family of curves is obtained from the given statement. Usually additional information is given which determines the constant k and hence some single curve in the family.

Before giving examples of this process, it should be pointed out that the units of measurement of the quantities involved are important and that k itself usually has units attached to its magnitude. Thus if, in Example 1, V is measured in cubic centimeters and p in grams per square centimeter, k is in gram-centimeters. On the other hand, if V is measured in cubic feet and p is in pounds per square foot, k is in foot-pounds. We need not concern ourselves with the units of k, however, *if, throughout the problem, we keep the measurements of the quantities involved in the same units.*

Example 3. It is known that the volume, V, of a gas is inversely proportional to the pressure, p, and that $V = 10$ cubic feet when $p = 2$ pounds per square foot. Find the volume when the pressure is 3 pounds per square foot.

SOLUTION: From Example 1, we have

$$V = \frac{k}{p}$$

and since $V = 10$ when $p = 2$, $k = 20$. Thus $V = 20/p$. Hence when $p = 3$ pounds per square foot, $V = 20/3$ cubic feet.

Example 4. It is known that the distance, s, that a ball travels in rolling down a certain inclined plane is directly proportional to the square of the time, t, and that $s = 15$ feet when $t = 3$ seconds. Find the distance the ball travels in the first 5 seconds and the time it takes for the ball to travel 3 yards.

SOLUTION: The equation of the family of curves is

$$s = kt^2$$

and since $s = 15$ when $t = 3$, we obtain $k = 15/9 = 5/3$. Hence $s = (5/3)t^2$. Then, when $t = 5$ seconds, $s = (5/3)(5)^2 = 125/3$ feet. On the other hand, to find t when $s = 3$ yards, we must first notice that 3 yards $= 9$ feet. Then we have

$$9 = \frac{5}{3} t^2$$

so that

$$t^2 = \frac{27}{5} \quad \text{and} \quad t = \sqrt{\frac{27}{5}} = \frac{3}{5} \sqrt{15} \text{ seconds}$$

3. $y = k/x^2$
   $9 = k/4$
   $k = 9/4$
   $y = 9/4 \cdot (-3/2)^2$
   $9/4 \cdot 9/4 =$
   $y = 81/16$

Exercise 9.3

1. If y is proportional to x and $y = 1$ when $x = 3$, what is y when $x = 5$?

2. If y is inversely proportional to x and $y = 3$ when $x = 2$, what is y when $x = 4$?

3. If y varies directly with the square of x and $y = 9$ when $x = 2$, what is y when $x = -\frac{3}{2}$?

4. If y is inversely proportional to the square of x and $y = 5$ when $x = 5$, what is y when $x = \frac{1}{10}$?

5. If s is proportional to the square root of t and $s = 3$ when $t = \frac{9}{4}$, what is s when $t = 16$?

6. If z is inversely proportional to the cube of w and $z = 40$ when $w = \frac{1}{2}$, what is z when $w = \frac{1}{3}$?

7. The distance s a body falls in time t starting from rest varies directly as the square of t. If $s = 256$ feet when $t = 4$ seconds, what is s when $t = 8$ seconds?

8. The amount a spring stretches under a force is proportional to the stretching force. If a force of 5 pounds stretches a particular spring one foot, how much force is required to stretch this spring 3 feet?

9. The attraction G between two objects is inversely proportional to the square of the distance d between them. If two particular objects are 1,000 feet apart the attraction between them is a force of 10 pounds. How far apart are these objects if the attraction between them is a force of 1,000 pounds?

10. If the current, i, in amperes in an electric circuit is inversely proportional to the resistance, R, in ohms and $i = 20$ amperes when $R = 10$ ohms, find i when $R = 5$.

11. The vibrating frequency (pitch) of a string is directly proportional to the square root of the tension. If the frequency is 30 times per second when the tension is 4 pounds, what is the frequency when the tension is 9 pounds?

12. Pressure at the bottom of a tank of water is directly proportional to the depth of the water in the tank. If the pressure is 1,000 pounds when the water is four feet deep, what is the pressure when the water is 5.5 feet deep?

13. The volume of a gas in a closed container is inversely proportional to the pressure. If the volume is 35 cubic inches when the pressure is 2 pounds per square inch, find the volume when the pressure is 7 pounds per square inch.

14. The weight of an object above the surface of the earth is inversely proportional to the square of its distance from the center of the earth. If an object weighs 150 pounds on the surface, how much does it weigh 100 miles above the surface? (Assume that the earth is a perfect sphere with a diameter of 8,000 miles.)

9.4 JOINT VARIATION

In applications, formulas and equations containing more than two variables frequently arise. The variables usually represent physical quantities or measurements and since such quantities or measurements usually vary with the others, the term **joint variation** is frequently used. Some of the equations associated with such terminology are given in the following examples.

Example 1. z varies directly as x and y: $z = kxy$

Example 2. z varies directly as x and inversely as y^2: $z = kx/y^2$

Example 3. z varies inversely as the product of x and y: $z = k/xy$

Example 4. z varies directly as x and y and inversely as the square of t: $z = kxy/t^2$

Although an intuitive geometrical interpretation (in three dimensions) can be made for some problems, no easy interpretation is possible for equations with more than three variables, and we shall not attempt any geometrical discussion of the case of three variables. The same algebraic and numerical procedures, however, may be used in determining k and using the resulting equations as in the case of equations with two variables.

Example 5. If z varies directly as x^2 and inversely as $\sqrt{y}$ and $z = 2$ when $x = 3$ and $y = 4$, find z when $x = 2$ and $y = 9$.

SOLUTION: $z = kx^2/\sqrt{y}$. Hence $2 = 9k/\sqrt{4} = 9k/2$ and so $k = 4/9$. Hence $z = \frac{4}{9}x^2/\sqrt{y}$, and when $x = 2$ and $y = 9$, it follows that $z = \frac{4}{9} \cdot \frac{4}{3} = \frac{16}{27}$.

Example 6. The elongation, E, of a steel wire when a mass, m, is hung from its free end varies jointly as m and the length, x, of the wire and inversely as the cross sectional area, A, of the wire. Given that $E = 0.001$ inches when $m = 20$ pounds, $x = 10$ inches, and $A = 0.01$ square inches, find E when $m = 40$ pounds, $x = 15.5$ inches, and $A = 0.015$ square inches.

SOLUTION: $E = kmx/A$ and so

$$0.001 = \frac{k(20)(10)}{.01} \quad \text{and} \quad k = \frac{(.001)(.01)}{200} = \frac{1}{20{,}000{,}000}$$

Hence

$$E = \frac{1}{20{,}000{,}000} \frac{mx}{A}$$

and when $m = 40$, $x = 15.5$, and $A = 0.015$, we have

$$E = \frac{1}{20{,}000{,}000} \frac{40(15.5)}{0.015} = 0.00207 \text{ inches (approximately)}$$

Example 7. The current, i, in amperes in an electric circuit varies directly as the electromotive force, E, in volts and inversely as the resistance, R, in ohms. If, in a certain circuit, $i = 30$ amperes when $R = 15$ ohms and $E = 450$ volts, find i when $R = 20$ ohms and $E = 200$ volts.

SOLUTION: $i = kE/R$ and so

$$30 = \frac{k \cdot 450}{15} \quad \text{and} \quad k = 1$$

Hence $i = E/R$ and when $R = 20$ and $E = 200$, we have $i = 200/20 = 10$ amperes.

Exercise 9.4

1. z varies directly as x and y, and $z = 60$ when $x = 2$ and $y = 3$. What is z when $x = 3$ and $y = 4$?

2. z varies directly as y and inversely as $\sqrt{w}$. If $z = 15$ when $y = 6$ and $w = 4$, find z when $y = 8$ and $w = 9$.

3. z varies directly as x^3 and inversely as y^2. If $z = 16/3$ when $x = 2$ and $y = 3$, find z when $x = 2$ and $y = 2$.

4. z varies directly as $\sqrt{x}$ and y^2 and inversely as $\sqrt[3]{w}$. If $z = 36$ when $x = 4$, $y = 3$, and $w = 8$, find z when $x = 9$, $y = 3$, and $w = 27$.

5. Boyle's law states that the pressure, P, of a gas varies directly as the absolute temperature, T, and inversely as the volume, V. If $P = 800$ pounds per square inch when $T = 80°$A. and $V = 1$ cubic foot, what is P when $T = 30°$A. and $V = 20$ cubic feet?

6. The kinetic energy, E, of a moving object varies directly as the mass, m, and the square of the velocity, v. If a 50-pound body, moving 20 feet per second has a kinetic energy of 10,000 foot-pounds, what is the kinetic energy of the same body when it is moving 30 feet per second?

7. The pitch, P, of a string of fixed length varies directly as the square root of the tension, T, and inversely as the square root of the mass, M. If $P = 16$ vibrations per second when $T = 64$ ounces and $M = 0.25$ ounces, what is the pitch of a string of the same length and a mass of 0.36 ounces under a tension of 81 ounces?

8. The safe load, L, of a horizontal beam varies directly as the breadth, b, and the depth, d, and inversely as the length, x. If a 3×5-inch beam of a certain length safely supports 600 pounds, what is the safe load for a 4×10-inch beam of the same length?

9. The resistance, R, of a wire varies directly as the length, x, and inversely as the square of its diameter, d. If 100 feet of a given wire of diameter 0.01 inches has a resistance of 6 ohms, what is the resistance of 50 feet of the same kind of wire but having a diameter of 0.015 inches?

10. Simple interest earned in a given time varies directly as the principal and the rate. If $600 earned $30 at 4 per cent, how much interest is earned by $1,000 at 5 per cent for the same period of time?

11. The mass of a steel rectangular sheet varies directly as the length, width, and thickness. If the mass of a 4×6-foot sheet 0.01 inch thick is 120 pounds, what is the mass of an 8×10-foot sheet which is 0.005 inch thick?

SEQUENCES, SERIES, AND
THE BINOMIAL THEOREM

10

10.1 SEQUENCES

Let us recall (Section 7.2) that a function is a rule which associates with each number, x, from a set of numbers called the domain of the function, a unique number, y. An important case where the domain of definition is the positive integers gives rise to the concept of a sequence.

Definition 10.1. A **sequence function** is a function with domain the set of positive integers.

It is customary to make some changes in the usual functional notation when we deal with sequence functions. In place of $f(x)$, $g(x)$, etc., we use $s(n)$ or s_n where s reminds us that we are dealing with a sequence function and n that the domain is the set of positive integers. Although the notation $s(n)$ is, in many respects, a preferable one, the notation s_n is much more common.

Example 1. $s(n) = s_n = \dfrac{1}{n}$. As n takes on the values $1, 2, 3, 4, \ldots$, we obtain $s(1) = s_1 = 1$, $s(2) = s_2 = \frac{1}{2}$, $s(3) = s_3 = \frac{1}{3}$, $s(4) = s_4 = \frac{1}{4}, \ldots$.

Definition 10.2. Let s be a sequence function. Then the set of functional values $s_1, s_2, s_3, \ldots$, is called a **sequence.** Each of $s_1, s_2, s_3, \ldots$, is called a **term** of the sequence. s_1 is called the first term, s_2 the

276

second term, . . . , s_n the nth term. If a sequence has a last term, it is a **finite sequence;** if there is no last term, it is an **infinite sequence.**

Example 2. If $s_n = (-1)^n$ defines the sequence function, the sequence is

$$-1, 1, -1, 1, \ldots$$

with first term -1, second term 1, third term -1, fourth term 1, and nth term $(-1)^n$.

Example 3. If $s_n = \sqrt{n}$, the sequence is

$$1, \sqrt{2}, \sqrt{3}, \sqrt{4}, \sqrt{5}, \ldots$$

with nth term $\sqrt{n}$.

Example 4. If $s_n = \dfrac{(n-1)(n-2)(n-3)}{n}$, the sequence is

$$0, 0, 0, \tfrac{3}{2}, \tfrac{24}{5}, 10, \ldots$$

with nth term $\dfrac{(n-1)(n-2)(n-3)}{n}$.

Sometimes the first few terms of a sequence are given as, for example, 1, 4, 9, 16, and the question is asked: What is the nth term of the sequence? While it is perfectly true that the "natural" answer to this question is $s_n = n^2$ there are, actually, an infinite number of correct answers possible. Thus, for example, we could have

$$s_n = n^4 - 10n^3 + 36n^2 - 50n + 24$$

which gives $s_1 = 1$, $s_2 = 4$, $s_3 = 9$, $s_4 = 16$, but $s_5 = 49$; or

$$s_n = 2n^2 - 10n + 35 - 2\left(\frac{25n - 12}{n^2}\right)$$

which gives s_1, s_2, s_3, s_4, as before, but $s_5 = 649/25$. In fact, we could have

$$s_n = n^2 + (n-1)(n-2)(n-3)(n-4)f(n)$$

where f is *any* function of n and the resulting sequence will obviously have 1, 4, 9, 16, as its first four terms!

Exercise 10.1

In problems 1–15, write the first five terms and the eleventh term of the sequence whose nth term is given.

1. $s_n = n^3$
2. $s_n = \dfrac{1}{n^2}$
3. $s_n = (2n - 1)^2$
4. $s_n = n^2 + n$
5. $s_n = n^2 - n$
6. $s_n = n^3 + 1$
7. $s_n = \dfrac{1}{n + 1}$
8. $s_n = \dfrac{1}{n^2 + 1}$
9. $s_n = n + (n - 1)$
10. $s_n = n + (n - 1)(n - 2)$
11. $s_n = (-1)^n n$
12. $s_n = (-1)^{n+1} n$
13. $s_n = (-1)^{n-1}$
14. $s_n = (-1)^{n^2 - 1}$
15. $s_n = (-1)^{n^2 + 1}$

In problems 16–20, write an algebraic expression for s_n.

16. $s_n =$ the nth odd positive integer.
17. $s_n =$ the nth even positive integer.
18. $s_n =$ the $(n + 1)$st odd positive integer.
19. $s_n =$ the $(n + 1)$st even positive integer.
20. $s_n =$ the $(n + 4)$th odd positive integer.

10.2   SERIES

Consider a sequence $s_1, s_2, s_3, \ldots, s_n, \ldots$, as defined in the preceding section and define S_n as the sum of the first n terms of the sequence. That is,

$$S_n = s_1 + s_2 + \cdots + s_n$$

Thus

$$S_1 = s_1$$
$$S_2 = s_1 + s_2$$
$$S_3 = s_1 + s_2 + s_3$$
$$\cdots \cdots \cdots \cdots$$

Notice that we again have a function, S, with domain the set of positive integers. For example, if $s_n = n^2$, $s_1 = 1$, $s_2 = 4$, $s_3 = 9$, and hence

$$S_1 = s_1 = 1$$
$$S_2 = s_1 + s_2 = 1 + 4 = 5$$
$$S_3 = s_1 + s_2 + s_3 = 1 + 4 + 9 = 14$$

The sequence of **partial sums** $S_1, S_2, S_3, \ldots$ is traditionally called a **series.** If the sequence $S_1, S_2, S_3, \ldots$ is a finite sequence, then we speak of a **finite** series; if the sequence $S_1, S_2, S_3, \ldots$ is an infinite sequence, then we speak of an **infinite** series. We shall consider only finite series in this chapter.

It is common practice to use the Greek capital letter Σ (*sigma*) to indicate a sum and to write, for example,

$$S_3 = s_1 + s_2 + s_3 = \sum_{n=1}^{3} s_n$$

Example 1. $\displaystyle\sum_{n=1}^{4} n = 1 + 2 + 3 + 4 = 10.$

Example 2. $\displaystyle\sum_{n=1}^{4} n + \sum_{n=5}^{10} n = (1 + 2 + 3 + 4) + $
$$(5 + 6 + 7 + 8 + 9 + 10)$$
$$= \sum_{n=1}^{10} n = 55.$$

Example 3. $\displaystyle\sum_{n=1}^{1} n = 1.$

Example 4. $\displaystyle\sum_{k=1}^{3} \frac{1}{k} = 1 + \frac{1}{2} + \frac{1}{3} = \frac{11}{6}.$

Example 5. $\displaystyle\sum_{m=2}^{4} m^2 = 2^2 + 3^2 + 4^2 = 29.$

Exercise 10.2

Find the numerical value of each of the following indicated sums.

1. $\displaystyle\sum_{n=1}^{5} n^3$

2. $\displaystyle\sum_{n=1}^{4} \frac{1}{n^2}$

3. $\displaystyle\sum_{n=1}^{6} (2n + 1)$

4. $\displaystyle\sum_{k=1}^{8} 2k$

5. $\displaystyle\sum_{k=1}^{5} (-1)^k k^2$

6. $\displaystyle\sum_{k=1}^{4} \frac{(-1)^{k+1}}{k + 1}$

7. $\displaystyle\sum_{k=3}^{5} \frac{1}{2k-1}$ 8. $\displaystyle\sum_{n=2}^{5} 2^n$ 9. $\displaystyle\sum_{n=2}^{5} (-1)^n 2^n$

10. $\displaystyle\sum_{n=1}^{4} n^2 + n$ 11. $\displaystyle\sum_{n=1}^{4} (-1)^n (n^2 + n)$

12. $\displaystyle\sum_{n=1}^{4} (-1)^{n+1}(n^2 + n)$ 13. $\displaystyle\sum_{n=1}^{4} [(-1)^n n^2 + (-1)^{n+1} n]$

14. $\displaystyle\sum_{k=0}^{5} (2 + 3k)$ 15. $\displaystyle\sum_{t=1}^{3} (3 + 4t)$ 16. $\displaystyle\sum_{t=3}^{5} (4 - 5t)$

17. $\displaystyle\sum_{p=2}^{5} \frac{p}{p+1}$ 18. $\displaystyle\sum_{k=1}^{4} \frac{k^2}{k+1}$ 19. $\displaystyle\sum_{s=2}^{5} (-1)^s \frac{s}{s+1}$

20. $\displaystyle\sum_{t=1}^{4} (-1)^{t+1} \frac{t^2}{t+1}$ 21. $\displaystyle\sum_{n=5}^{10} (-1)^n$

10.3 ARITHMETIC PROGRESSIONS

Consider the sequence

$$5, 8, 11, 14, \ldots$$

where each term is obtained from the preceding one by adding 3. Thus $8 = 5 + 3$, $11 = 8 + 3$, $14 = 11 + 3$. Here, even though we have given only four terms of the sequence, the sequence is uniquely defined since we have given, in addition to the first four terms, a rule for proceeding from one term to the next. Thus the fifth term is $14 + 3 = 17$ and the sixth term is $17 + 3 = 20$ or in symbols, $s_2 = s_1 + 3$, $s_3 = s_2 + 3$, $s_4 = s_3 + 3$, and, in general, $s_1 = 5$ and

$$s_n = s_{n-1} + 3 \qquad (n = 2, 3, 4, \ldots)$$

Such a formula, where the nth term of a sequence is given by an expression involving one or more of the preceding terms, is called a **recursion formula.** We now make the following definition.

Definition 10.3. Let $s_1, s_2, s_3, \ldots$, be a sequence in which

$$s_n = s_{n-1} + d \qquad (n = 2, 3, 4, \ldots)$$

Then the sequence is called an **arithmetic sequence** or **arithmetic progression.** The number d is called the **common difference** of the sequence (since $s_n - s_{n-1} = d$).

Notice that an arithmetic progression is uniquely defined if s_1 and d are given.

Example 1. Given an arithmetic progression in which $s_1 = -4$ and $d = \frac{1}{2}$, the first four terms of the progression are

$$s_1 = -4, \quad s_2 = -4 + \tfrac{1}{2} = -3\tfrac{1}{2}, \quad s_3 = -3\tfrac{1}{2} + \tfrac{1}{2} = -3,$$
$$s_4 = -3 + \tfrac{1}{2} = -2\tfrac{1}{2}$$

Example 2. Given an arithmetic progression in which the first term is 5 and the common difference is -2, the first four terms of the progression are

$$s_1 = 5, \, s_2 = 5 + (-2) = 3, \, s_3 = 3 + (-2) = 1, \, s_4 = 1 + (-2) = -1$$

Example 3. The arithmetic progression $2, 2, 2, 2, \ldots$, has first term 2 and common difference 0.

Since we have

$$s_1 = s_1, \quad s_2 = s_1 + d, \quad s_3 = s_2 + d = (s_1 + d) + d = s_1 + 2d,$$
$$s_4 = s_3 + d = (s_1 + 2d) + d = s_1 + 3d$$

it is easy to see that, in general,

$$s_n = s_1 + (n - 1)d \tag{1}$$

The following examples illustrate the use of formula (1).

Example 4. Find the fifteenth term of the arithmetic progression whose first term is 2 and whose common difference is $\frac{1}{2}$.

SOLUTION: $s_{15} = s_1 + (n - 1)d = 2 + (15 - 1)\tfrac{1}{2} = 9.$

Example 5. Find the first term of an arithmetic progression if the fifth term is 29 and d is 3.

SOLUTION: $s_5 = 29, d = 3$. Hence, by (1),

$$29 = s_1 + (5 - 1)3$$

and

$$s_1 = 29 - 12 = 17$$

Example 6. Given that the first term of an arithmetic sequence is 56 and the seventeenth term is 32, find the tenth term and the twenty-fifth term.

SOLUTION: First we find d by using (1):

$$s_{17} = s_1 + (17 - 1)d,$$
$$32 - 56 = 16d$$
$$16d = -24$$
$$d = -\tfrac{3}{2}$$

Now we can find the tenth term, s_{10}, as

$$s_{10} = 56 + (10 - 1)(-\tfrac{3}{2}) = 56 - \tfrac{27}{2} = \tfrac{85}{2}$$

and the twenty-fifth term, s_{25}, as

$$s_{25} = 56 + (25 - 1)(-\tfrac{3}{2}) = 56 - 36 = 20$$

We now wish to find a formula for the sum, S_n, of the first n terms of an arithmetic progression $s_1, s_2, \ldots, s_n, \ldots$. To do this we note that the first n terms of an arithmetic progression are given not only by

$$s_1, s_1 + d, s_1 + 2d, \ldots, s_1 + (n - 1)d$$

but also, in reverse order, by

$$s_n, s_n - d, s_n - 2d, \ldots, s_n - (n - 1)d$$

Hence $S_n = \sum_{k=0}^{n-1} (s_1 + kd)$ or, expanding,

$$S_n = s_1 + (s_1 + d) + (s_1 + 2d) + \cdots + [s_1 + (n - 1)d]$$

and, also, $S_n = \sum_{k=0}^{n-1} (s_n - kd)$ or, expanding,

$$S_n = s_n + (s_n - d) + (s_n - 2d) + \cdots + [s_n - (n - 1)d]$$

Adding together both sides of the two equalities we have

$$2S_n = ns_1 + ns_n$$

so that

$$S_n = \frac{n}{2}(s_1 + s_n) \tag{2}$$

Formula (2) together with formula (1) may be used to solve a wide variety of problems involving arithmetic progressions. If any three of the five quantities s_1, s_n, n, d, and S_n are given, we can always find the other two by the use of these two formulas. The following three examples are typical.

Example 7. An arithmetic progression has $d = -2$ and $s_5 = 4$. Find the sum of the first ten terms.

SOLUTION: From (1) we have

$$4 = s_1 + (5 - 1)(-2)$$

and hence that $s_1 = 12$. Then, again by (1),

$$s_{10} = 12 + (10 - 1)(-2) = -6$$

and, finally, by (2)

$$S_{10} = \tfrac{10}{2}[12 + (-6)] = 30$$

Example 8. The fifth term of an arithmetic progression is -8 and the twelfth term is 6. Find the sum of the first twenty-five terms.

SOLUTION: From $s_5 = -8$ and $s_{12} = 6$ we have, by (1),

$$-8 = s_1 + 4d \qquad \text{and} \qquad 6 = s_1 + 11d$$

Solving this system of equations for s_1 and d, we obtain $d = 2$ and $s_1 = -16$. Again using (1) we have

$$s_{25} = -16 + (25 - 1)2 = 32$$

and, finally, by (2),

$$S_{25} = \tfrac{25}{2}(-16 + 32) = 200$$

Example 9. An arithmetic progression has $d = \tfrac{1}{2}$, $s_n = 12$, and $S_n = 147$. Find s_1 and n.

SOLUTION: From (1) we have

$$12 = s_1 + (n - 1)\tfrac{1}{2}$$

while from (2) we have

$$147 = \frac{n}{2}(s_1 + 12)$$

Multiplying both sides of each equation by 2 we have the system of equations

$$24 = 2s_1 + n - 1 \quad \text{and} \quad 294 = n(s_1 + 12)$$

From the first of these equations, $n = 25 - 2s_1$. Substituting this in the second, we have

$$294 = (25 - 2s_1)(s_1 + 12) = 300 + s_1 - 2s_1{}^2$$

or, simplifying, $2s_1{}^2 - s_1 - 6 = 0$. Since $2s_1{}^2 - s_1 - 6 = (2s_1 + 3)(s_1 - 2)$ we have $s_1 = -3/2$ or $s_1 = 2$. But $n = 25 - 2s_1$ and hence either

$$s_1 = -\tfrac{3}{2} \quad \text{and} \quad n = 25 - 2(-\tfrac{3}{2}) = 28$$

or

$$s_1 = 2 \quad \text{and} \quad n = 25 - 2(2) = 21$$

Hence there are two arithmetic progressions satisfying the given conditions.

Exercise 10.3

In problems 1–12, evaluate the indicated sum by using the formula for the sum of an arithmetic progression.

1. $\displaystyle\sum_{k=0}^{20}(2 + 3k)$

2. $\displaystyle\sum_{n=0}^{25}(3 + 5n)$

3. $\displaystyle\sum_{n=5}^{20}\left(\frac{1}{2} + \frac{n}{2}\right)$

4. $\displaystyle\sum_{n=3}^{30}(2 - 7n)$

5. $\displaystyle\sum_{k=0}^{10}(3 - k)$

6. $\displaystyle\sum_{k=0}^{20}(-2 + 4k)$

7. $\displaystyle\sum_{t=3}^{12}\left(-3 - \frac{t}{2}\right)$

8. $\displaystyle\sum_{k=0}^{10}(\sqrt{3} - \sqrt{3}k)$

9. $\displaystyle\sum_{n=1}^{10}\left(\frac{1}{3} - \frac{n}{5}\right)$

10. $\displaystyle\sum_{k=2}^{11}\left(\frac{3}{4} - \frac{k}{7}\right)$

11. $\displaystyle\sum_{k=3}^{17}\left(\frac{11}{19} + \frac{5k}{7}\right)$

12. $\displaystyle\sum_{k=1}^{10}\left(\frac{7}{3} - \frac{5k}{12}\right)$

In problems 13–18, the given data refer to arithmetic progressions.

13. If $s_1 = 3$ and $d = 4$, find s_{12} and S_{12}.

14. If $s_1 = 2$ and $d = 5$, find s_{16} and S_{16}.

15. If $s_1 = 5$ and $d = 7$, find s_{10} and S_{10}.

16. If $s_5 = 6$, $d = 4$, and $s_n = 86$, find n and S_n.

17. If $s_3 = 5$, $d = 8$, and $s_n = 165$, find n and S_n.

18. If $s_4 = -3$, $d = 2$, and $s_n = 25$, find n and S_n.

19. The sum of the first seven terms of an arithmetic progression is 105 and the seventh term is 27. What is the first term?

20. The fifth and tenth terms of an arithmetic progression are 21 and 41, respectively. Find the first term and the common difference.

21. Find the twentieth term of an arithmetic progression if the first term is 4 and the common difference is 2.

22. Find the fifteenth term of an arithmetic progression if the third term is 6 and the common difference is 5.

23. Find the sum of the first fifteen terms of an arithmetic progression if the sixth term is 4 and the common difference is −3.

24. Find the sum of the first twenty terms of an arithmetic progression if the third term is −5 and the common difference is $\frac{1}{5}$.

*25. If the formula for the sum of a particular arithmetic progression is

$$S_n = \tfrac{5}{2}n^2 + \tfrac{7}{2}n$$

find the first term and the common difference.

In problems 26–37, the given data refer to arithmetic progressions.

26. If $s_1 = 4$, $d = 4$, and $s_n = 56$, find n and S_n.

27. If $s_1 = 3$, and $s_8 = 45$, find d and S_8.

28. If $d = 4$ and $s_8 = 109$, find s_1 and S_8.

29. If $s_1 = 13$, $s_n = 29$, and $S_n = 84$, find d and n.

30. If $s_1 = 3$ and $S_9 = 66$, find s_9 and d.

31. If $s_1 = 5$, $S_n = 275$, and $d = 5$, find n and s_n.

32. If $s_1 = \frac{1}{2}$ and $d = \frac{1}{3}$, find s_{10} and S_{10}.

33. If $s_3 = \frac{7}{8}$ and $d = -\frac{1}{2}$, find s_1 and S_{15}.

*34. If $S_3 = 0$ and $S_5 = 35$, find S_{10} and s_2.

*35. If $S_2 = -4$ and $S_4 = 0$, find S_{20} and s_3.

*36. If $s_2 = 10$ and $S_5 = 55$, find S_{20}.

*37. If $s_5 = \frac{3}{2}$ and $S_{10} = \frac{65}{4}$, find S_{13}.

38. If $1,000 plus 4 per cent interest on the outstanding principal is paid at the end of each year on a $10,000 loan until the loan is paid off, what is the total interest paid?

39. A person earns 3 dollars on the first day of the month, 6 dollars on the second day, and $3n$ dollars on the nth day. How much does he earn in a month of 30 days?

40. In drilling an oil well 500 feet deep the cost is $50 for the first foot and increases $40 for each additional foot. What is the cost of drilling the well?

41. A man saved \$5 the first week of the year and increased his weekly savings by \$1.75 until he was saving \$20.75 a week. He then increased his weekly savings by \$2 a week for the remaining weeks of the year. At the end of 52 weeks, how much had he saved?

42. A ball rolling down an inclined plane travels 8 feet during the first second and in the nth second it travels 6 feet farther than it did in the $(n-1)$th second. How far does it roll in 15 seconds? How long would it take to roll 1,300 feet?

43. In a lottery with ticket numbers from 1 to 1,000 each ticket holder received twice as many cents as the number of his ticket. What should a ticket cost to make the organization holding the lottery show a profit of \$1,000?

10.4 GEOMETRIC PROGRESSIONS

Another important type of sequence is the **geometric progression** in which each term is found by multiplying the preceding term by a fixed number $r \neq 0$. That is,

$$s_n = s_{n-1} r$$

The number r is called the **common ratio** since the ratio of each term to its preceding term is r:

$$\frac{s_n}{s_{n-1}} = \frac{s_{n-1} r}{s_{n-1}} = r \text{ if } s_{n-1} \neq 0$$

Now $s_2 = s_1 r$, $s_3 = s_2 r = (s_1 r)r = s_1 r^2$, $s_4 = s_2 r = (s_1 r^2)r = s_1 r^3$, and, in general,

$$s_n = s_1 r^{n-1} \tag{1}$$

Example 1. If the first term of a geometric progression is 9 and the common ratio is $-\frac{2}{3}$, find the first five terms.

SOLUTION:
$$s_1 = 9$$
$$s_2 = 9(-\tfrac{2}{3}) = -6$$
$$s_3 = (-6)(-\tfrac{2}{3}) = 4$$
$$s_4 = 4(-\tfrac{2}{3}) = -\tfrac{8}{3}$$
$$s_5 = (-\tfrac{8}{3})(-\tfrac{2}{3}) = \tfrac{16}{9}$$

Example 2. The seventh term of a geometric progression is 192 and $r = 2$. Find the first four terms.

SOLUTION: First we find the first term from formula (1):

$$192 = s_1(2)^{7-1} = 2^6 s_1 = 64 s_1$$
$$s_1 = \tfrac{192}{64} = 3$$

Then

$$s_1 = 3, \qquad s_2 = 3 \cdot 2 = 6, \qquad s_3 = 6 \cdot 2 = 12, \qquad \text{and} \qquad s_4 = 12 \cdot 2 = 24$$

To find the sum, S_n, of the first n terms of a geometric progression, we write $S_n = \sum_{k=0}^{n-1} s_1 r^k$ or, expanding,

$$S_n = s_1 + s_1 r + s_1 r^2 + \cdots + s_1 r^{n-1} \tag{2}$$

We then multiply both sides of this equality by r to obtain

$$S_n r = r \sum_{k=0}^{n-1} s_1 r^k = \sum_{k=1}^{n} s_1 r^k,$$

or, expanding,

$$S_n r = s_1 r + s_1 r^2 + s_1 r^3 + \cdots + s_1 r^n \tag{3}$$

Now we subtract (2) from (3) to obtain

$$S_n r - S_n = s_1 r^n - s_1$$

Hence

$$S_n(r - 1) = s_1(r^n - 1)$$

and thus

$$S_n = s_1 \frac{r^n - 1}{r - 1} \qquad (r \neq 1) \tag{4}$$

(What is the situation if $r = 1$?)

As was the case in arithmetic progressions, we have five quantities, s_1, s_n, n, r, and S_n, and if any three are given we can find the other two by use of formulas (1) and (4). The following three examples illustrate this procedure.

Example 3. Find the sum of the first six terms of a geometric progression whose first term is $\tfrac{1}{3}$ and whose second term is -1.

SOLUTION: First we use (1) with $n = 2$ to obtain

$$-1 = \tfrac{1}{3}r, \quad r = -3$$

Then, by (4),

$$S_6 = \frac{1}{3}\frac{(-3)^6 - 1}{-3 - 1} = \frac{1}{3} \cdot \frac{729 - 1}{-4} = -\frac{182}{3}$$

Example 4. The fourth term of a geometric progression is $\tfrac{1}{2}$ and the sixth term is $\tfrac{1}{8}$. Find the first term and the common ratio.

SOLUTION: By (1), first with $n = 4$, and then with $n = 6$, we have

$$\tfrac{1}{2} = s_1 r^3 \quad \text{and} \quad \tfrac{1}{8} = s_1 r^5$$

Multiplying the second equation by 4 we have $\tfrac{1}{2} = 4s_1 r^5$. Hence

$$\tfrac{1}{2} = s_1 r^3 = 4s_1 r^5$$

and since $s_1 \neq 0$ (why?), $r^3 = 4r^5$. But then, since $r \neq 0$ (why?) we have $r^2 = \tfrac{1}{4}$; $r = \tfrac{1}{2}$ or $r = -\tfrac{1}{2}$. If $r = \tfrac{1}{2}$, $s_1 = 1/2r^3 = 4$, while if $r = -\tfrac{1}{2}$, $s_1 = -4$. Hence there are two geometric progressions satisfying the given conditions.

Example 5. The first term of a geometric progression is 27, the nth term is 32/9, and the sum of n terms is 665/9. Find n and r.

SOLUTION: From (1) we have

$$\tfrac{32}{9} = 27r^{n-1}$$

while from (4) we have

$$\frac{665}{9} = 27\frac{r^n - 1}{r - 1} = \frac{27r^n - 27}{r - 1}$$

Multiplying through the first equation by r we have

$$\frac{32r}{9} = 27r^n$$

Substituting $32r/9$ for $27r^n$ in the second equation we have

$$\frac{665}{9} = \frac{32r/9 - 27}{r - 1}$$

Multiplying both sides of this equation by $9(r-1)$ we obtain

$$665(r-1) = 32r - 243$$

and hence

$$665r - 665 = 32r - 243$$

Thus $633r = 422$ and $r = \frac{422}{633} = \frac{2}{3}$. Finally,

$$\tfrac{32}{9} = 27(\tfrac{2}{3})^{n-1}$$

so that

$$(\tfrac{2}{3})^{n-1} = \tfrac{32}{243} = (\tfrac{2}{3})^5$$

Thus $n-1 = 5$ so that $n = 6$.

Exercise 10.4

In problems 1–12, evaluate the indicated sum using the formula for the sum of a geometric progression.

1. $\displaystyle\sum_{n=1}^{6} 2^n$
2. $\displaystyle\sum_{n=1}^{5} 3^n$
3. $\displaystyle\sum_{n=1}^{7} \frac{1}{2^n}$

4. $\displaystyle\sum_{n=1}^{8} 3 \cdot 2^n$
5. $\displaystyle\sum_{n=0}^{4} 5 \cdot 7^n$
6. $\displaystyle\sum_{n=0}^{5} \frac{4}{3^n}$

7. $\displaystyle\sum_{n=1}^{8} (-1)^n 2^n$
8. $\displaystyle\sum_{k=2}^{7} (-1)^k 3^k$
9. $\displaystyle\sum_{k=3}^{8} \frac{(-1)^{k+1}}{2^k}$

10. $\displaystyle\sum_{k=0}^{4} \frac{(-1)^k}{3^k}$
11. $\displaystyle\sum_{n=2}^{6} \left(\frac{3}{4}\right)^n$
12. $\displaystyle\sum_{n=2}^{5} \left(\frac{7}{6}\right)^n$

In problems 13–31, the given data refer to geometric progressions.

13. If $s_1 = 3$ and $r = 5$, find S_6.
14. If $s_1 = \frac{1}{2}$ and $r = \frac{1}{3}$, find S_6.
15. If $s_1 = -\frac{1}{5}$ and $r = -\frac{1}{2}$, find S_8.
16. If $s_1 = -5$ and $r = -\frac{3}{4}$, find S_4.
17. If $s_1 = 4$ and $s_6 = 4/243$, find S_5.
18. If $s_3 = 40/27$ and $s_6 = 320/729$, find r and s_1.
19. If $s_3 = -189/8$ and $s_5 = -1{,}701/32$, find r and s_1.
20. If $s_2 = 9/5$ and $s_5 = -243/5$, find r and s_1.
21. If $s_4 = 54$ and $r = 3$, find s_8.
22. If $r = 2/3$ and $s_5 = 4/9$, find s_7.

23. If $r = 4/7$ and $s_5 = -16/49$, find s_4.

24. If $r = -3$, $s_n = -189$, and $S_n = -140$, find n and s_1.

25. If $S_1 = \frac{1}{3}$ and $S_2 = \frac{1}{2}$, find r and s_1.

26. If $S_1 = \frac{2}{3}$ and $S_2 = \frac{5}{6}$, find r and s_1.

27. If $S_1 = \frac{4}{5}$ and $S_2 = \frac{3}{5}$, find r and s_1.

28. If $S_1 = 3/2$ and $S_2 = -2$, find r and s_1.

29. If $r = \frac{3}{4}$, $s_n = 81$, and $S_n = 781$, find s_1 and n.

30. If $r = 3/2$, $s_1 = 32$, and $S_n = 665$, find s_n and n.

31. If $s_1 = 7/4$, $s_n = 112$, and $S_n = 889/4$, find r and n.

32. Find the sum of the first five terms of a geometric progression if the first term is 5 and the fifth term is $16/125$.

33. The third and sixth terms of a geometric progression are $1/12$ and $-1/96$ respectively. Find the first term and the common ratio.

34. The first term of a geometric progression is 2 and the common ratio is $-\frac{1}{3}$. Find the seventh term.

35. The first term of a geometric progression is 8 and the common ratio is $3/2$. Which term is equal to $729/8$?

36. The first term of a geometric progression is $1/3$ and the second term is $4/5$. Which term is equal to $6,912/625$?

37. A ball is dropped from a height of 21 feet. On each rebound it reaches a height $\frac{2}{3}$ of that from which it fell. How far has it traveled when it hits the ground for the sixth time?

38. A man earns 1 cent the first day of the month, 2 cents the second day, 4 cents the third day, . . . , 2^{n-1} cents the nth day. What are his total earnings after the fifteenth day's work is done?

39. A population in a city is increasing at the rate of 8 per cent per year. If the city now has a population of 1,000,000, what will be the population 11 years later? (Use logarithms.)

40. The population of a city is 50,000 and it increases 5 per cent per year. What is the population after 9 years? (Use logarithms.)

41. The starting salary for a job is \$4,000 per year. What is the salary after 11 years if there is a 5 per cent increase each year? (Use logarithms.)

*42. Two lines intersect at a 45° angle. From a point four inches from the intersection and on one of the lines, a perpendicular is drawn to a point, P, on the other line. From P a perpendicular is drawn back to a point, Q, on the first line. From Q a perpendicular is drawn back to the second line, etc. Find the sum of the lengths of the first ten perpendiculars.

*43. A person earns $1 the first day, $2 the second day, $3 the third day, and $n the nth day. What is the first day at which the total earnings under this rate of pay are less than the earnings of the person described in problem 38?

10.5 THE BINOMIAL THEOREM

An important sequence of terms arises in the expansion of $(a + b)^n$:

$$(a + b)^1 = a + b$$
$$(a + b)^2 = a^2 + 2ab + b^2$$
$$(a + b)^3 = a^3 + 3a^2b + 3ab^2 + b^3 \tag{1}$$
$$(a + b)^4 = a^4 + 4a^3b + 6a^2b^2 + 4ab^3 + b^4$$
$$(a + b)^5 = a^5 + 5a^4b + 10a^3b^2 + 10a^2b^3 + 5ab^4 + b^5$$

. . . .

We see that the third term in the expansion of $(a + b)^3$ is $3ab^2$ and that the third term in the expansion of $(a + b)^4$ is $6a^2b^2$. But what, for example, is the seventh term in the expansion of $(a + b)^{10}$? Of course, it is possible to multiply out $(a + b)^{10}$ and look at the seventh term, but for n a large number this becomes very tedious. Thus it is to our advantage to have a formula for the kth term in the expansion of $(a + b)^n$. Before giving such a formula, it is desirable to make the following definition.

Definition 10.4. $n! = n(n - 1)(n - 2) \cdots 3 \cdot 2 \cdot 1$ for n a positive whole number. The symbol $n!$ is read **n factorial** and means the product of all the positive integers up to and including n.

Examples.

$$4! = 4 \cdot 3 \cdot 2 \cdot 1 = 24$$
$$\frac{7!}{10!} = \frac{7 \cdot 6 \cdot 5 \cdot 4 \cdot 3 \cdot 2 \cdot 1}{10 \cdot 9 \cdot 8 \cdot 7 \cdot 6 \cdot 5 \cdot 4 \cdot 3 \cdot 2 \cdot 1} = \frac{1}{10 \cdot 9 \cdot 8} = \frac{1}{720}$$

Now let us look at just the coefficients of each term in the expansions (1) and arrange them as follows:

$$
\begin{array}{ccccccccccc}
 & & & & 1 & & 1 & & & & \\
 & & & 1 & & 2 & & 1 & & & \\
 & & 1 & & 3 & & 3 & & 1 & & \\
 & 1 & & 4 & & 6 & & 4 & & 1 & \\
1 & & 5 & & 10 & & 10 & & 5 & & 1
\end{array}
\qquad (2)
$$

. . . .

This arrangement is known as **Pascal's** [1] **triangle,** since it was he who first noticed the relationships in this array of numbers. We see that we can find any number (except the first and the last, which are both 1) in any row by adding the two numbers diagonally above it. For example, in the fifth row, $5 = 1 + 4$, $10 = 4 + 6$, etc. Now we can immediately write the sixth row:

$$1 \quad 6 \quad 15 \quad 20 \quad 15 \quad 6 \quad 1$$

and, in fact, these are the coefficients in the expansion of $(a + b)^6$. We can thus write the expansion without direct multiplication [2] if we notice in (1) that the first term is always a^n and that the exponent of a drops by one in each succeeding term, while b starts in the second term to the first power and the exponent increases by one in each succeeding term:

$$(a + b)^6 = a^6 + 6a^5b + 15a^4b^2 + 20a^3b^3 + 15a^2b^4 + 6ab^5 + b^6$$

This process can be continued, of course, but to find the seventh term in the expansion of $(a + b)^{10}$ we would have to write the first ten rows of Pascal's triangle, and again, for n large, this becomes tedious.

Now there is a very definite relationship between the number n and the coefficient of the kth term in the expansion of $(a + b)^n$ and, in fact, it can be shown that the kth term (for $k > 1$) in such an expansion is given by the formula:

$$k\text{th term} = \frac{n(n - 1)(n - 2) \ldots (n - k + 2)}{(k - 1)!} a^{n-k+1}b^{k-1}$$

Example 1. Find the third term in the expansion of $(a + b)^6$.

SOLUTION: Here $k = 3$, $n = 6$, so that $n - k + 2 = 5$, $n - k + 1 = 4$, and $k - 1 = 2$. Hence the third term is given by

[1] Blaise Pascal, French mathematician-philosopher, 1623–1662.

[2] Providing, of course, that we accept without proof the validity of the Pascal triangle for all expansions. A proof may be found in any college algebra text.

$$\frac{6 \cdot 5}{2!} a^4b^2 = 15a^4b^2$$

Example 2. Find the seventh term in the expansion of $(a + b)^{10}$.

SOLUTION: Here $k = 7$, $n = 10$, so that $n - k + 2 = 5$, $n - k + 1 = 4$, and $k - 1 = 6$. Hence the seventh term is given by

$$\frac{10 \cdot 9 \cdot 8 \cdot 7 \cdot 6 \cdot 5}{6!} a^4b^6 = 210a^4b^6$$

Example 3. Find the sixth term in the expansion of $(x^2 - \frac{1}{2})^{12}$.

SOLUTION: Here we take $a = x^2$, $b = -\frac{1}{2}$, $n = 12$, $k = 6$ and note that $n - k + 2 = 8$, $n - k + 1 = 7$, and $k - 1 = 5$. Hence the sixth term is given by

$$\frac{12 \cdot 11 \cdot 10 \cdot 9 \cdot 8}{5!} (x^2)^7(-\tfrac{1}{2})^5 = -\tfrac{99}{4}x^{14}$$

The general expansion of $(a + b)^n$, known as the **binomial expansion,** can now be written as

$$(a + b)^n = a^n + na^{n-1}b + \frac{n(n - 1)}{2!} a^{n-2}b^2 + \frac{n(n - 1)(n - 2)}{3!} a^{n-3}b^3$$
$$+ \cdots + \frac{n(n - 1)(n - 2) \ldots (n - k + 2)}{(k - 1)!} a^{n-k+1}b^{k-1}$$
$$+ \cdots + nab^{n-1} + b^n$$

From this expansion, we may observe that the exponents of a descend from n to 0; the exponents of b rise from 0 to n; while each coefficient (after the first) is obtained from the preceding one by multiplying by the exponent of a and dividing by one more than the exponent of b. Thus $na^{n-1}b$ gives us

$$\frac{n(n - 1)}{2}$$

for the next coefficient when we multiply n by $n - 1$ and divide by $1 + 1 = 2$. Then from

$$\frac{n(n - 1)}{2} a^{n-2}b^2$$

we get

$$\frac{n(n-1)}{2} \cdot \frac{n-2}{3} = \frac{n(n-1)(n-2)}{2 \cdot 3}$$

and so on.

The coefficients n, $\dfrac{n(n-1)}{2!}$, $\dfrac{n(n-1)(n-2)}{3!}$, etc., are known as **binomial coefficients** and are frequently written as

$$\binom{n}{1} = n, \quad \binom{n}{2} = \frac{n(n-1)}{2}, \quad \binom{n}{3} = \frac{n(n-1)(n-2)}{3!}$$

and, in general, as

$$\binom{n}{k} = \frac{n(n-1)(n-2) \ldots (n-k+1)}{k!}$$

Example 4. Find the first five terms in the binomial expansion of $(a+b)^8$.

SOLUTION: $(a+b)^8 = a^8 + 8a^7b + \dfrac{8 \cdot 7}{2} a^6b^2 + \dfrac{28 \cdot 6}{3} a^5b^3$

$$+ \frac{56 \cdot 5}{4} a^4b^4 + \cdots$$

$$= a^8 + 8a^7b + 28a^6b^2 + 56a^5b^3 + 70a^4b^4 + \cdots$$

(The student should note that, by symmetry, the last four terms are $56a^3b^5$, $28a^2b^6$, $8ab^7$, and b^8.)

Example 5. Find the binomial expansion of $(2x-5)^4$.

SOLUTION: Here we take $a = 2x$ and $b = -5$ to obtain

$$(2x-5)^4 = (2x)^4 + 4(2x)^3(-5) + \frac{4 \cdot 3}{2} (2x)^2(-5)^2$$

$$+ \frac{6 \cdot 2}{3} (2x)(-5)^3 + \frac{4 \cdot 1}{4} (-5)^4$$

$$= 16x^4 - 160x^3 + 600x^2 - 1{,}000x + 625$$

Exercise 10.5

In problems 1–12, evaluate each of the given symbols.

1. $5!$ **2.** $6!$ **3.** $7!$ **4.** $9!$

5. $10!$ **6.** $\dfrac{3!}{2!1!}$ **7.** $\dfrac{8!}{4!4!}$ **8.** $\dfrac{9!1!}{3!6!}$

9. $\dfrac{15!}{10!5!}$ **10.** $\dfrac{16!}{12!6!}$ **11.** $\dfrac{12!1!}{5!7!3!}$ **12.** $\dfrac{18!}{5!12!4!3!}$

In problems 13–21, use Pascal's triangle to expand each of the given expressions.

13. $(x + y)^7$ **14.** $(a + 2b)^5$ **15.** $(3x + 2y)^4$

16. $(x - y)^6$ **17.** $(2x - 3y)^5$ **18.** $(x + y)^8$

19. $(x + y)^9$ **20.** $(2x + y)^6$ **21.** $(3x - 2y)^5$

In problems 22–33, obtain the first four terms in the binomial expansion of each of the given expressions.

22. $(x + y)^{15}$ **23.** $(x + y)^{12}$ **24.** $(2x + y)^{11}$

25. $(x + 3y)^{13}$ **26.** $(x - 3y)^{11}$ **27.** $(a + 4b)^{12}$

28. $(a - 2b)^{14}$ **29.** $(2a + 3xy)^{10}$ **30.** $(a^2 + z^3)^{12}$

31. $(2a^2b - 3z^2w)^{11}$ **32.** $(a^3 - x^2y^2)^{14}$ **33.** $(2a - b^2)^{11}$

In problems 34–42, find the indicated term without expanding.

34. $(x + y)^{15}$; 8th   **35.** $(x + y)^{12}$; 8th   **36.** $(2x + y)^{11}$; 8th

37. $(x - y)^{15}$; 9th   **38.** $(2x - y)^{16}$; 4th   **39.** $(x - 3y)^{15}$; 4th

40. $(2a - b)^{12}$; 6th   **41.** $(3a + 2b)^{10}$; 4th   **42.** $(a - 2b)^{16}$; 10th

COMPLEX NUMBERS

11

11.1 INTRODUCTION

The historical development of the concept of number has origins which precede written records. It is fairly certain, however, that the first numbers considered were the natural numbers, 1, 2, 3, 4, ...

By 2000 B.C., positive rational numbers such as 1/2, 3/5, 5, 77/15, were being used extensively by the Babylonian and some other contemporary civilizations. The Greeks recognized the existence of irrational numbers such as $\sqrt{2}$, $\sqrt{3}$, and π by 200 B.C. and by A.D. 1000 both negative numbers and zero had been incorporated into European mathematics.

Now it is easy to characterize the positive integers as being members of the infinite sequence 1, 2, 3, . . . , and rational numbers as the quotient of one integer by another. But irrational numbers are defined in a negative fashion as numbers that are *not* rational and, in fact, as pointed out in Section 1.11, it was a long time before the existence of irrational numbers was discovered. Thus when we say that the system of real numbers is the set of rational numbers together with the set of all irrational numbers, our definition necessarily is somewhat vague.

To define a real number precisely, however, is not an easy task and, in fact, this was not done until the latter part of the nineteenth century. At this level, about all we can say is that the totality of all measurements of distance on a number line can be regarded as the set of real numbers.

11.2 EXTENSION AND SOLVABILITY

Even before mathematicians had a clear understanding of the nature of a real number they had gone on to invent still other types of numbers. One way of portraying this development of the number system is to study the kinds of numbers that are needed to solve equations. Thus, for example, to solve equations such as $x + 2 = 5$ only positive integers are needed. On the other hand, to solve equations such as $x + 5 = 3$ it is clear that negative numbers are needed. So from one point of view the desire to solve equations of this type necessitates the invention of negative numbers, or as mathematicians say, necessitates "extending the system".

Similarly, an equation such as $5x = 3$ has no solution if only integers are available but the invention of rational numbers such as $\frac{3}{5}$ changes the status of such equations from "unsolvable" to "solvable." Thus solvability is a *relative* concept; that is, whether or not an equation has a solution depends on what numbers are available or allowed as solutions. Likewise, the equation $x^2 = 2$ has a solution if $\sqrt{2}$ is allowed, but if only rational numbers are allowed, then this equation has no solution. Since the system of real numbers includes all the irrational numbers, the equation $x^2 = 2$ has a solution *after* the system of rational numbers has been extended to the system of real numbers.

Exercise 11.2

1. If only positive integers are available, which of the following equations are solvable?

(a) $x - 3 = 0$ (b) $2x - 4 = 0$ (c) $x + 3 = 0$
(d) $2x + 4 = 0$ (e) $2x = 6$ (f) $3x = 10$
(g) $3x = 12$ (h) $x + 1 = 1$ (i) $x + 2 = 2$

2. If only integers are available, which of the following equations are solvable?

(a) $3x + 6 = 2$ (b) $5x + 4 = 9$ (c) $x + 2 = 2$
(d) $x - 3 = -3$ (e) $3x + 9 = 30$ (f) $3x - 1 = -4$
(g) $x^2 - 4 = 0$ (h) $x - 1 = -1$ (i) $2x + 7 = 6$

3. If only rational numbers are available, which of the following equations are solvable?

(a) $x - 2 = 0$ (b) $x^3 + 8 = 0$ (c) $x^3 - 9 = 0$
(d) $x^2 - 4x - 28 = 0$ (e) $x^2 + 1 = 0$ (f) $5x - 7 = 0$
(g) $x^2 + x - 1 = 0$ (h) $7x + 9 = 0$ (i) $x^2 - 3 = 0$

4. Which of the following statements are true and which are false?

(a) Every integer is a real number.
(b) Every rational number is a real number.
(c) Every real number is an irrational number.
(d) Every integer is a rational number.

*5. Prove that, if only rational numbers are available, the equation $2^x = 5$ has no solution.

11.3 DEFINITION AND PROPERTIES OF COMPLEX NUMBERS

We now consider the equation $x^2 = -1$ and note that, if only real numbers are available, this equation has no solution since the square of any real number is positive. Just as before this situation leads to the invention of new numbers, in this case to the extension of the *system of real numbers* to the *system of complex numbers.*

Definition 11.1. The system of **complex numbers** consists of all polynomials with real number coefficients in the complex unit i together with the following agreements:

(1) Equality: $a + bi = c + di$ if and only if $a = c$ and $b = d$;
(2) Operations: all the ordinary laws governing operations on polynomials or quotients of polynomials are permitted;
(3) $i^2 = -1$; that is, -1 may be substituted for i^2 whenever i^2 occurs.

It follows from agreements (1)–(3), that any complex number (polynomial in i) can be written as $a + bi$ where a and b are real numbers. The following examples illustrate this fact.

Example 1. $(2 + 3i) + (4 + i) = (2 + 4) + (3 + 1)i = 6 + 4i$.

Example 2. $5i^4 = 5i^2 \cdot i^2 = 5(-1)(-1) = 5.$

Example 3. $6i^3 + 3i^6 + 5i - 2 = 6i^2i + 3(i^2)^3 + 5i - 2$
$$= 6(-1)i + 3(-1)^3 + 5i - 2$$
$$= -6i - 3 + 5i - 2$$
$$= -5 - i = -5 + (-1)i$$

Example 4. $(2 + 3i)(4 + i) = 8 + 14i + 3i^2$
$$= 8 + 14i - 3$$
$$= 5 + 14i$$

An expression of the form

$$\frac{a + bi}{c + di}$$

may be "rationalized" by multiplying numerator and denominator by $c - di$.

Example 5. $\dfrac{5 + 2i}{3 + 4i} = \dfrac{(5 + 2i)(3 - 4i)}{(3 + 4i)(3 - 4i)} = \dfrac{15 - 14i - 8i^2}{3^2 - 16i^2}$
$$= \dfrac{15 - 14i - 8(-1)}{9 - 16(-1)} = \dfrac{23 - 14i}{25} = \dfrac{23}{25} - \dfrac{14i}{25}$$

Example 6. $\dfrac{5 + 2i}{i} = \dfrac{(5 + 2i)(-i)}{i(-i)} = \dfrac{-5i - 2i^2}{-i^2} = \dfrac{-5i + 2}{1}$
$$= 2 - 5i$$

Example 7. Expand $(2 + 3i)^3$.

SOLUTION: The identity $(a + b)^3 = a^3 + 3a^2b + 3ab^2 + b^3$ is still valid. In this case $a = 2$, $b = 3i$, so we have

$$(2 + 3i)^3 = 2^3 + 3 \cdot 2^2(3i) + 3 \cdot 2(3i)^2 + (3i)^3$$
$$= 8 + 36i + 54i^2 + 27i^3 = 8 + 36i - 54 - 27i = -46 + 9i$$

Example 8. Given $f(x) = x^3 + x + 1$, evaluate $f(1 + i)$.

SOLUTION:

$$f(1 + i) = (1 + i)^3 + (1 + i) + 1 = 1 + 3i + 3i^2 + i^3 + 1 + i + 1$$
$$= 1 + 3i + 3(-1) + (-1)i + 1 + i + 1 = 3i$$

Example 9. Show that $\left(-\dfrac{1}{2} + \dfrac{\sqrt{3}\,i}{2}\right)^3 = 1.$

SOLUTION: We first note that

$$\left(-\frac{1}{2} + \frac{\sqrt{3}\,i}{2}\right)^3 = \left(\frac{1}{2}\right)^3 (-1 + \sqrt{3}\,i)^3 = \frac{1}{8}(-1 + \sqrt{3}\,i)^3$$

Then we again apply the identity $(a + b)^3 = a^3 + 3a^2b + 3ab^2 + b^3$ to obtain

$$
\begin{aligned}
(-1 + \sqrt{3}\,i)^3 &= (-1)^3 + 3(-1)^2 \sqrt{3}\,i + 3(-1)(\sqrt{3}\,i)^2 + (\sqrt{3}\,i)^3 \\
&= -1 + 3\sqrt{3}\,i - 9i^2 + 3\sqrt{3}\,i^3 \\
&= -1 + 3\sqrt{3}\,i + 9 - 3\sqrt{3}\,i = 8
\end{aligned}
$$

Hence

$$\left(-\frac{1}{2} + \frac{\sqrt{3}\,i}{2}\right)^3 = \frac{1}{8} \cdot 8 = 1$$

Example 10. Show that $\left(\dfrac{1}{\sqrt{2}} + \dfrac{1}{\sqrt{2}}\,i\right)^4 = -1.$

SOLUTION: We first note that

$$\left(\frac{1}{\sqrt{2}} + \frac{1}{\sqrt{2}}\,i\right)^4 = \left(\frac{1}{\sqrt{2}}\right)^4 (1 + i)^4 = \frac{1}{4}(1 + i)^4$$

Now we apply the identity $(a + b)^4 = a^4 + 4a^3b + 6a^2b^2 + 4ab^3 + b^4$ to obtain

$$
\begin{aligned}
(1 + i)^4 &= 1^4 + 4i + 6i^2 + 4i^3 + i^4 \\
&= 1 + 4i - 6 - 4i + 1 = -4
\end{aligned}
$$

Hence

$$\left(\frac{1}{\sqrt{2}} + \frac{1}{\sqrt{2}}\,i\right)^4 = \frac{1}{4}(-4) = -1$$

The system of complex numbers contains all the real numbers (and so really is an extension) since real numbers such as 5, $\sqrt{3}$, and π can be written as $5 = 5 + 0 \cdot i$, $\sqrt{3} = \sqrt{3} + 0 \cdot i$, $\pi = \pi + 0 \cdot i$ respectively, and, in general, $a = a + 0 \cdot i$.

Numbers of the form bi where b is a nonzero real number are called "pure imaginary." For example, i, $-5i$, $\frac{1}{2}i$, $\sqrt{2}\,i$, etc., are pure imaginary numbers, while $2 + 3i$, $4 - \sqrt{2}\,i$, etc., are not pure imaginary. The use of the word "imaginary" reflects the historical fact that at one time these numbers were regarded with suspicion. (Actually, negative numbers were once called "fictitious" and were not accepted by all mathematicians for some time after their introduction.)

For any complex number $a + bi$, we call a the **real part** and bi the **imaginary part**. For example, if $z = 5 + 3i$, the real part of z is 5 and the imaginary part is $3i$. So it is sometimes said that two complex numbers are equal if and only if their real parts are equal and their imaginary parts are equal.

The use of the word "imaginary" is unfortunate in that it may cause students to feel that, in some sense, these numbers are not quite proper. Actually, all numbers are inventions of the human mind, and as complex numbers have proved their usefulness in, for example, the study of alternating currents, there is no reason whatsoever not to accept them.

Exercise 11.3

In problems 1–30, reduce the given expression to a complex number of the form $a + bi$ where a and b are real numbers.

1. $2i^2$
2. $(2i)^2$
3. $1 + 2i^2$
4. $(1 + 2i)^2$
5. $4i^3$
6. $1 - 2i^7$
7. $1 - i^6$
8. $2 - (3i)^3$
9. $(1 + i)^2$
10. $1 + 2i - (3 + i)$
11. $3 - 2i + (1 - 2i)^2$
12. $1 - i - (1 + i)$
13. $4 + (2 - i)^2$
14. $1 - (1 + i)^2$
15. $i^4 - (2i)^2 + 5i^6$
16. $(2 + 5i) - (4 + i)$
17. i^{4k} (k a positive integer)
18. i^{4k+1} (k a positive integer)
19. i^{-1}
20. i^{-2}
21. i^{4k+2} (k a positive integer)
22. i^{4k+3} (k a positive integer)
23. i^{19}
24. i^{-15}
25. i^{-12}
26. $(\sqrt{3} + i) + (2 - \sqrt{2}\,i)$
27. $\left(1 - \dfrac{\sqrt{2}\,i}{2}\right) - \left(\dfrac{1}{2} - \sqrt{2}\,i\right)$
28. $(\sqrt{3} - \sqrt{2}\,i) + (\sqrt{2} + \sqrt{3}\,i)$
29. $a + bi + c + di$
30. $a + bi - (c - di)$

In problems 31–58, perform the indicated operations and reduce the answer to a complex number of the form $a + bi$ where a and b are real numbers.

31. $(1 + 2i)(2 - i)$

32. $(2 - 3i)(2 + i)$

33. $(1 - 3i)(1 + 3i)$

34. $(2 - i)(2 + i)$

35. $(-2 + i)(-3 + 2i)$

36. $(-3 - i)(3 - 2i)$

37. $(2 - i)(-2 + i)$

38. $(2 - i)^2$

39. $(1 + 2i)^2$

40. $(2 - 3i)^2$

41. $(2 - 3i)^3$

42. $(1 + i)^3$

43. $\left(\frac{3}{2} - 2i\right)\left(\frac{1}{3} + \frac{i}{2}\right)$

44. $\left(\frac{5}{2} - 5i\right)\left(\frac{5}{2} + 5i\right)$

45. $(\frac{1}{2} - \sqrt{2}\,i)(\sqrt{2} + i)$

46. $(\sqrt{3} - 5i)(2 + \sqrt{3}\,i)$

47. $\left(\frac{\sqrt{2}}{2} + \frac{\sqrt{2}\,i}{2}\right)^2$

48. $\left(\frac{\sqrt{2}}{2} - \frac{\sqrt{2}\,i}{2}\right)^2$

49. $\left(\frac{\sqrt{3}}{2} - \frac{i}{2}\right)^3$

50. $\left(\frac{\sqrt{3}}{2} + \frac{i}{2}\right)^3$

51. $\left(-\frac{\sqrt{2}}{2} + \frac{\sqrt{2}\,i}{2}\right)^2$

52. $\left(\frac{\sqrt{2}}{2} - \frac{\sqrt{2}\,i}{2}\right)^3$

53. $\dfrac{2 + 3i}{1 - i}$

54. $\dfrac{-2 + 3i}{2 + i}$

55. $\dfrac{-2 + 3i}{-3 + 2i}$

56. $\dfrac{\sqrt{2} + 3i}{1 - \sqrt{2}\,i}$

57. $\dfrac{2 + \sqrt{3}\,i}{-\sqrt{3} + 2i}$

58. $\dfrac{\sqrt{5} - \sqrt{2}\,i}{\sqrt{3} - \sqrt{5}\,i}$

In problems 59–67, show that the given complex number satisfies the given equation.

59. $x^2 - 2x + 5 = 0$; $1 + 2i$

60. $x^2 - x + 1 = 0$; $\dfrac{1}{2} - \dfrac{\sqrt{3}\,i}{2}$

61. $2x^2 + 3x + 2 = 0$; $-\dfrac{3}{4} + \dfrac{\sqrt{7}\,i}{4}$

62. $2x^2 + 2 = x$; $\dfrac{1}{2} - \dfrac{\sqrt{15}\,i}{2}$

63. $x^3 - 3x^2 + 3x - 2 = 0$; $\dfrac{1}{2} + \dfrac{\sqrt{3}\,i}{2}$

64. $x^3 + 2 = x^2$; $1 - i$

65. $2x^3 + 3 = 2x^2 + x$; $1 + \dfrac{\sqrt{2}\,i}{2}$

66. $x^3 + x^2 = x^4 + 4x - 2$; $1 + i$

67. $\dfrac{2}{x-1} - \dfrac{1}{x-2} - 1 = 0$; $2 - i$

68. Show that $\dfrac{\sqrt{2}}{2} + \dfrac{\sqrt{2}\,i}{2}$ is a square root of i.

69. Show that $-\dfrac{\sqrt{2}}{2} + \dfrac{\sqrt{2}\,i}{2}$ is a square root of $-i$.

70. Show that $\dfrac{\sqrt{3}}{2} + \dfrac{i}{2}$ is a cube root of i.

11.4 THE SQUARE ROOT OF −1

With complex numbers available, the equation $x^2 = -1$ now has two solutions, $x = i$ and $x = -i$. In other words, we have two numbers whose square is −1. We will *agree* however that $\sqrt{-1} = i$ or, in other words, that the principal square root of −1 is i. (Recall that in the case of a number such as 4 when we had to make a choice we *agreed* that $\sqrt{4} = 2$ and $\sqrt{4} \neq -2$.)

The student should also recall at this point that we stated our law $\sqrt[n]{a}\,\sqrt[n]{b} = \sqrt[n]{ab}$ (Section 8.2) under the conditions that either n be odd or that a and b be nonnegative. We can now see the reason for this restriction since if $\sqrt{a}\,\sqrt{b} = \sqrt{ab}$ for all numbers a and b, we would have $\sqrt{-1}\,\sqrt{-1} = \sqrt{(-1)(-1)} = \sqrt{1} = 1$, whereas, actually, $\sqrt{-1}\,\sqrt{-1} = i \cdot i = i^2 = -1$.

We can, however, make the following useful definition.

Definition 11.2. If a is a real number and $a > 0$, then $\sqrt{-a} = \sqrt{a}\,i$. Or, in other words, if a is positive, the principal square root of $-a$ is defined to be $\sqrt{a}\,i$.

Since $(\sqrt{a}\,i)^2 = \sqrt{a}\,\sqrt{a}\,i^2 = -a$ we see that $\sqrt{a}\,i$ is indeed a square root of $-a$. But $-\sqrt{a}\,i$ is also a square root of $-a$ since $(-\sqrt{a}\,i)^2 = (-\sqrt{a})(-\sqrt{a})i^2 = -a$. So when we write $\sqrt{-a} = \sqrt{a}\,i$ for $a > 0$ this is by the definition of the principal square root of a negative real number given above.

Example 1. $\sqrt{-4} = \sqrt{4}\ i = 2i$. (Check: $(2i)^2 = 4i^2 = -4$.)

Example 2. $\sqrt{-27} = \sqrt{27}\ i = 3\sqrt{3}\ i$. (Check: $(3\sqrt{3}\ i)^2 = 27i^2 = -27$.)

Exercise 11.4

Express each of the following numbers as a complex number in the form $a + bi$ where a and b are real numbers.

1. $\sqrt{-4}$ 2. $\sqrt{-25}$ 3. $\sqrt{-24}$
4. $\sqrt{-18}$ 5. $\sqrt{-2}$ 6. $\sqrt{-45}$
7. $(\sqrt{-3})^2$ 8. $(\sqrt{-8})^2$ 9. $(\sqrt{-35})^2$
10. $(\sqrt{-4})^3$ 11. $(\sqrt{-12})^3$ 12. $(\sqrt{-18})^3$
13. $1 - \sqrt{-4}$ 14. $2 + \sqrt{-5}$ 15. $10 + \sqrt{-12}$
16. $(1 - \sqrt{-9})^2$ 17. $(1 + \sqrt{-15})^2$ 18. $(2 - \sqrt{-27})^2$
19. $(3 - \sqrt{-3})^3$ 20. $(2 - \sqrt{-50})^3$ 21. $(1 + \sqrt{-18})^3$

11.5   THE QUADRATIC EQUATION

Before the invention of complex numbers, an equation such as $x^2 + x + 1 = 0$ had no solution because the quadratic formula yielded

$$x = \frac{-1 \pm \sqrt{1 - 4}}{2}$$

and $\sqrt{-3}$ was not meaningful. Now this equation has two solutions:

$$x_1 = \frac{-1 + \sqrt{-3}}{2} \quad \text{and} \quad x_2 = \frac{-1 - \sqrt{-3}}{2}$$

or

$$x_1 = -\frac{1}{2} + \frac{\sqrt{3}\ i}{2}, \quad x_2 = -\frac{1}{2} - \frac{\sqrt{3}\ i}{2}$$

In general, we have the following theorem.

Theorem 11.1. The equation $ax^2 + bx + c = 0$, with a, b, and c real numbers, and $a \neq 0$ has the solutions

$$x_1 = \frac{-b + \sqrt{b^2 - 4ac}}{2a} \quad \text{and} \quad x_2 = \frac{-b - \sqrt{b^2 - 4ac}}{2a}$$

PROOF: The proof of this theorem given previously in Section 6.6 is now valid *without* the restriction previously imposed, namely that $b^2 - 4ac \geqq 0$.

It is now possible to classify the solutions of any quadratic equation $ax^2 + bx + c = 0$ (a, b, c real numbers) by examining the number $b^2 - 4ac$. This number, $b^2 - 4ac$, is called the **discriminant** and since we must take its square root it follows that:

(1) If $b^2 - 4ac > 0$, there are two (unequal) real numbers which satisfy the equation.

(2) If $b^2 - 4ac < 0$, there are two (unequal) complex numbers which satisfy the equation and these complex numbers are not real numbers.

(3) If $b^2 - 4ac = 0$, there is only one real number which satisfies the equation.

In practice, of course, by the time the discriminant has been found, the solutions can be written immediately. Example 4 that follows, however, shows how the concept of the discriminant can be useful in cases where the actual solution is not the important thing. Furthermore, the concept of a discriminant carries over into equations of higher degree where the actual solutions are not so easily obtained.

Example 1. Determine the nature of the solutions of the quadratic equation $x^2 + 2x - 5 = 0$.

SOLUTION: Here $a = 1$, $b = 2$, $c = -5$, $b^2 - 4ac = 4 + 20 = 24 > 0$, so that the discriminant is positive and hence the solution set of this equation consists of two real numbers.

Example 2. Determine the nature of the solutions of the quadratic equation $x^2 + 2x + 5 = 0$.

SOLUTION: Here $b^2 - 4ac = 4 - 20 = -16 < 0$, so that the solution set of this equation consists of two complex numbers that are not real numbers.

Example 3. Determine the nature of the solutions of the quadratic equation $x^2 - 4x + 4 = 0$.

SOLUTION: Here $b^2 - 4ac = 0$, and so there is exactly one (real) solution of this quadratic.

Example 4. If the equation $6x^2 + 9y^2 + x - 6y = 0$ is regarded as a quadratic equation in y, find the values of x which give real solutions for y.

SOLUTION: This is essentially the same problem as Example 2 of Section 7.8. There we actually solved for y, whereas here we note that $a = 9$, $b = -6$, $c = 6x^2 + x$. Hence the discriminant is

$$(-6)^2 - 4 \cdot 9(6x^2 + x) = 36(1 - 6x^2 - x)$$

Setting $1 - 6x^2 - x \geq 0$ we complete the problem as before.

Exercise 11.5

In problems 1–20, compute the discriminant to determine the nature of the solutions of the following quadratic equations.

1. $x^2 - 2x + 1 = 0$
2. $2x^2 + 3x - 1 = 0$
3. $x^2 + 5x - 2 = 0$
4. $x^2 + x + 1 = 0$
5. $x^2 - x + 1 = 0$
6. $4x^2 + 4x + 1 = 0$
7. $x^2 - 5 = 0$
8. $x^2 + 2 = 0$
9. $-2x^2 + 2x - 1 = 0$
10. $5x^2 = 3x - 2$
11. $6x^2 = 2x - 3$
12. $4x^2 - x + 3 = 0$
13. $x^2 + 2x + 10 = 0$
14. $x^2 + x = 30$
15. $16x^2 + 9 = 24x$
16. $5x^2 - 7x + 3 = 0$
17. $6x^2 - 6x + 5 = 0$
18. $7x^2 = 9x + 3$
19. $9x^2 + 7 = 2x$
20. $6x^2 + 11x - 5 = 0$

Problems 21–30: In problems 1–10, find the solutions and check them by substitution in the original equation.

In problems 31–40, treat the following equations as quadratics in y and determine the range for x (that is, the restrictions, if any, on x in order that y be real).

31. $y^2 - 6y + 2x^2 - 9 = 0$
32. $9y^2 + x^2 - 25 = 0$
33. $y^2 - 2xy - 2y - 4x^2 - 8x - 4 = 0$
34. $x^2 + 24y = 4y^2 + 4x + 36$
35. $x^2 - 4xy + y^2 + 3 = 0$
36. $x^2 + 2x - 4y = 23 - 4y^2$
37. $x^2 - 5xy + 6y^2 = 10$

38. $(x + y)^2 + x + y = 12$ **39.** $(2x + y)(x - y - 1) = 0$
40. $x^2 + 2xy + 2y^2 - 1 = 0$

11.6 EQUATIONS OF HIGHER DEGREE AND FACTORING

There is a very close connection between the *roots* of a polynomial equation $f(x) = a_n x^n + a_{n-1} x^{n-1} + \cdots + a_1 x + a_0 = 0$ and the *factors* of the polynomial $f(x)$ as given by the following theorem.

Theorem 11.2. The polynomial equation

$$f(x) = a_n x^n + a_{n-1} x^{n-1} + \cdots + a_1 x + a_0 = 0$$

has a root a if the polynomial $f(x)$ has the factor $x - a$ and, conversely, $f(x)$ has a factor $x - a$ if $f(x) = 0$ has a root a.

PROOF: We first note that, for every positive integer k, $x^k - a^k$ has a factor $x - a$. For example,

$$x^2 - a^2 = (x - a)(x + a),$$
$$x^3 - a^3 = (x - a)(x^2 + ax + a^2),$$
$$x^4 - a^4 = (x - a)(x^3 + a^2 x + ax^2 + a^3)$$

as may be verified by direct multiplication and, in general,[1]

$$x^k - a^k = (x - a)(x^k + x^{k-1}a + x^{k-2}a^2 + \cdots + xa^{k-1} + a^k) \qquad (1)$$

Then since

$$f(x) - f(a) = a_n(x^n - a^n) + a_{n-1}(x^{n-1} - a^{n-1}) + \cdots + a_1(x - a)$$

and since each of $x^n - a^n$, $x^{n-1} - a^{n-1}, \ldots, x - a$ have a factor $x - a$ by (1), we have

$$f(x) - f(a) = (x - a)q(x) \qquad (2)$$

where $q(x)$ is a polynomial in x and a.

Now if $f(x) = 0$ has a root a, we have $f(a) = 0$ so that (2) becomes

$$f(x) = (x - a)q(x) \qquad (3)$$

and $x - a$ is a factor of $f(x)$ as desired.

[1] The general result will certainly appear plausible to the student. A proof involves "mathematical induction" as discussed in any textbook on college algebra.

Conversely, if $x - a$ is a factor of $f(x)$, then

$$f(x) = (x - a)q(x)$$

and

$$f(a) = (a - a)q(x) = 0$$

so that a is a root of $f(x) = 0$.

Example 1. The factors of $x^2 - x - 6$ are $x - 3$ and $x + 2 = x - (-2)$, and the roots of $x^2 - x - 6 = 0$ are 3 and -2.

Example 2. The roots of the quadratic equation $x^2 - x - 3 = 0$ are

$$x_1 = \frac{1 + \sqrt{13}}{2} \quad \text{and} \quad x_2 = \frac{1 - \sqrt{13}}{2}$$

Hence

$$x^2 - x - 3 = \left(x - \frac{1 + \sqrt{13}}{2}\right)\left(x - \frac{1 - \sqrt{13}}{2}\right)$$

Example 3. The roots of the quadratic equation $x^2 + x + 3 = 0$ are

$$x_1 = \frac{-1 + \sqrt{11}\, i}{2} \quad \text{and} \quad x_2 = \frac{-1 - \sqrt{11}\, i}{2}$$

Hence

$$x^2 + x + 3 = \left(x - \frac{-1 + \sqrt{11}\, i}{2}\right)\left(x - \frac{-1 - \sqrt{11}\, i}{2}\right)$$

Thus any quadratic polynomial $ax^2 + bx + c$ $(a, b, c,$ real) is factorable into linear factors if we allow the use of complex numbers. In fact, a result due to Gauss (mentioned in Section 5.8) tells us that, with the use of complex numbers, any polynomial

$$x^n + a_{n-1}x^{n-1} + \cdots + a_1x + a_0$$

with complex coefficients $a_{n-1}, \ldots, a_0$ can be factored into n linear factors, $(x - c_1), (x - c_2), \ldots, (x - c_n)$. In particular, polynomials of the form $x^n - a$ have n distinct factors and, for some polynomials of this form, we can find these factors by methods we have studied thus far. The following example illustrates this point.

Example 4. The equation $x^3 - 1 = 0$ which had only one solution, $x = 1$, when only real numbers were available, now has three solutions. To find these solutions we write

$$x^3 - 1 = (x - 1)(x^2 + x + 1) = 0$$

Hence $x = 1$ or $x^2 + x + 1 = 0$ and thus

$$x = 1, \quad -\frac{1}{2} + \frac{\sqrt{3}\,i}{2}, \quad \text{or} \quad -\frac{1}{2} - \frac{\sqrt{3}\,i}{2}$$

Each of these numbers has its cube equal to 1 and so each of these numbers is a cube root of 1.

Exercise 11.6

In problems 1–14, factor the given quadratic polynomial by first finding the roots of the corresponding quadratic equation.

1. $x^2 + 4x + 5$
2. $x^2 - 5x + 3$
3. $2x^2 - x - 4$
4. $2x^2 - 5x - 2$
5. $x^2 + 2x - 2$
6. $2x^2 - x - 2$
7. $3x^2 - 3x - 4$
8. $x^2 + 2x + 3$
9. $x^2 + x + 5$
10. $2x^2 - 7x + 8$
11. $x^2 - kx + 2$
12. $3bx^2 + cx + 1$
13. $6ax^2 - 7ax - 3$
14. $ax^2 - 3ax + 5$

In problems 15–24, find all the roots of the given equation by first finding a rational root by the methods of Section 6.10 and then expressing the given polynomial as a product of a linear factor and a quadratic.

15. $x^3 - 4x^2 - 7x + 10 = 0$
16. $x^3 - 5x^2 + 2x + 8 = 0$
17. $x^3 - 5x^2 - 19x - 10 = 0$
18. $2x^3 - 6x^2 - 25x + 25 = 0$
19. $2x^3 + 3x^2 - 11x - 6 = 0$
20. $x^3 + x^2 - 2 = 0$
21. $2x^3 - 3x^2 + 3x - 1 = 0$
22. $4x^3 - 5x^2 + 4x - 3 = 0$
23. $15x^3 - 11x^2 - 5x + 6 = 0$
24. $2x^3 - x^2 - 1 = 0$

11.7 MATRICES

A two by two (2×2) **matrix** is a square array of four numbers written in the form

$$\begin{pmatrix} a_{11} & a_{12} \\ a_{21} & a_{22} \end{pmatrix}$$

In general, the entry a_{ij} denotes the number in the ith **row** and jth **column.** Thus a_{11} is in the first row and first column, a_{12} is in the first row and second column, etc.[2]

Some of the most frequently studied systems in all of mathematics and mathematical physics are systems of matrices. They are also used in engineering, economics, psychology, etc. In this section we will define equality of matrices, matrix addition, and matrix multiplication.

Let

$$A = \begin{pmatrix} a_{11} & a_{12} \\ a_{21} & a_{22} \end{pmatrix} \quad \text{and} \quad B = \begin{pmatrix} b_{11} & b_{12} \\ b_{21} & b_{22} \end{pmatrix}$$

Definition 11.3. $A = B$ if and only if $a_{11} = b_{11}, a_{12} = b_{12}, a_{21} = b_{21},$ and $a_{22} = b_{22}.$

Definition 11.4. $A + B = \begin{pmatrix} a_{11} + b_{11} & a_{12} + b_{12} \\ a_{21} + b_{21} & a_{22} + b_{22} \end{pmatrix}.$

Definition 11.5. $A \cdot B = \begin{pmatrix} a_{11}b_{11} + a_{12}b_{21} & a_{11}b_{12} + a_{12}b_{22} \\ a_{21}b_{11} + a_{22}b_{21} & a_{21}b_{12} + a_{22}b_{22} \end{pmatrix}.$

Example 1.

$$\begin{pmatrix} 1 & 2 \\ -1 & 3 \end{pmatrix} + \begin{pmatrix} 0 & -2 \\ 1 & 3 \end{pmatrix} = \begin{pmatrix} 1+0 & 2+(-2) \\ -1+1 & 3+3 \end{pmatrix} = \begin{pmatrix} 1 & 0 \\ 0 & 6 \end{pmatrix}$$

[2] Although our notation for a matrix resembles our notation for a determinant (Section 5.12), a matrix is *not* a determinant. (A determinant is actually a symbol for a number; for example, $\begin{vmatrix} 1 & 2 \\ 3 & 5 \end{vmatrix} = -1$.) Compare, also, the 2×3 and 3×4 matrices used in Sections 5.9 and 5.10.

Example 2.

$$\begin{pmatrix} 1 & 4 \\ 0 & 7 \end{pmatrix} + \begin{pmatrix} 0 & 0 \\ 0 & 0 \end{pmatrix} = \begin{pmatrix} 1 & 4 \\ 0 & 7 \end{pmatrix}$$

Example 3.

$$\begin{pmatrix} 1 & -2 \\ 3 & 4 \end{pmatrix} \cdot \begin{pmatrix} 4 & 2 \\ 0 & -5 \end{pmatrix} = \begin{pmatrix} 1 \cdot 4 + (-2) \cdot 0 & 1 \cdot 2 + (-2)(-5) \\ 3 \cdot 4 + 4 \cdot 0 & 3 \cdot 2 + 4 \cdot (-5) \end{pmatrix}$$
$$= \begin{pmatrix} 4 & 12 \\ 12 & -14 \end{pmatrix}$$

Example 4.

$$\begin{pmatrix} 0 & 1 \\ 3 & 2 \end{pmatrix} \cdot \begin{pmatrix} 1 & 1 \\ 4 & 2 \end{pmatrix} = \begin{pmatrix} 0 \cdot 1 + 1 \cdot 4 & 0 \cdot 1 + 1 \cdot 2 \\ 3 \cdot 1 + 2 \cdot 4 & 3 \cdot 1 + 2 \cdot 2 \end{pmatrix} = \begin{pmatrix} 4 & 2 \\ 11 & 7 \end{pmatrix}$$

The examples of matrix multiplication merit a closer inspection. In Example 3, if we write

$$\begin{pmatrix} 1 & -2 \end{pmatrix}\begin{pmatrix} 4 \\ 0 \end{pmatrix}$$

that is, consider only the first row and first column of the respective matrices, we multiply corresponding elements and add,

$$\begin{pmatrix} 1 & -2 \end{pmatrix}\begin{pmatrix} 4 \\ 0 \end{pmatrix} \text{ gives } \begin{pmatrix} 1 \cdot 4 + (-2) \, 0 \end{pmatrix}$$

to obtain the entry in the first row and first column. Next

$$\begin{pmatrix} 1 & -2 \end{pmatrix}\begin{pmatrix} 2 \\ -5 \end{pmatrix} \text{ gives } \begin{pmatrix} 1 \cdot 2 + (-2)(-5) \end{pmatrix}$$

and we obtain the entry in the first row and second column. Then

$$\begin{pmatrix} 3 & 4 \end{pmatrix}\begin{pmatrix} 4 \\ 0 \end{pmatrix} \text{ gives } \begin{pmatrix} 3 \cdot 4 + 4 \cdot 0 \end{pmatrix}$$

and finally

$$\begin{pmatrix} 3 & 4 \end{pmatrix}\begin{pmatrix} 2 \\ -5 \end{pmatrix} \text{ gives } \begin{pmatrix} 3 \cdot 2 + 4 \cdot (-5) \end{pmatrix}$$

Exercise 11.7

In problems 1–8, perform the indicated addition.

1. $\begin{pmatrix} 1 & 1 \\ 0 & 1 \end{pmatrix} + \begin{pmatrix} 1 & -1 \\ 0 & 0 \end{pmatrix}$

2. $\begin{pmatrix} 1 & 0 \\ 0 & 1 \end{pmatrix} + \begin{pmatrix} 1 & 0 \\ 0 & 1 \end{pmatrix}$

3. $\begin{pmatrix} 2 & -3 \\ 1 & 2 \end{pmatrix} + \begin{pmatrix} 1 & 2 \\ -2 & -1 \end{pmatrix}$

4. $\begin{pmatrix} 1 & 2 \\ -1 & 3 \end{pmatrix} + \begin{pmatrix} 0 & -2 \\ 1 & -3 \end{pmatrix}$

5. $\begin{pmatrix} 1 & 2 \\ 3 & 4 \end{pmatrix} + \begin{pmatrix} -1 & -2 \\ -3 & -4 \end{pmatrix}$

6. $\begin{pmatrix} 1 & 2 \\ a & 0 \end{pmatrix} + \begin{pmatrix} 1 & a \\ a & 0 \end{pmatrix}$

7. $\begin{pmatrix} a & 0 \\ b & 0 \end{pmatrix} + \begin{pmatrix} 1 & a \\ 2 & 0 \end{pmatrix}$

8. $\begin{pmatrix} a & 0 \\ 0 & -a \end{pmatrix} + \begin{pmatrix} a & -b \\ b & -a \end{pmatrix}$

In problems 9–16, perform the indicated multiplication.

9. $\begin{pmatrix} 1 & 2 \\ -1 & 0 \end{pmatrix} \cdot \begin{pmatrix} 1 & 1 \\ 4 & 2 \end{pmatrix}$

10. $\begin{pmatrix} 1 & 0 \\ 0 & -1 \end{pmatrix} \cdot \begin{pmatrix} -2 & 1 \\ 0 & -3 \end{pmatrix}$

11. $\begin{pmatrix} 0 & 1 \\ 1 & 0 \end{pmatrix} \cdot \begin{pmatrix} 0 & 1 \\ 1 & 0 \end{pmatrix}$

12. $\begin{pmatrix} 1 & 2 \\ 0 & 3 \end{pmatrix} \cdot \begin{pmatrix} 1 & 2 \\ 0 & 3 \end{pmatrix}$

13. $\begin{pmatrix} 1 & -2 \\ 3 & 0 \end{pmatrix} \cdot \begin{pmatrix} 2 & -1 \\ 0 & 1 \end{pmatrix}$

14. $\begin{pmatrix} 2 & 1 \\ 0 & 1 \end{pmatrix} \cdot \begin{pmatrix} 1 & -2 \\ 3 & 0 \end{pmatrix}$

15. $\begin{pmatrix} 2 & a \\ b & 0 \end{pmatrix} \cdot \begin{pmatrix} a & 2 \\ 1 & -1 \end{pmatrix}$

16. $\begin{pmatrix} a & -a \\ 0 & 1 \end{pmatrix} \cdot \begin{pmatrix} 0 & -1 \\ b & b \end{pmatrix}$

17. Verify that the matrix $x = \begin{pmatrix} -3 & 1 \\ 2 & 0 \end{pmatrix}$ satisfies the matrix equation

$$\begin{pmatrix} 2 & -1 \\ 0 & 1 \end{pmatrix} x + \begin{pmatrix} 8 & -2 \\ -2 & 0 \end{pmatrix} = \begin{pmatrix} 0 & 0 \\ 0 & 0 \end{pmatrix}$$

18. Verify that the matrix $x = \begin{pmatrix} 2 & 1 \\ -1 & 1 \end{pmatrix}$ satisfies the matrix equation

$$\begin{pmatrix} 1 & -1 \\ -1 & 0 \end{pmatrix} x + \begin{pmatrix} -3 & 0 \\ 2 & 1 \end{pmatrix} = \begin{pmatrix} 0 & 0 \\ 0 & 0 \end{pmatrix}$$

19. Verify that the matrix $x = \begin{pmatrix} -2 & 1 \\ 0 & 1 \end{pmatrix}$ satisfies the matrix equation

$$x^2 + \begin{pmatrix} -4 & 1 \\ 0 & -1 \end{pmatrix} = \begin{pmatrix} 0 & 0 \\ 0 & 0 \end{pmatrix}$$

20. Verify that the matrix $x = \begin{pmatrix} 1 & 0 \\ -1 & 2 \end{pmatrix}$ satisfies the matrix equation

$$x^2 + \begin{pmatrix} 2 & 1 \\ 0 & -1 \end{pmatrix} x + \begin{pmatrix} -2 & -2 \\ 2 & -2 \end{pmatrix} = \begin{pmatrix} 0 & 0 \\ 0 & 0 \end{pmatrix}$$

11.8 THE ALGEBRA OF MATRICES

The system consisting of all 2×2 matrices whose entries are real numbers, with the operations of addition and multiplication as defined in Section 11.7, has the following properties:

If A, B, and C are any 2×2 matrices and $I = \begin{pmatrix} 1 & 0 \\ 0 & 1 \end{pmatrix}$ and $\Theta = \begin{pmatrix} 0 & 0 \\ 0 & 0 \end{pmatrix}$ then:

(1) $A + B = B + A$ The commutative property for addition.

(2) $(A + B) + C = A + (B + C)$ The associative property for addition.

(3) $(A \cdot B) \cdot C = A \cdot (B \cdot C)$ The associative property for multiplication.

(4) $A \cdot (B + C) = A \cdot B + A \cdot C$ The distributive properties.
$(A + B) \cdot C = A \cdot C + B \cdot C$

(5) $\Theta + A = A$, $\Theta \cdot A = \Theta$ Properties of the matrix Θ.

(6) $I \cdot A = A \cdot I = A$ Properties of the matrix I.

(7) If $A = \begin{pmatrix} a_{11} & a_{12} \\ a_{21} & a_{22} \end{pmatrix}$ and $-A = \begin{pmatrix} -a_{11} & -a_{12} \\ -a_{21} & -a_{22} \end{pmatrix}$, then $A + (-A) = \Theta$.

These properties all follow from the basic properties of real numbers given in Chapter 1 and Definitions 11.3, 11.4, and 11.5 of Section 11.7. Properties 1, 2, 5, and 7 follow almost immediately from Definitions 11.3 and 11.4 and the corresponding properties of real numbers, whereas property 6 is an immediate consequence of Definitions 11.3 and 11.5 and the fact that $1 \cdot a = a$ for all real numbers a. Properties 3 and 4, on the other hand, involve a certain amount of computation, although it is easy to verify special cases as in Examples 1 and 2.

Example 1. If $A = \begin{pmatrix} 1 & 1 \\ -1 & 1 \end{pmatrix}$, $B = \begin{pmatrix} 0 & -2 \\ 1 & 1 \end{pmatrix}$, and $C = \begin{pmatrix} 1 & 0 \\ -1 & 2 \end{pmatrix}$, then

$$(A \cdot B) \cdot C = \left[\begin{pmatrix} 1 & 1 \\ -1 & 1 \end{pmatrix} \cdot \begin{pmatrix} 0 & -2 \\ 1 & 1 \end{pmatrix} \right] \cdot \begin{pmatrix} 1 & 0 \\ -1 & 2 \end{pmatrix}$$

$$= \begin{pmatrix} 1 & -1 \\ 1 & 3 \end{pmatrix} \cdot \begin{pmatrix} 1 & 0 \\ -1 & 2 \end{pmatrix} = \begin{pmatrix} 2 & -2 \\ -2 & 6 \end{pmatrix}$$

while, also,

$$A \cdot (B \cdot C) = \begin{pmatrix} 1 & 1 \\ -1 & 1 \end{pmatrix} \cdot \left[\begin{pmatrix} 0 & -2 \\ 1 & 1 \end{pmatrix} \cdot \begin{pmatrix} 1 & 0 \\ -1 & 2 \end{pmatrix} \right]$$

$$= \begin{pmatrix} 1 & 1 \\ -1 & 1 \end{pmatrix} \cdot \begin{pmatrix} 2 & -4 \\ 0 & 2 \end{pmatrix} = \begin{pmatrix} 2 & -2 \\ -2 & 6 \end{pmatrix}$$

Example 2. If A, B, and C are as in Example 1, then

$$A \cdot (B + C) = \begin{pmatrix} 1 & 1 \\ -1 & 1 \end{pmatrix} \cdot \left[\begin{pmatrix} 0 & -2 \\ 1 & 1 \end{pmatrix} + \begin{pmatrix} 1 & 0 \\ -1 & 2 \end{pmatrix} \right]$$

$$= \begin{pmatrix} 1 & 1 \\ -1 & 1 \end{pmatrix} \cdot \begin{pmatrix} 1 & -2 \\ 0 & 3 \end{pmatrix} = \begin{pmatrix} 1 & 1 \\ -1 & 5 \end{pmatrix}$$

while, also,

$$A \cdot B + A \cdot C = \begin{pmatrix} 1 & 1 \\ -1 & 1 \end{pmatrix} \cdot \begin{pmatrix} 0 & -2 \\ 1 & 1 \end{pmatrix} + \begin{pmatrix} 1 & 1 \\ -1 & 1 \end{pmatrix} \cdot \begin{pmatrix} 1 & 0 \\ -1 & 2 \end{pmatrix}$$

$$= \begin{pmatrix} 1 & -1 \\ 1 & 3 \end{pmatrix} + \begin{pmatrix} 0 & 2 \\ -2 & 2 \end{pmatrix} = \begin{pmatrix} 1 & 1 \\ -1 & 5 \end{pmatrix}$$

The student may have noticed that the commutative property for multiplication, $A \cdot B = B \cdot A$, does not appear in the foregoing list. To see why it should not, suppose that

$$A = \begin{pmatrix} 1 & 2 \\ 0 & 3 \end{pmatrix} \quad \text{and} \quad B = \begin{pmatrix} 1 & 0 \\ 1 & 2 \end{pmatrix}$$

Then

$$A \cdot B = \begin{pmatrix} 3 & 4 \\ 3 & 6 \end{pmatrix} \quad \text{while} \quad B \cdot A = \begin{pmatrix} 1 & 2 \\ 1 & 8 \end{pmatrix}$$

and hence $A \cdot B \neq B \cdot A$. This example shows that the commutative property for matrix multiplication simply does not hold; that is, it is *not* the case that for *any* two matrices C and D, $C \cdot D = D \cdot C$. On the other hand, it is *sometimes* true that $C \cdot D = D \cdot C$. For example, if

$$C = \begin{pmatrix} 0 & 1 \\ -1 & 0 \end{pmatrix} \quad \text{and} \quad D = \begin{pmatrix} 1 & 1 \\ -1 & 1 \end{pmatrix}, \text{ then } C \cdot D = \begin{pmatrix} -1 & 1 \\ -1 & -1 \end{pmatrix} = D \cdot C$$

Exercise 11.8

1. Show that $I^2 = I$; $I^3 = I$; $\Theta^2 = \Theta$; where $I^2 = I \cdot I$, $I^3 = I \cdot I \cdot I$, $\Theta^2 = \Theta \cdot \Theta$.

2. Show that $(-I)^2 = I$; $(-A)^2 = A^2$; $A(-A) = -A^2$ where $(-I)^2 = (-I) \cdot (-I)$, $(-A)^2 = (-A) \cdot (-A)$, and $A^2 = A \cdot A$.

3. If $A = \begin{pmatrix} 1 & 0 \\ -1 & 2 \end{pmatrix}$ and $B = \begin{pmatrix} 0 & 1 \\ 1 & -1 \end{pmatrix}$, show that $A \cdot B \neq B \cdot A$.

4. If A and B are as in problem 3 and $C = \begin{pmatrix} 1 & 2 \\ -1 & 0 \end{pmatrix}$, show that $(A \cdot B) \cdot C = A \cdot (B \cdot C)$.

5. If A, B, and C are as given in problems 3 and 4, show that $A \cdot (B + C) = A \cdot B + A \cdot C$.

6. If A, B, and C are as given in problems 3 and 4, show that $(A + B) \cdot C = A \cdot C + B \cdot C$.

7. If $A = \begin{pmatrix} 2 & -1 \\ 1 & 2 \end{pmatrix}$, $B = \begin{pmatrix} -1 & 3 \\ -3 & -1 \end{pmatrix}$, and $C = \begin{pmatrix} 1 & 0 \\ 2 & -1 \end{pmatrix}$, show that although $A \cdot B = B \cdot A$, $A \cdot C \neq C \cdot A$ and $B \cdot C \neq C \cdot B$.

*8. If $A = \begin{pmatrix} 1 & -2 \\ 0 & 1 \end{pmatrix}$ and $B = \begin{pmatrix} -1 & 1 \\ 1 & 0 \end{pmatrix}$, show that if X is a matrix such that $A \cdot X = A \cdot B$ then $X = B$.

*9. Prove property 3 for all 2×2 matrices.

*10. Prove property 4 for all 2×2 matrices.

11.9 COMPLEX NUMBERS AS MATRICES

Instead of defining a complex number to be a polynomial in i (Definition 11.1), we can define a complex number to be a certain type of matrix. For example, instead of the complex number $3 + 5i$, let us write

$$\begin{pmatrix} 3 & -5 \\ 5 & 3 \end{pmatrix};$$

instead of the complex number $1 + 6i$, let us write

$$\begin{pmatrix} 1 & -6 \\ 6 & 1 \end{pmatrix};$$

and instead of the complex number $i = 0 + 1i$, let us write

$$\begin{pmatrix} 0 & -1 \\ 1 & 0 \end{pmatrix}$$

In general, instead of the complex number $a + bi$ we will write

$$\begin{pmatrix} a & -b \\ b & a \end{pmatrix}$$

Now the matrices of the form

$$\begin{pmatrix} a & -b \\ b & a \end{pmatrix}$$

behave under matrix operations just like complex numbers as is illustrated in Examples 1 and 2.

Example 1. $i^2 = -1$, and the matrix $\begin{pmatrix} 0 & -1 \\ 1 & 0 \end{pmatrix}$ which represents i has the analogous property

$$\begin{pmatrix} 0 & -1 \\ 1 & 0 \end{pmatrix}^2 = \begin{pmatrix} -1 & 0 \\ 0 & -1 \end{pmatrix} = -I$$

Example 2. $(5 + 3i)(2 + 4i) = -2 + 26i$ and

$$\begin{pmatrix} 5 & -3 \\ 3 & 5 \end{pmatrix} \cdot \begin{pmatrix} 2 & -4 \\ 4 & 2 \end{pmatrix} = \begin{pmatrix} -2 & -26 \\ 26 & -2 \end{pmatrix}$$

Exercise 11.9

In problems 1–10, write the 2×2 matrix corresponding to the given complex number

1. $1 + i$ 2. $1 - i$
3. $\sqrt{2} - 3i$ 4. $\sqrt{2} + i$
5. $-i$ 6. -1
7. $\sqrt{2} - \sqrt{3}\, i$ 8. $5 - \sqrt{3}\, i$
9. $\sqrt{3}$ 10. $\sqrt{2}\, i$

In problems 11–20, write the complex number corresponding to the given matrix

11. $\begin{pmatrix} 1 & -2 \\ 2 & 1 \end{pmatrix}$ 12. $\begin{pmatrix} -1 & -3 \\ 3 & -1 \end{pmatrix}$

13. $\begin{pmatrix} 1 & -1 \\ 1 & 1 \end{pmatrix}$ 14. $\begin{pmatrix} 1 & \sqrt{2} \\ -\sqrt{2} & 1 \end{pmatrix}$

15. $\begin{pmatrix} \sqrt{2} & 1 \\ -1 & \sqrt{2} \end{pmatrix}$ 16. $\begin{pmatrix} \sqrt{3} & -\sqrt{5} \\ \sqrt{5} & \sqrt{3} \end{pmatrix}$

17. $\begin{pmatrix} 0 & 3 \\ -3 & 0 \end{pmatrix}$ 18. $\begin{pmatrix} -2 & 0 \\ 0 & -2 \end{pmatrix}$

19. $\begin{pmatrix} -\sqrt{3} & 0 \\ 0 & -\sqrt{3} \end{pmatrix}$ 20. $\begin{pmatrix} 0 & \sqrt{5} \\ -\sqrt{5} & 0 \end{pmatrix}$

21. Show that if we restrict ourselves to the 2×2 matrices which represent complex numbers, multiplication is commutative.

Hint: Let $A = \begin{pmatrix} a & -b \\ b & a \end{pmatrix}$ and $B = \begin{pmatrix} c & -d \\ d & c \end{pmatrix}$, and show that $A \cdot B = B \cdot A$.

In problems 22–33, perform the indicated operations by first representing the complex numbers as 2×2 matrices and using the matrix operations of addition and multiplication. Then replace the resulting matrix by the corresponding complex number.

22. $(2 + i) + (1 - 2i)$ 23. $(\sqrt{2} + i) + (-3 + 2i)$
24. $(\sqrt{2} - \sqrt{3}\,i) + (\sqrt{2} - i)$ 25. $(2 + i)(1 - i)$
26. $(3 + 2i)^2$ 27. $(1 - 3i)(2 + \sqrt{2}\,i)$
28. $(3 - 2i)(-i)$ 29. $\sqrt{2}\,(4 + \sqrt{2}\,i)$
30. $(-1 - 2i)(-1 + 2i)$ 31. $(\sqrt{2} + i)(\sqrt{2} - i)$
32. $(1 - \sqrt{2}\,i)(1 + \sqrt{2}\,i)$ 33. $i(1 - i)$

References for Further Study

Bell, E. T., *Men of Mathematics*, 2d ed. New York: McGraw-Hill Book Company, 1945.

Bell, E. T., *Mathematics, Queen and Servant of Science*. New York: McGraw-Hill Book Company, 1951.

Boyer, C. B., *A History of Mathematics*. New York: John Wiley and Sons, 1968.

Cajori, F., *A History of Mathematics*, 2d ed. New York: The Macmillan Company, 1919.

Dantzig, T., *Number, the Language of Science*, 3d ed. New York: The Macmillan Company, 1946.

Dickson, L. E., *New First Course in the Theory of Equations*. New York: John Wiley and Sons, 1939.

Dubisch, R., *The Nature of Number*. New York: The Ronald Press Company, 1952.

Eves, H., *An Introduction to the History of Mathematics*. New York: Rinehart and Company, 1953.

Freund, J. E., *A Modern Introduction to Mathematics*. Englewood Cliffs, New Jersey: Prentice-Hall, 1956.

Griffiths, L. W., *Introduction to the Theory of Equations*, 2d ed. New York: John Wiley and Sons, 1947.

Howes, V. E., *Pre-Calculus Mathematics Series*, Vol. 1: *Algebra* and Vol. 2: *Functions*. New York: John Wiley and Sons, 1967.

Kasner, E. and J. R. Newman, *Mathematics and the Imagination*. New York: Simon and Schuster, 1940.

Kline, M., *Mathematics in Western Culture*. New York: Oxford University Press, 1953.

Kramer, E. E., *The Main Stream of Mathematics*. New York: Oxford University Press, 1951.

Meserve, B. E., *Fundamental Concepts of Algebra*. Cambridge, Massachusetts: Addison-Wesley Publishing Company, 1953.

Newman, J. R., ed., *The World of Mathematics*, 4 volumes, New York: Simon and Schuster, 1956.

Newsom, C. V. and H. Eves, *Introduction to College Mathematics*, 2d ed. New York: Prentice-Hall, 1954.

Polya, G., *How to Solve It*. Princeton, New Jersey: Princeton University Press, 1945.

Richardson, M., *College Algebra*, alt. ed. Englewood Cliffs, New Jersey: Prentice-Hall, 1958.

Schaaf, W. L., *Basic Concepts of Mathematics*, 2d ed. New York: John Wiley and Sons, 1965.

Stabler, E. R., *An Introduction to Mathematical Thought*. Cambridge, Massachusetts: Addison-Wesley Publishing Company, 1953.

Stein, S. K., *Mathematics: The Man-made Universe*. San Francisco: W. H. Freeman and Co., 1963.

Struik, D. J., *A Concise History of Mathematics*, 2 volumes, New York: Dover Publications, 1948.

Young, F. H., *The Nature of Mathematics*. New York: John Wiley and Sons, 1968.

MATHEMATICAL TABLES

TABLE 1. TABLE OF POWERS AND ROOTS

n	n^2	$\sqrt{n}$	n	n^2	$\sqrt{n}$
1	1	1.000	51	2,601	7.141
2	4	1.414	52	2,704	7.211
3	9	1.732	53	2,809	7.280
4	16	2.000	54	2,916	7.348
5	25	2.236	55	3,025	7.416
6	36	2.449	56	3,136	7.483
7	49	2.645	57	3,249	7.549
8	64	2.828	58	3,364	7.615
9	81	3.000	59	3,481	7.681
10	100	3.162	60	3,600	7.745
11	121	3.316	61	3,721	7.810
12	144	3.464	62	3,844	7.874
13	169	3.605	63	3,969	7.937
14	196	3.741	64	4,096	8.000
15	225	3.872	65	4,225	8.062
16	256	4.000	66	4,356	8.124
17	289	4.123	67	4,489	8.185
18	324	4.242	68	4,624	8.246
19	361	4.358	69	4,761	8.306
20	400	4.472	70	4,900	8.366
21	441	4.582	71	5,041	8.426
22	484	4.690	72	5,184	8.485
23	529	4.795	73	5,329	8.544
24	576	4.898	74	5,476	8.602
25	625	5.000	75	5,625	8.660
26	676	5.099	76	5,776	8.717
27	729	5.196	77	5,929	8.774
28	784	5.291	78	6,084	8.831
29	841	5.385	79	6,241	8.888
30	900	5.477	80	6,400	8.944
31	961	5.567	81	6,561	9.000
32	1,024	5.656	82	6,724	9.055
33	1,089	5.744	83	6,889	9.110
34	1,156	5.830	84	7,056	9.165
35	1,225	5.916	85	7,225	9.219
36	1,296	6.000	86	7,396	9.273
37	1,369	6.082	87	7,569	9.327
38	1,444	6.164	88	7,744	9.380
39	1,521	6.244	89	7,921	9.433
40	1,600	6.324	90	8,100	9.486
41	1,681	6.403	91	8,281	9.539
42	1,764	6.480	92	8,464	9.591
43	1,849	6.557	93	8,649	9.643
44	1,936	6.633	94	8,836	9.695
45	2,025	6.708	95	9,025	9.746
46	2,116	6.782	96	9,216	9.797
47	2,209	6.855	97	9,409	9.848
48	2,304	6.928	98	9,604	9.899
49	2,401	7.000	99	9,801	9.949
50	2,500	7.071	100	10,000	10.000

TABLE 2. FOUR-PLACE LOGARITHMS

	0	1	2	3	4	5	6	7	8	9
10	0000	0043	0086	0128	0170	0212	0253	0294	0334	0374
11	0414	0453	0492	0531	0569	0607	0645	0682	0719	0755
12	0792	0828	0864	0899	0934	0969	1004	1038	1072	1106
13	1139	1173	1206	1239	1271	1303	1335	1367	1399	1430
14	1461	1492	1523	1553	1584	1614	1644	1673	1703	1732
15	1761	1790	1818	1847	1875	1903	1931	1959	1987	2014
16	2041	2068	2095	2122	2148	2175	2201	2227	2253	2279
17	2304	2330	2355	2380	2405	2430	2455	2480	2504	2529
18	2553	2577	2601	2625	2648	2672	2695	2718	2742	2765
19	2788	2810	2833	2856	2878	2900	2923	2945	2967	2989
20	3010	3032	3054	3075	3096	3118	3139	3160	3181	3201
21	3222	3243	3263	3284	3304	3324	3345	3365	3385	3404
22	3424	3444	3464	3483	3502	3522	3541	3560	3579	3598
23	3617	3636	3655	3674	3692	3711	3729	3747	3766	3784
24	3802	3820.	3838	3856	3874	3892	3909	3927	3945	3962
25	3979	3997	4014	4031	4048	4065	4082	4099	4116	4133
26	4150	4166	4183	4200	4216	4232	4249	4265	4281	4298
27	4314	4330	4346	4362	4378	4393	4409	4425	4440	4456
28	4472	4487	4502	4518	4533	4548	4564	4579	4594	4609
29	4624	4639	4654	4669	4683	4698	4713	4728	4742	4757
30	4771	4786	4800	4814	4829	4843	4857	4871	4886	4900
31	4914	4928	4942	4955	4969	4983	4997	5011	5024	5038
32	5051	5065	5079	5092	5105	5119	5132	5145	5159	5172
33	5185	5198	5211	5224	5237	5250	5263	5276	5289	5302
34	5315	5328	5340	5353	5366	5378	5391	5403	5416	5428
35	5441	5453	5465	5478	5490	5502	5514	5527	5539	5551
36	5563	5575	5587	5599	5611	5623	5635	5647	5658	5670
37	5682	5694	5705	5717	5729	5740	5752	5763	5775	5786
38	5798	5809	5821	5832	5843	5855	5866	5877	5888	5899
39	5911	5922	5933	5944	5955	5966	5977	5988	5999	6010
40	6021	6031	6042	6053	6064	6075	6085	6096	6107	6117
41	6128	6138	6149	6160	6170	6180	6191	6201	6212	6222
42	6232	6243	6253	6263	6274	6284	6294	6304	6314	6325
43	6335	6345	6355	6365	6375	6385	6395	6405	6415	6425
44	6435	6444	6454	6464	6474	6484	6493	6503	6513	6522
45	6532	6542	6551	6561	6571	6580	6590	6599	6609	6618
46	6628	6637	6646	6656	6665	6675	6684	6693	6702	6712
47	6721	6730	6739	6749	6758	6767	6776	6785	6794	6803
48	6812	6821	6830	6839	6848	6857	6866	6875	6884	6893
49	6902	6911	6920	6928	6937	6946	6955	6964	6972	6981
50	6990	6998	7007	7016	7024	7033	7042	7050	7059	7067
51	7076	7084	7093	7101	7110	7118	7126	7135	7143	7152
52	7160	7168	7177	7185	7193	7202	7210	7218	7226	7235
53	7243	7251	7259	7267	7275	7284	7292	7300	7308	7316
54	7324	7332	7340	7348	7356	7364	7372	7380	7388	7396

TABLE 2. FOUR-PLACE LOGARITHMS (Continued)

	0	1	2	3	4	5	6	7	8	9
55	7404	7412	7419	7427	7435	7443	7451	7459	7466	7474
56	7482	7490	7497	7505	7513	7520	7528	7536	7543	7551
57	7559	7566	7574	7582	7589	7597	7604	7612	7619	7627
58	7634	7642	7649	7657	7664	7672	7679	7686	7694	7701
59	7709	7716	7723	7731	7738	7745	7752	7760	7767	7774
60	7782	7789	7796	7803	7810	7818	7825	7832	7839	7846
61	7853	7860	7868	7875	7882	7889	7896	7903	7910	7917
62	7924	7931	7938	7945	7952	7959	7966	7973	7980	7987
63	7993	8000	8007	8014	8021	8028	8035	8041	8048	8055
64	8062	8069	8075	8082	8089	8096	8102	8109	8116	8122
65	8129	8136	8142	8149	8156	8162	8169	8176	8182	8189
66	8195	8202	8209	8215	8222	8228	8235	8241	8248	8254
67	8261	8267	8274	8280	8287	8293	8299	8306	8312	8319
68	8325	8331	8338	8344	8351	8357	8363	8370	8376	8382
69	8388	8395	8401	8407	8414	8420	8426	8432	8439	8445
70	8451	8457	8463	8470	8476	8482	8488	8494	8500	8506
71	8513	8519	8525	8531	8537	8543	8549	8555	8561	8567
72	8573	8579	8585	8591	8597	8603	8609	8615	8621	8627
73	8633	8639	8645	8651	8657	8663	8669	8675	8681	8686
74	8692	8698	8704	8710	8716	8722	8727	8733	8739	8745
75	8751	8756	8762	8768	8774	8779	8785	8791	8797	8802
76	8808	8814	8820	8825	8831	8837	8842	8848	8854	8859
77	8865	8871	8876	8882	8887	8893	8899	8904	8910	8915
78	8921	8927	8932	8938	8943	8949	8954	8960	8965	8971
79	8976	8982	8987	8993	8998	9004	9009	9015	9020	9025
80	9031	9036	9042	9047	9053	9058	9063	9069	9074	9079
81	9085	9090	9096	9101	9106	9112	9117	9122	9128	9133
82	9138	9143	9149	9154	9159	9165	9170	9175	9180	9186
83	9191	9196	9201	9206	9212	9217	9222	9227	9232	9238
84	9243	9248	9253	9258	9263	9269	9274	9279	9284	9289
85	9294	9299	9304	9309	9315	9320	9325	9330	9335	9340
86	9345	9350	9355	9360	9365	9370	9375	9380	9385	9390
87	9395	9400	9405	9410	9415	9420	9425	9430	9435	9440
88	9445	9450	9455	9460	9465	9469	9474	9479	9484	9489
89	9494	9499	9504	9509	9513	9518	9523	9528	9533	9538
90	9542	9547	9552	9557	9562	9566	9571	9576	9581	9586
91	9590	9595	9600	9605	9609	9614	9619	9624	9628	9633
92	9638	9643	9647	9652	9657	9661	9666	9671	9675	9680
93	9685	9689	9694	9699	9703	9708	9713	9717	9722	9727
94	9731	9736	9741	9745	9750	9754	9759	9763	9768	9773
95	9777	9782	9786	9791	9795	9800	9805	9809	9814	9818
96	9823	9827	9832	9836	9841	9845	9850	9854	9859	9863
97	9868	9872	9877	9881	9886	9890	9894	9899	9903	9908
98	9912	9917	9921	9926	9930	9934	9939	9943	9948	9952
99	9956	9961	9965	9969	9974	9978	9983	9987	9991	9996

Answers to Odd-Numbered Problems

Exercise 1.2

1. (a) $\{4\}$ (b) $\varnothing$ (c) $\{0, 1, 2, 3, 4, 5, 6, 7, 8, 9\}$ (d) $\{7\}$ (e) $\varnothing$ (f) $\{4, 3\}$ (g) $\varnothing$
3. $\{a, b\}, \{a, c\}, \{b, c\}, \{a, b, c\}$ **5.** 8 **7.** (a) Yes (b) Yes (c) Yes (d) Yes

Exercise 1.3

1. Associative property of addition; 10
3. Associative property of multiplication; 30
5. Left-hand distributive property; 32
7. Right-hand distributive property; 46
9. Right-hand distributive property; 25
11. Commutative property of multiplication; 30 **13.** Closed **15.** Closed
17. Not closed **19.** Closed **21.** Not closed

Exercise 1.4

1. $\{1, 2, 3, 4, 6, 7\}$ **3.** $\{1, 3, 4, 5, 6, 7\}$ **5.** $\{3, 7\}$ **7.** $\{1, 3, 4, 6, 7\}$
9. $A \cap B = \{x | x \in A \text{ and } x \in B\}, A \cup B = \{x | x \in A \text{ or } x \in B\}$
11. (a) $A = \varnothing$ (b) $A = \varnothing$ and $B = \varnothing$ (c) No **13.** $A' = \{x | x \in U \text{ and } x \notin A\}$

Exercise 1.6

1. (a) 2 (b) 1 (c) -2 (d) 98 (e) -1 (f) -98 (g) 0 (h) -2
3. (a) $5 - 3$ (b) $5 - (-3)$ (c) $-5 - 3$ (d) $-5 - (-3)$ (e) $0 - 0$ (f) $0 - (-2)$
5. (a) 5 (b) 5 (c) 0 (d) 1 (e) 0 (f) 25 (g) 50 (h) 101

Exercise 1.7

1. 2   **3.** 1   **5.** 11   **7.** −3   **9.** 27   **11.** −11   **13.** 1   **15.** −22   **17.** 46
19. −40

Exercise 1.8

1. −15   **3.** −20   **5.** 27   **7.** 105   **9.** −30   **11.** 280   **13.** −6   **15.** −3   **17.** 5
19. −9   **21.** −2   **23.** −5

Exercise 1.9

1. All of the numbers listed except −100/0, $\sqrt{3}$, π, and $\sqrt[3]{2}$ are rational
numbers.   **3.** $\frac{1}{2}$ **5.** 8   **7.** $\frac{5}{7}$ **9.** $\frac{1}{2}$ **11.** $\frac{5}{2}$ **13.** 6   **15.** $\frac{1}{6}$ **17.** $\frac{2}{3}$ **19.** $-\frac{4}{3}$
21. $\frac{28}{5}$ **23.** 1   **25.** $\frac{6}{35}$ **27.** −1

Exercise 1.10

1. $\frac{13}{12}$ **3.** −1   **5.** $\frac{1}{12}$ **7.** $\frac{46}{45}$ **9.** $\frac{47}{12}$ **11.** $\frac{19}{12}$ **13.** $\frac{7}{12}$ **15.** $\frac{3}{2}$ **17.** $\frac{51}{50}$ **19.** $\frac{6}{25}$
21. $-\frac{1}{12}$ **23.** $-\frac{5}{26}$ **25.** $\frac{1}{4}$ **27.** $-\frac{35}{36}$ **29.** $\frac{125}{3}$ **31.** $\frac{1}{3}$ **33.** $\frac{1}{12}$ **35.** −15
37. $\frac{243}{20}$

Exercise 1.13

1. 5   **3.** −8   **5.** $-\frac{5}{4}$ **7.** −2   **9.** $\frac{13}{4}$ **11.** $\frac{7}{2}$ **13.** $7a - 7$ **15.** $a + 3b$
17. $-a - 12$ **19.** $\frac{1}{3}x + y$ **21.** $9x + 6y$ **23.** $\frac{21}{8}x - 3y$ **25.** Term   **27.** Term
29. Term   **31.** Factor   **33.** Factor   **35.** Factor   **37.** $9b$ **39.** d **41.** y
43. 7   **45.** $4a$ **47.** $9(x + y)$

Exercise 2.1

1. Second-degree monomial   **3.** First-degree binomial
5. Second-degree trinomial   **7.** Fifth-degree binomial
9. $x + 1$; $a_1 = 1$, $a_0 = 1$
11. $-3x^3 + x^2 - 3x + 1$; $a_3 = -3$, $a_2 = 1$, $a_1 = -3$, $a_0 = 1$ **13.** $\frac{1}{2}$; $a_0 = \frac{1}{2}$
15. $4x^4 + 0x^3 - 4x^2 + 0x + 1$; $a_4 = 4$, $a_3 = 0$, $a_2 = -4$, $a_1 = 0$, $a_0 = 1$
17. $\frac{7}{2}x^3 + 0x^2 - \frac{1}{2}x + 1$; $a_3 = \frac{7}{2}$, $a_2 = 0$, $a_1 = -\frac{1}{2}$ $a_0 = 1$
19. $yx^2 + x - 3y$; $a_2 = y$, $a_1 = 1$, $a_0 = -3y$
21. $-2x^2 + ax + \frac{1}{2}$; $a_2 = -2$, $a_1 = a$, $a_0 = \frac{1}{2}$
23. $(y^2 + y)x^3 - (y + 1)x^2 + yx + (-y^2 + 2y - 1)$; $a_3 = y^2 + y$, $a_2 = -(y + 1)$,
$a_1 = y$, $a_0 = -y^2 + 2y - 1$

Exercise 2.2

1. $8x^2 + 3x - 1$ **3.** $-10y^2 + y + 5$ **5.** $-2x + 5y - 13$

7. $-7a - 3b - 5c$ **9.** $-13a + 8b - 2c$ **11.** $11a^2 - 7a + 1$

13. $8x^2 - 3x - 3$ **15.** $-10y^2 - y + 5$ **17.** $10a + 2c$

19. $13x + 6y - 4z$ **21.** $8a^2 + 3a + 7$ **23.** $7x^3 + 4x^2 + 2x$

25. $-3x + 3$

Exercise 2.3

1. -27 **3.** $8a^3$ **5.** 324 **7.** x^6y^4 **9.** $-3x^2y^3$ **11.** $-12a^3b^3$ **13.** $-9v^7s^{19}$

15. $-x^4y^4$ **17.** $-30x^3y^2z^2$ **19.** $-12x^{a+4}y^{b+4}$

Exercise 2.4

1. $12x - 3a$ **3.** $x^3 - x^4$

5. $-12xy + 21y^2$ **7.** $3a^2bc + 2ab^2c - abc^2$

9. $5m^2n^2 + 4mn^3 - m^2n^3$ **11.** $6a^2 - 11ab + 4b^2$

13. $12x^2 - 32x + 5$ **15.** $a^2 - b^2$

17. $25a^4 - 16$ **19.** $2a^4 - 9a^2 + 10$

21. $a^3 - 8$ **23.** $2x^3 - 9x^2 + 5x - 4$

25. $x^3 + 4x^2 + 7x + 6$ **27.** $a^3 + 2a^2b + 2ab^2 + b^3$

29. $x^4 + x^3 - 4x^2 + 16x - 8$ **31.** $35a^2 - 29ab + 6b^2$

33. $a^3 + b^3$ **35.** $x^3 + 4x^2y + 4xy^2 + y^3$

37. $8x^5 + 12x^4y - 18x^3y^2 - 31x^2y^3 - 8xy^4 + 5y^5$

39. $4x^2 + y^2 + 16z^2 - 4xy + 16xz - 8yz$

41. $6a^2 - 8a^3$ **43.** $x^2y + x^2z + y^2x + y^2z$

45. $3x^2 + 4x + 2$ **47.** $a^3 + 6a^2 + 3a$

49. $5x - 10$ **51.** $2x^2 + 8x + 8$

53. $4xy$ **55.** $1 \cdot 10^2 + 8 \cdot 10 + 0 \cdot 1 = 180$

57. $2 \cdot 10^2 + 9 \cdot 10 + 4 = 294$ **59.** $6 \cdot 10^2 + 7 \cdot 10 + 2 = 672$

Exercise 2.6

1. $3 \cdot 5$ **3.** $2 \cdot 2 \cdot 2 \cdot 3$ **5.** $2 \cdot 2 \cdot 5 \cdot 5$

7. $2 \cdot 7 \cdot 13$ **9.** $2 \cdot 71$ **11.** -48

13. 126 **15.** $5x$ **17.** 0

19. $4y$ **21.** $8(3x - 2)$ **23.** $3(x^2 - x + 1)$

25. $2a(4a - 3)$ **27.** $5(a - b)$ **29.** $3x(7y - 1)$

31. $x^2(x^2 - 2x - 5)$ **33.** $a^2(3b^3c^4 - 5b - 7c^2)$ **35.** $2bx^2(2bx - 3 + 4c)$

37. $9ab(3a + 1)$ **39.** $6xy(4x - 2y - 1)$ **41.** $(x + y)(x - 3)$

43. $(x - 2)(2h - 3k)$ **45.** $(y - 2)(7x + 10z)$ **47.** $(x - 2y)(7 - 4a)$

49. $(x - y)(a^2 + b^2 - 3)$
53. $(x + 2y)(3x^2 - 9x + 17)$
51. $(x^2 - x + 1)(a + b)$
55. $(x^2 - 3x + 5)(x^2 - 7)$

Exercise 2.7

1. $(r + s)(r - s)$
3. $2(3x + 2y)(3x - 2y)$
5. $(2c + 3d)(2c - 3d)$
7. $(1 + 7xy)(1 - 7xy)$
9. $2(2a + b)(2a - b)$
11. $(10a + x^2)(10a - x^2)$
13. $3(3ax + 1)(3ax - 1)$
15. $(m^2n + r^3s^4)(m^2n - r^3s^4)$
17. $(a + 2b + c)(a + 2b - c)$
19. $(x + 3y + 5z)(x + 3y - 5z)$
21. $(a + b + c + d)(a + b - c - d)$
23. $a^2(a + b + 1)(a + b - 1)$
25. $(2xy + z^2 + 1)(2xy - z^2 - 1)$
27. $(x + y)(x - y)(36a^2x^2 + 25b^2y^2)$
29. $(6x - 7y - 1)^2(1 + x)(1 - x)$
31. $(r - s)(r^2 + rs + s^2)$
33. $(3x - 2y)(9x^2 + 6xy + 4y^2)$
35. $(m + 4)(m^2 - 4m + 16)$
37. $(1 - 3y^2)(1 + 3y^2 + 9y^4)$
39. $3(2x - 1)(4x^2 + 2x + 1)$
41. $(3xy + z^2 + 1)[9x^2y^2 - 3xy(z^2 + 1) + (z^2 + 1)^2]$
43. $(a - b + c)[a^2 + a(b - c) + (b - c)^2]$
45. $(x - 3)^2(a + b)(a^2 - ab + b^2)$
47. $(xy^2 - xy + z^2)[x^2y^4 + xy^2(xy - z^2) + (xy - z^2)^2]$
49. $a^3(a - b + 1)[(a - b)^2 - (a - b) + 1]$

Exercise 2.8

1. $(x - 3)(x - 1)$
3. $a(x - 2)^2$
5. $2(x - 5)(x + 1)$
7. $(a + 4)^2$
9. $(x - 2)^2$
11. $(x - 3)^2$
13. $x(x + 4)(x - 1)$
15. $(a + 2b)(a + 5b)$
17. $(3x + 2y)^2$
19. $(1 - 7x)^2$
21. $3(2x - 1)(x + 3)$
23. $2(x - 2y)(x + 3y)$
25. $a(x - 5)^2$
27. $(6x - 7)(x + 1)$
29. $(5 + y)(3 - y)$
31. $(5x + 3y)^2$
33. $(5a + 7b)(2a - 3b)$
35. $m(m + 2)^2$
37. $2xy(3x + 2y)(x - 5y)$
39. $(a^3 - 3b)(a^3 - 4b)$
41. $(x + y - 5)(x + y + 4)$
43. $(z - 2)^2$
45. $2(a + b)(x - 6)(x + 5)$
47. $x(2x - 3)(x - 4)$
49. $(2x + 2y - 1)^2$

Exercise 2.9

1. $(x - y)(a + b)$
3. $(a + 1)(5a^2 - 1)$
5. $(x - y)(4z - 3)$
7. $(x - 3)(x^2 + 1)$
9. $(a - b)(a^2 + ab + b^2 - a)$
11. $(x + y)(x - y + 3)$
13. $(x - 3)(x^2 - 2)$
15. $(2a + 2b + c)(2a - 2b - c)$
17. $(4a + 3x - 1)(4a - 3x + 1)$
19. $(4x + 3y)(5x + 2z)$
21. $(2x + y)(2x + y - 7)$
23. $(x + 3)(x^2 - 2y)$

Exercise 2.10

1. $2(4 + x)(4 - x)$ **3.** $2(7a + 1)(4a - 1)$ **5.** $(2x + 3y)(x + 2y)$
7. $(2a + 3b)(2a - 3b)$ **9.** $2a(x + 3a - 2y)$ **11.** $(2x - y)(2x + 3y)$
13. $3(x + 1)^2(x - 1)$ **15.** $5(a + 1)(3a + 7)$ **17.** $(x - 2y)(x + 2y - 1)$
19. $-3a^2y(2x + 3y)$ **21.** $(m + 1)(m - 1)(n + 1)(n - 1)$
23. $(9x - 4y)^2$ **25.** $(x - 2y + z)(x - 2y - z)$
27. $3a(a - 1)(a^2 + a + 1)$ **29.** $(x - 2y)(x + y - 2)$
31. $(r + 2)(r - 2)(r^2 - r + 4)$ **33.** $4(2x + 3yz)(2x - 3yz)$
35. $(x - 5 + 2y)(x - 5 - 2y)$ **37.** $(x + \frac{1}{2}y^2)(x - \frac{1}{2}y^2)$
39. $x(x + 2y)$ **41.** $(a + b - c)(a - b + c)$
43. $x(2y + x)(2y - x)$ **45.** $2(x - 2)^2$
47. $(2x + y + 4)(2x + y - 5)$ **49.** $(y + 5)(x - 2)$
51. $3x(x - 4)$ **53.** $5(x - 1)^3(x - 5)$
55. $(x - 2)^2(x^2 + 4x + 6)$

Exercise 3.1

1. 0   **3.** 1   **5.** 0, 3   **7.** x/y **9.** $-3x/y$ **11.** $5a^2$ **13.** $-2xy^4/z$
15. $-2x(a + b)/(a - b)$ **17.** $(3 + 5b)/2$ **19.** $-x/y$ **21.** $x/(x - y)$ **23.** $-5x/2$
25. $-(2a + 3)/(a + 4)$ **27.** $x/(x + 3)$

Exercise 3.2

1. $2 + \frac{1}{2}$ **3.** $1 + \frac{7}{8}$ **5.** $3x^2 + x$
7. $2a^2 - 3a$ **9.** $2y + 4y^3$ **11.** $4y^3 + 2z - 2$
13. $2s^3 - 3s^4t + 4t^4$ **15.** $-2x + \frac{9x + 4}{6x^2}$ **17.** $-x^2 + \frac{3}{2}x - \frac{5}{2} + \frac{1}{x}$
19. $-3y + 2 + \frac{3}{5x^2}$ **21.** $x - 3$ **23.** $a + 1$
25. $3x + 8$ **27.** $x^2 - x + 2$ **29.** $x^2 + 3x + 2$
31. $2x + 21 + \frac{144}{x - 7}$ **33.** $x + 5 + \frac{22}{x - 4}$ **35.** $x^2 - x + 3$
37. $2x^2 - x - 6 - \frac{11}{2x - 3}$ **39.** $2x^2 + xy - 2y^2$
41. $2x - 3$ **43.** $\frac{5}{2}x - \frac{1}{4} + \frac{3}{4(2x - 1)}$
45. $\frac{3}{2}x - \frac{13}{4} + \frac{43}{4(2x + 3)}$ **47.** $\frac{2}{3}x^2 - \frac{4}{9}x + \frac{2}{27} + \frac{35}{27(3x - 4)}$
49. $x - \frac{x}{x^3 + 1}$ **51.** $x - 1 + \frac{1}{x^2 + x + 1}$
53. $x - \frac{12}{x^2 + 6x + 5}$

Exercise 3.3

1. $x^2 + 5x - 2$, $R = 0$ **3.** $2x^2 + 4x - 6$, $R = -2$
5. $x^3 - 4x^2 + 21x - 88$, $R = 350$ **7.** $3x^3 + 3x^2 + 2x + 2$, $R = 12$
9. $2x^2 + 4x + 2$, $R = 0$ **11.** $2x^2 + \frac{11}{3}x + \frac{11}{9}$, $R = -\frac{16}{27}$
13. $x^3 - 2x^2 + 4x - 8$, $R = 11$ **15.** $x^2 + ax + a^2$, $R = 2a^3$

Exercise 3.4

1. $\dfrac{a^3b}{2}$ **3.** $\dfrac{15x^4}{2y^5}$ **5.** $\dfrac{3(a-2)}{a+4}$ **7.** $\dfrac{x-3}{(x-2)^2}$ **9.** $\dfrac{a+1}{a-2}$ **11.** $-\dfrac{(x+1)(x+3)}{(x+2)^2}$
13. 1   **15.** $-\dfrac{2x+3y}{(3x-2y)(x+y)^2}$

Exercise 3.5

1. 30   **3.** 1,008   **5.** $18abc$ **7.** $84ab^3$ **9.** $x^2y^2(x-y)$ **11.** $(a+b)^2(a-b)$
13. $(2x-1)(x+2)$ **15.** $(2x+5)(x+1)(x-1)$

Exercise 3.6

1. $5x/12$

3. $77a/60$

5. $(9-2a)/15$

7. $5x/(2x-1)(x+2)$

9. $\dfrac{6x^2 + 5x - 23}{(2x-3)(2x+3)(x-2)}$

11. $(x-5)/x(x+3)$

13. $(-5x+5)/(x-2)(x-3)$

15. $20x^2/(x-1)(x-2)(x+3)$

17. $(2x-3)/(x+2)(x-3)$

19. $(3a+5)/(a+1)$

21. $(a-2)/(a+2)$

23. $\dfrac{2x^2 - 9ax + 21a^2}{(x-3a)^2(x+3a)}$

25. $3(y-x)/y$

27. $\dfrac{6x^3 - 21x^2 - 11x + 2}{12}$

29. $5/(x+3)$

31. $\dfrac{9x^2 + 18x + 4y^2 + 24y + 10}{36}$

33. $\dfrac{4x^2 - 8x - 9y^2 + 36y - 32}{36}$

35. $-\dfrac{8x(x^2+1)}{(x^2-1)^3}$

37. $\dfrac{x(5x^3 + 9x + 4)}{(x^3 + 2)(x^2 + 3)}$

Exercise 3.7

1. $\frac{4}{7}$

3. $\frac{6}{19}$

5. x

7. $3x$

9. $x+2$

11. $\dfrac{x(x^2 - xy + 1)}{x^2 + x + 1}$

13. -1 **15.** $1 - x$ **17.** $(x + 2)/x^2$

19. $2xy/(y^2 - x^2)$ **21.** $(x^2 - y^2)/x^2$

23. $(x^2 + y^2 + 1)/(x^2 + 1)$ **25.** $\dfrac{-(2x + h)}{x^2(x + h)^2}$

27. $(y^2 - x)/y^3$ **29.** $(2x - 1 - x^2)/(1 + x^2)$

31. $\dfrac{y^2 + 2xy - x^2}{xy(y + x)}$

Exercise 4.2 (*Note:* Different answers may be obtained here depending on what the letters represent, whether one equation or more than one equation is used, etc.)

1. $x + 3 = 15$ **3.** $y = x + 3,\ x + y = 25$

5. $3x + x = 28$ **7.** $x + 3x = 20$

9. $2x + 10 = 44$ **11.** $2x + 3 = 3x - 8$

13. $x + (x - 3) = 27$ **15.** $y = x - 3,\ x + y = 27$

17. $x + y = 20,\ y = 3x$ **19.** $11{,}425 + 750x = 1{,}600$

21. $x + 5x = 90$ **23.** $x(x - 5) = 150$

25. $x(x + 2)(x - 1) = 30$ **27.** $5x + 10(2x) + 25(4x) = 500$

29. $x = 4z,\ y = \frac{1}{3}x,\ x + y + z + t = 1{,}000,\ 2x + 4y + 10z + 15t = 4{,}250$

Exercise 4.3 (*Note:* Different answers may be obtained here depending on what the letters represent, whether one equation or more than one equation is used, etc.)

1. $x + (x + 1) = 47$ **3.** $(x + 1)^2 - x^2 = 9$

5. $x^2 + (x + 2)^2 = 10[(x + 2)^2 - x^2] + 2$

7. $\dfrac{10}{15} + \dfrac{10}{x} = 1$

9. $5/x + 5/y + 5/z = 1,\ 6/x + 6/y = 1,\ 10/x + 10/z = 1$

11. $x/6 - x/8 = 1$ **13.** $x + y = 100,\ 71x + 96y = 7{,}600$

15. $300 \cdot 50 + 60x = 58(300 + x)$ **17.** $x + y = 100,\ 3x + 4y = 320$

19. $6 \cdot 0.3 + 0.6x = 0.4(x + 6)$ **21.** $x + 20 \cdot 0.1 = 0.15(20 + x)$

23. $x + y = 10,\ 0.60x + 0.40y = 0.48 \cdot 10$

25. $5v = 3(v + 20)$ **27.** $150t + 210t = 1{,}080$

29. $5v + 5(2v - 20) = 350$ **31.** $20 = 5(x - y),\ 20 = \frac{5}{2}(x + y)$

33. $\dfrac{d}{v + 40} = 1,\ \dfrac{d}{v - 40} = \dfrac{7}{6}$ **35.** $x/50 + y/8 = 1,\ x/50 + y/12 = 3/4$

Exercise 5.1

1. Conditional   **3.** Conditional   **5.** Identity

7. Identity   **9.** Conditional

11. $a_5x^5 + a_4x^4 + a_3x^3 + a_2x^2 + a_1x + a_0 = 0$

13. 3   **15.** 5; 2 in x; 3 in y

17. 6; 5 in x; 3 in y **19.** 7; 5 in x; 4 in y

21. $a_0 = -1$, $a_1 = 5$, $a_2 = -3$, $a_3 = 2$

23. $a_0 = -1$, $a_1 = 1$, $a_2 = -5$, $a_3 = 0$, $a_4 = 1$

25. $a_0 = 0$, $a_1 = 2$, $a_2 = -5$, $a_3 = a_4 = a_5 = 0$, $a_6 = 1$

Exercise 5.2

1. $x = 2$ **3.** $x = 5$ **5.** $x = 14/3$ **7.** $x = 7/5$

9. $x = 1$ **11.** Equivalent   **13.** Equivalent

15. Equivalent   **17.** Not equivalent   **19.** Not equivalent

Exercise 5.3

1. $\frac{13}{7}$ **3.** $-\frac{5}{3}$ **5.** $\frac{17}{5}$

7. $\frac{10}{21}$ **9.** $\frac{1}{6}$ **11.** -2

13. 11   **15.** $(2 + 3b - 5a)/8$ **17.** $a(p - r)/p$ if $p \neq 0$

19. $(a + b)/(2a - 1)$ if $a \neq \frac{1}{2}$

21. 12   **23.** 15   **25.** 11

27. 13   **29.** 10

31. 5 inches and 15 inches   **33.** 15° and 75°

35. 23 and 24   **37.** 80 gallons   **39.** 30 mph

41. 3 hours

Exercise 5.6

1. 2   **3.** $\frac{8}{3}$ **5.** -1 **7.** $\frac{4}{3}$ **9.** 0

11. No slope   **13.** 3   **15.** 1   **17.** 4   **19.** 5

21. $-\frac{4}{3}$ **23.** $-\frac{4}{5}$ **25.** 0   **27.** No slope

29. 6, 9, and 10; 11 and 12

Exercise 5.7

1. $x = -1$, $y = 1$ **3.** $x = -4$, $y = 3$ **5.** $x = 2$, $y = 3$

7. $x = -5$, $y = 2$ **9.** $x = -2$, $y = 4$ **11.** $x = \frac{1}{2}$, $y = \frac{3}{2}$

Exercise 5.8

1. $x = -1$, $y = 1$ **3.** $x = -2$, $y = 1$ **5.** $x = 3$, $y = -2$

7. $x = 1$, $y = -3$ **9.** $x = \frac{7}{19}$, $y = -\frac{4}{19}$ **11.** $x = \frac{3}{4}$, $y = \frac{1}{2}$

13. $x = \frac{3}{2}$, $y = 1$ **15.** $x = y = 0$ **17.** $x = \frac{1}{2}$, $y = \frac{1}{4}$

19. $x = \frac{6}{17}$, $y = -\frac{25}{17}$ **21.** 11 and 14

23. 12 and 15 years   **25.** Jack, $15; Bill, $5

27. Both receive 300 acres

29. 80 pounds of 71¢ and 20 pounds of 96¢

31. 80 $3 shares and·20 $4 shares   **33.** Boat, 6 mph; current, 2 mph

35. 62.5, 62.5, and 75 feet   **37.** $5,000 in each

39. 74   **41.** 62

43. $y = (\frac{2}{3})x^2 + \frac{4}{3}$

Exercise 5.10

1. $x = -6$, $y = 3$, $z = 0$ **3.** $x = -8$, $y = 3$, $z = 1$

5. $x_1 = \frac{7}{23}$, $x_2 = -\frac{6}{23}$, $x_3 = -\frac{87}{23}$ **7.** $x = -2$, $y = 5$, $z = -3$

9. $x = \frac{2}{9}$, $y = -\frac{5}{6}$, $z = \frac{1}{2}$, $w = \frac{5}{9}$ **11.** $x = \frac{1}{5}$, $y = \frac{1}{3}$, $z = -\frac{1}{2}$

13. 600, 2 cents; 200, 4 cents; 150, 10 cents; 50, 5 cents   **15.** 6 square feet

17. 274   **19.** $12,000 at 4 per cent, $8,000 at 5 per cent, $10,000 at 6 per

cent   **21.** $x^2 + y^2 - 7x - 9y + 20 = 0$ **23.** $x^2 + 3y^2 - 4x = 0$

Exercise 5.11

1. $a_{11} = 1$, $a_{12} = 2$, $a_{21} = 2$, $a_{22} = 5$, $b_1 = 1$, $b_2 = 3$; $x = -1$, $y = 1$

3. $a_{11} = 2$, $a_{12} = -1$, $a_{21} = -5$, $a_{22} = 2$, $b_1 = 4$, $b_2 = 11$; $x = -19$, $y = -42$

5. $a_{11} = 2$, $a_{12} = -3$, $a_{21} = -2$, $a_{22} = -5$, $b_1 = 0$, $b_2 = -1$; $x = \frac{3}{16}$, $y = \frac{1}{8}$

7. $a_{11} = \dfrac{1}{a}$, $a_{12} = \dfrac{1}{b}$, $a_{21} = \dfrac{1}{b}$, $a_{22} = -\dfrac{1}{a}$, $b_1 = c$, $b_2 = c$; $x = \dfrac{abc(a + b)}{a^2 + b^2}$,

$y = \dfrac{abc(a - b)}{a^2 + b^2}$ **9.** $a_{11} = 2$, $a_{12} = 3$, $a_{21} = 0$, $a_{22} = 1$, $b_1 = 0$, $b_2 = 5 - a$;

$x = \dfrac{3(a - 5)}{2}$, $y = 5 - a$

Exercise 5.12

1. -2 **3.** -7 **5.** -5 **7.** -37 **9.** 0   **11.** $x = \frac{3}{4}$, $y = \frac{1}{2}$ **13.** $x = \frac{3}{2}$, $y = 1$

15. $x = \frac{1}{2}$, $y = \frac{1}{4}$ **17.** $x = -1$, $y = 1$ **19.** $x = \frac{6}{17}$, $y = -\frac{25}{17}$

21. $X = \{(x, y) \mid x = 0 \text{ and } y = 1\} = \{(0, 1)\}$; independent and consistent

23. $X = \{(x, y) \mid x = -9 \text{ and } y = 14\} = \{(-9, 14)\}$; independent and consistent

25. $X = \varnothing$; inconsistent   **27.** $X = \{(x, y) \mid y = x - 3\}$; dependent

29. $X = \varnothing$; inconsistent

Exercise 5.13

1. -21 **3.** 78   **5.** -24 **7.** 141   **13.** $x = 15$, $y = 5$, $z = -11$

15. $a = 19$, $b = -13$, $c = -4$ **17.** $r = \frac{29}{4}$, $s = -\frac{49}{4}$, $t = -\frac{11}{4}$

19. $x_1 = -3$, $x_2 = 5$, $x_3 = -1$

Exercise 6.2

1. $\{3, -2\}$ **3.** $\{-5, -2\}$ **5.** $\{-\frac{1}{2}, 3\}$ **7.** $\{\frac{3}{4}, -\frac{3}{2}\}$ **9.** $\{0, 1\}$ **11.** $\{0, \frac{4}{9}\}$
13. $\{0, 2\}$ **15.** $\{0, 3, -2\}$ **17.** $\{0, 3, 5\}$ **19.** $\{0, -\frac{1}{3}, 5\}$ **21.** $\{a, b\}$
23. $\{-c - 5a, -c + 5a\}$ **25.** $\{3b + k, 3b - k\}$

Exercise 6.3

1. $3\sqrt{3}$ **3.** $3\sqrt{2}$ **5.** $4\sqrt{6}$ **7.** $\frac{1}{5}\sqrt{5}$ **9.** $\frac{1}{6}\sqrt{15}$ **11.** $\frac{1}{9}\sqrt{6}$ **13.** $1 + \sqrt{2}$
15. $1 + \sqrt{3}$ **17.** $\dfrac{1 + \sqrt{5}}{3}$ **19.** $\dfrac{-2 - \sqrt{3}}{2}$

Exercise 6.4

1. $x = -1$ or 3 **3.** $x = 3$ or 4

5. $x = \dfrac{-3 + \sqrt{3}}{3}$ or $\dfrac{-3 - \sqrt{3}}{3}$ **7.** $x = \dfrac{1 + \sqrt{5}}{2}$ or $\dfrac{1 - \sqrt{5}}{2}$

9. $x = \dfrac{3\sqrt{3} + \sqrt{15}}{6}$ or $\dfrac{3\sqrt{3} - \sqrt{15}}{6}$ **11.** $x = \dfrac{-\sqrt{3} + 2\sqrt{2}}{5}$ or $\dfrac{-\sqrt{3} - 2\sqrt{2}}{5}$

13. $x = \dfrac{p + \sqrt{p^2 - 4q}}{2}$ or $\dfrac{p - \sqrt{p^2 - 4q}}{2}$

Exercise 6.5

1. $(1, 2), 4$ **3.** $(2, 3), 5$ **5.** $(-1, -2), 5$
7. $(\frac{5}{4}, 1), \sqrt{41}/4$ **9.** $(-\frac{5}{4}, 0), \frac{5}{4}$ **11.** $(x + 5)^2 + 5$
13. $(x + 1)^2 + 4$ **15.** $36 - (4 - x)^2$ **17.** $9 - (2 + x)^2$

Exercise 6.6

1. $x = 3$ or -1 **3.** $x = 3$ or 4 **5.** $w = -1 \pm \sqrt{6}$

7. $u = \dfrac{1 \pm \sqrt{11}}{5}$ **9.** $v = \dfrac{-4 \pm \sqrt{6}}{5}$ **11.** $x = \dfrac{-1 \pm \sqrt{5}}{2}$

13. $p = \dfrac{-\sqrt{3} \pm \sqrt{7}}{2}$ **15.** $r = \dfrac{5 \pm \sqrt{5}}{2}$ **17.** $x = \dfrac{-3 \pm \sqrt{969}}{16}$

19. $y = \dfrac{(1 \pm \sqrt{5})x}{4}$ **21.** $y = \dfrac{3 + x \pm \sqrt{9 + 6x + 9x^2}}{4}$

23. $y = \dfrac{-x \pm \sqrt{9x^2 - 16x + 16}}{4}$ **25.** 5 and 12 inches

27. 19 and 21 or -1 and 1   **29.** 16 and 17 or -17 and -16
31. 54 square inches   **33.** $\frac{6}{5}$ or -1
35. 4 by 6 inches   **37.** 3, 4, and 5 or -5, -4, and -3
39. $\frac{3}{2}$ or $\frac{1}{2}$ seconds

Exercise 6.8

1. $\{12\}$ 3. $\{1\}$ 5. $\{-2\}$ 7. $\{-1, 3\}$ 9. $\{4\}$ 11. $\{27\}$ 13. $\{2\}$ 15. $\{2\}$
17. $\{4\}$ 19. $\{-\frac{1}{3}, -\frac{5}{3}\}$

Exercise 6.9

1. $\{-1, 1\}$
5. $\{\sqrt{\frac{4}{2}}, -\sqrt{\frac{18}{3}}\}$
9. $\{\frac{8}{5}\}$
13. $\{\frac{3}{4}, 1\}$
17. $\{-3, 3\}$

3. $\{-\sqrt{5}, \sqrt{5}\}$
7. $\{2\}$
11. $\{-\frac{49}{75}\}$
15. $\{\frac{7}{3}, \frac{4}{9}\}$

Exercise 6.10

1. $-2, -1, 1, 3$ 3. 2 5. None
7. $\frac{1}{2}$ 9. $\frac{5}{2}$ 11. $\frac{5}{2}$

Exercise 6.11

1. $x = 5$ 3. $x = 6$ 5. $x = -2$
7. $x = 2$ 9. $x = -\frac{3}{2}$ 11. $x = -3$
13. $x = 8$ 15. $x = -\frac{1}{2}$ 17. $x = \frac{5}{3}$
19. $x = -3$ 21. $x = -3$ or 6 23. no solution
25. $x = -\frac{15}{2}$ or $\frac{7}{2}$ 27. $x = -4$ or 2 29. $x = 2$ or 4
31. $x = 7$ 33. no solution   35. $x = 0$
37. $x = \frac{3}{2}$ 39. $x = 18$
41. $x = \frac{72}{25}, y = \frac{192}{25}$ 43. $x = 2, y = 6$
45. $x = -2, y = 1$ 47. $3\frac{3}{7}$ days   49. 6 hours
51. 10 gallons   53. $3\frac{1}{3}$ gallons   55. 10 mph and 15 mph
57. 520 mph, 560 miles   59. 12.5 miles, 6 miles   61. 48 and 72
63. 6 hours   65. 900 miles   67. $2\frac{1}{2}$ quarts
69. 38 pounds   71. 7.5 and 22.5 pounds

73. $x = \frac{1}{2}$ 75. $x = \dfrac{-6 \pm \sqrt{111}}{15}$

77. $y' = \dfrac{4 - x}{2x^2 y}$ 79. $y' = \dfrac{y(3y^2 - 1)}{2y^3 - 3xy^2 - x}$

Exercise 7.2

1. $f(x) = x + 5$ 3. $f(x) = \sqrt{x - 5}$ 5. $f(x) = 1/(x^2 - 1)$
7. $f(x, y) = (x + y)^2$ 9. $f(x) = |x|$

11. f is the rule which associates with every number, 3 more than the given number.

13. f is the rule which associates with every number, the square of 1 more than the number.

15. f is the rule which associates with every number, the absolute value of one more than the number.

17. f is the rule which associates with every pair of numbers, the square of the first minus the square of the second.

19. f is the rule which associates with every number, the given number if the given number is positive, the number 1 if the given number is 0, and the number -1 if the given number is negative.

21. Domain: all real numbers; range: all real numbers.

23. Domain: all real numbers greater than or equal to -3; range; all non-negative real numbers.

25. Domain: all real numbers; range: the number 1.

27. Domain: all real numbers; range: all non-negative real numbers.

29. Domain: all real numbers except 2 and -2 and numbers lying between 2 and -2; range: all positive real numbers.

Exercise 7.3

1. $3, 0, 2, 4$ **3.** $0, 1, 1, \frac{1}{4}, 16$

5. $5, 4, 10, 2a^2 - 3a + 5, 2b^2 - 3b + 5$

7. $0, -1, x - 2, (a^2 - 1)/(a + 1), x^2 - 1$

9. $-\dfrac{1}{5}, \dfrac{(a + 1)(2a - 1)}{a(2a - 3)}, \dfrac{(x + 1)(2x - 1)}{(x - 1)(2x - 5)}$

Exercise 7.4

17. 4.2 and -1.2 **19.** 3.4 and 0.6 **21.** No real root

23. 2.3 **25.** $h(2x + 3 + h)$ **27.** $\dfrac{h}{h + 1}$

29. $-\dfrac{h}{x(x + h)}$ **31.** $y = x^2 - 2x + 4$ **33.** $y = \dfrac{1}{1 - x}$

Exercise 7.5

21. $0 < x^2 < \frac{1}{4}; \ 0 < x^3 < \frac{1}{8}$

23. $0 < x^2 < \frac{1}{4}; \ -\frac{1}{8} < x^3 < \frac{1}{8}$

25. $x^2 > 0; \ x^3 < 1{,}000$

Exercise 7.6

1. $x < 8$ 3. $x \leqslant -2$ 5. $x \geqslant -2$ 7. $x < -2$ 9. $x \leqslant -2$ 11. $x < 1$
13. $x \geqslant 0$ 15. $x > \frac{8}{3}$ 17. $\{x | 2 < x < 5\}$ 19. $\{x | x \leqslant -1\}$ 21. $\varnothing$
23. $\{x | x = -3\}$ 25. R° 27. $\{x | -3 < x < \frac{1}{2}$ and $x \neq 0\}$
29. $\{x | x < \frac{1}{2}$ or $x > \frac{7}{2}\}$ 31. $\{x | -2 \leqslant x \leqslant 7\}$

Exercise 7.7

1. $\{x | x > \frac{5}{2}\}$ 3. $\{x | x \geqslant -\frac{1}{2}\}$ 5. $\{x | -1 < x < \frac{1}{4}\}$ 7. $\{x | x \geqslant 3$ or $x \leqslant 0\}$
9. $\{x | x > 3$ or $-1 < x < 0\}$ 11. $\{x | x \geqslant 0$ or $x \leqslant -3\}$ 13. $\{x | -\frac{2}{3} < x < 1\}$
15. $\varnothing$ 17. R° 19. $\{x | x \geqslant 3$ or $0 \leqslant x \leqslant 1\}$ 21. $\{x | x > 3$ or $-1 < x < 2\}$
23. $\{x | x < \frac{1}{3}$ or $\frac{4}{5} < x < \frac{3}{2}\}$ 25. $\{x | x \geqslant 5$ or $x \leqslant -\frac{2}{5}$ or $\frac{1}{2} \leqslant x \leqslant \frac{4}{3}\}$

Exercise 7.8

1. $\{x | x \neq 0\}$ 3. $\{x | x \neq 1$ and $x \neq -1\}$ 5. $\{x | -5 \leqslant x \leqslant 5\}$
7. $\{x | x \geqslant 0\}$ 9. $\{x | x \leqslant 2\}$ 11. $\{x | -\frac{5}{2} \leqslant x \leqslant -\frac{3}{2}\}$ 13. $\{x | -\frac{3}{2} \leqslant x \leqslant \frac{3}{2}\}$
15. $\{x | x \leqslant -2$ or $x \geqslant 4\}$ 17. $\{x | -2(\sqrt{2} + 1) \leqslant x \leqslant 2(\sqrt{2} - 1)\}$
19. None

Exercise 7.10

1. $(0, 4)$ and $(4, 0)$ 3. Do not intersect   5. $(1, 2)$ and $(4, -13)$
7. $(-1, 2\sqrt{6}), (-1, -2\sqrt{6})$ 9. $(2, \sqrt{5}), (2, -\sqrt{5})$
11. $x = 5, y = 3; x = -5, y = -3$ 13. $x = 1, y = 2$
15. $x = 2, y = 1; x = 1, y = 2; x = -1, y = -2; x = -2, y = -1$
17. $x = 1, y = 1; x = -1, y = 1$ 19. $x = 1, y = 1; x = 1, y = -1$

Exercise 8.1

1. 32	3. 1	5. 729	7. a^8	9. y^6
11. $a^3 b^3$	13. $1{,}728$	15. 2^{3+k}	17. $16a^6$	19. a^3/b^3
21. a^{6k}	23. $\frac{1}{27}$	25. -32	27. b/a	29. $1/a^2 b^2$

31. y/x^2 33. $3y/16x$ 35. $16a^4/b^6$
37. $a^{mx} k^{mz}/b^{2m}$ 39. $a^2/25$ 41. $1/a^n$
43. $-8x^3 y^6$ 45. $a^{2k} x^{k-1}$ 47. $-x^9/a^6$ 49. b/a
51. $p^2 q^2/8r$ 53. $-27a^2/8b^3 cd^2$ 55. y
57. $\frac{5}{4}$ 59. 4 61. $5/a^n$
63. $1/(x + y)$ 65. $b^3/a(a + b)$ 67. $(2 + x^2 - 6x^4)/x^4$
69. x/y 71. $ab/(b - a)$ 73. $ab/(a + b)^2$
75. $xy/(x + y)$

Exercise 8.2

1. $5\sqrt{3}$ **3.** $2\sqrt[3]{2}$ **5.** 8

7. $3a^2b$ **9.** -1 **11.** $2a^3|b|\sqrt{2a}$

13. x^k **15.** $(|x^3|\sqrt{2})/3y^2$ **17.** $(-2\sqrt[3]{2})/xy^3$

19. $(|y^2 + x|\sqrt{3})/y^2$ **21.** $5\sqrt{2}$ **23.** $2\sqrt{2x}$

25. 0 **27.** $(4a + |a|)x\sqrt{x}$

29. $(3b + 4|a|c^2 + 2b^2|c|)\sqrt{b}$ **31.** $\sqrt[3]{6}/3$

33. $2\sqrt{5}/5$ **35.** $\sqrt{2xy}/xy$ **37.** $4\sqrt{3}/3$

39. $\sqrt{2}$ **41.** $a\sqrt{10abc}/5bc^3$ **43.** $2\sqrt{a(x+y)}|x+y|$

45. $2\sqrt{x-y}/(x-y)$ **47.** $\sqrt{3b(ab+15)}/3|b|$ **49.** $\sqrt{7x(112x^3+3)}/7x^2$

51. $-17\sqrt{2x}/4x$ **53.** $\sqrt[3]{3a}$

55. $\dfrac{(4a^2 - 4|b| + 3a^3b)\sqrt{2ab}}{2a^2|b|}$ **57.** $6\sqrt{2}$

59. 5 **61.** $x\sqrt[3]{x^2}$ **63.** $3x\sqrt{2x}$

65. $-2\sqrt[3]{9}$ **67.** 1 **69.** 23

71. $x + 6\sqrt{xy} + 9y$ **73.** $3x - 7\sqrt{xy} - 6y$

75. $a^3x\sqrt{x} - 3a^2bx\sqrt{y} + 3ab^2y\sqrt{x} - b^3y\sqrt{y}$

77. $(5\sqrt{2} - 2)/46$ **79.** $\sqrt{2} + 1$ **81.** $3 + 2\sqrt{2}$

83. $\dfrac{a^2 - 4a\sqrt{x} + 4x}{a^2 - 4x^2}$ **85.** $\dfrac{x\sqrt{y} + y\sqrt{x}}{x - y}$ **87.** $(a + \sqrt{a^2 - b^2})/b$

89. $-(5\sqrt{2} + 3\sqrt{6} + 4\sqrt{3} + 8)/2$

91. $\dfrac{a(\sqrt{a} - 1 - \sqrt{b})(a - b + 1 + 2\sqrt{a})}{(a - b + 1)^2 - 4a}$

93. $\dfrac{2(\sqrt[3]{25} - \sqrt[3]{10} + \sqrt[3]{4})}{7}$ **95.** $\dfrac{1}{5(\sqrt{3} - \sqrt{2})}$

97. $\dfrac{-1}{3(1 - \sqrt{3})}$ **99.** $\dfrac{1}{\sqrt{x+h} + \sqrt{x}}$

Exercise 8.3

1. 2 **3.** 8 **5.** $a\sqrt[3]{a^2}$ **7.** $\frac{1}{4}$ **9.** 2

11. $\sqrt{|a|}$ **13.** $\sqrt{3}$ **15.** $\sqrt{2a}$ **17.** $|x|\sqrt{2|ax|}$ **19.** $\sqrt[3]{9ax^2}/3$

21. x **23.** $\sqrt{a}/a$ **25.** $x\sqrt[12]{x}$ **27.** $\sqrt[4]{x}$ **29.** $4/|a|b^2$

31. $2\sqrt[6]{2}$ **33.** $\sqrt[6]{x^5}$ **35.** $\sqrt[6]{x}$ **37.** $\sqrt[16]{5}$ **39.** $|a|\sqrt[6]{|a|}$

41. $16a\sqrt[3]{a}$ **43.** $2a^3\sqrt[5]{2}$ **45.** $\sqrt[4]{2}$

47. $3\sqrt[3]{x^2y}/|x|y$ **49.** $|a^3|b^2\sqrt{3}$ **53.** $(9x^2 + 5)\sqrt[5]{x^4}/5x^2$

55. $\dfrac{(x+1)\sqrt[3]{(x-1)^2}\sqrt[3]{x+2}}{(x-1)(x+2)}$ **57.** $-4\sqrt[3]{(1-x)(1-x^2)}/3(1-x^2)^2$

Exercise 8.6

1. $\log_2 4 = 2$ **3.** $\log_2 8 = 3$ **5.** $\log_{64} 4 = \frac{1}{3}$ **7.** $\log_2 \frac{1}{8} = -3$
9. $\log_5 125 = 3$ **11.** $\log_3 18 = p$ **13.** $\log_\pi 1 = 0$ **15.** $\log_b a = \log_b a$
17. $2^4 = 16$ **19.** $5^3 = 125$ **21.** $(\frac{1}{2})^3 = \frac{1}{8}$ **23.** $2^{-4} = \frac{1}{16}$
25. $(\frac{1}{125})^{-1/3} = 5$ **27.** $10^{-1} = \frac{1}{10}$ **29.** $10^0 = 1$ **31.** $7^p = 49$
33. 3   **35.** 3   **37.** 1   **39.** 4
41. 12   **43.** 5   **45.** -3 **47.** 32
49. 8   **51.** $\frac{1}{216}$ **53.** $\frac{1}{343}$ **55.** 25
57. 343   **59.** 32   **61.** 1.25   **63.** -1

Exercise 8.8

1. 1   **3.** 3.322   **5.** -1 **7.** -3.322
9. 0.5   **11.** 1.292   **13.** 3.907   **15.** -1.322
17. 6.3702   **19.** 0.7078   **21.** -4.2468 **23.** $\log_a 7$

25. $\log_a 24$ **27.** $\log_a x^2$ **29.** $\log_a \dfrac{5x}{3}$ **31.** 1.3980

33. 1.6990   **35.** 0.6020   **37.** 0.0970

Exercise 8.9

1. $x = 2$ **3.** $x = 3$ or -1 **5.** $x = \dfrac{\log_{10} 28}{2 \log_{10} 5}$

7. $x = \pm \sqrt{\dfrac{\log_{10} 15}{\log_{10} 18}}$ **9.** $x = \frac{3}{7}$ **11.** $x = 1,000$

13. $x = \frac{1}{2}$ or $-\frac{1}{2}$ **15.** $x = 4$ or -2 **17.** $x = 5$

Exercise 8.10

1. 1.36×10^2 **3.** 2.768×10 **5.** 6.721×10
7. 5.232×10^2 **9.** 3.0103×10^{-2} **11.** 3.75×10^{-4}
13. 2.65×10^{-1} **15.** 10^{-3} **17.** 1.6990   **19.** 0.7050
21. 1.7316   **23.** $0.7076 - 3$ **25.** -1.7185 **27.** 1.7160
29. $0.7160 - 1$ **31.** 1.3592   **33.** 1.7295   **35.** $0.7035 - 2$
37. 0.7280   **39.** 2.7241   **41.** $0.9499 - 1$ **43.** 0.2637
45. 0.9024   **47.** $0.9011 - 3$ **49.** $0.3008 - 1$ **51.** 1.7784

Exercise 8.11

1. 32   **3.** 100   **5.** 1   **7.** 4^{x^2+3} **9.** 10
11. 2   **13.** 200   **15.** 27.8   **17.** 0.0871   **19.** 0.0187

21. 6.78   **23.** 1.269   **25.** 3,645   **27.** 1.029   **29.** 53.95
31. 2.076   **33.** 141.1   **35.** 0.8562   **37.** 1,398

Exercise 8.12

1. 29,550   **3.** −0.07057   **5.** −0.00004760   **7.** −0.01382
9. 0.01228   **11.** 5,704   **13.** 2.364   **15.** 1.812
17. 37.95   **19.** 4.074   **21.** 4.644   **23.** 3.299
25. −3.995   **27.** 179.9 seconds **29.** 11.54 years

Exercise 8.13

1. 0.01826   **3.** 0.6665   **5.** 0.05012
7. 0.000002991   **9.** 0.0002866

Exercise 9.2

1. $y = -x + 3$ **3.** $y = x^2 + 1$
5. $y = 4/x^2$ **7.** $y = 2x$
9. $y = 3(x + 1)$ or $y = -3(x + 1)$ **11.** $y = x + 1$ or $y = -x - 1$
13. $f(x) = x^2 + 1$ **15.** $f(x) = 1/(x - 1)$ **17.** $f(z) = (\frac{1}{2})\sqrt{z}$
19. $f(3) = \frac{9}{4}$ **21.** $f(-\frac{1}{3}) = \frac{10}{3}$ **23.** $f(16) = 3$
25. $y = 59$ **27.** $y = 4x^{2/3} + 1$

Exercise 9.3

1. $y = \frac{5}{3}$ **3.** $y = \frac{81}{16}$ **5.** $x = 8$
7. $s = 1,024$ feet   **9.** 100 feet   **11.** 45 times per second
13. 10 cubic inches

Exercise 9.4

1. 120   **3.** 12
5. 15 pounds per square inch   **7.** 15 vibrations per second
9. $\frac{4}{3}$ ohms   **11.** 200 pounds

Exercise 10.1

1. 1, 8, 27, 64, 125; 1,331   **3.** 1, 9, 25, 49, 81; 441
5. 0, 2, 6, 12, 20; 110   **7.** $\frac{1}{2}, \frac{1}{3}, \frac{1}{4}, \frac{1}{5}, \frac{1}{6}; \frac{1}{12}$
9. 1, 3, 5, 7, 9; 21   **11.** −1, 2, −3, 4, −5; −11

13. $1, -1, 1, -1, 1; 1$

15. $1, -1, 1, -1, 1; 1$

17. $2n$

19. $2n + 2$

Exercise 10.2

1. 225   3. 48   5. -15 7. $\frac{143}{315}$

9. -20 11. 12   13. 8   15. 33

17. $\frac{61}{20}$ 19. $-\frac{7}{60}$ 21. 0

Exercise 10.3

1. 672   3. 108   5. -22 7. $-\frac{135}{2}$

9. $-\frac{23}{3}$ 11. $\frac{15,405}{133}$ 13. 47 and 300   15. 68 and 365

17. 23 and 1,771   19. 3   21. 42   23. -30

25. 6 and 5   27. 6 and 192   29. $\frac{16}{3}$ and 4   31. 10 and 50

33. $\frac{15}{8}$ and $-\frac{195}{8}$ 35. 320 and 1   37. 26

39. $1,395   41. $2,806.25   43. $11.01

Exercise 10.4

1. 126   3. $\frac{127}{128}$ 5. 14,005   7. 170

9. $\frac{21}{256}$ 11. $\frac{7,029}{4,096}$ 13. 11,718   15. $-\frac{17}{128}$

17. $\frac{484}{81}$ 19. $\frac{3}{2}$ or $-\frac{3}{2}$ and $-\frac{21}{2}$

21. 4,374   23. $-\frac{4}{7}$ 25. $\frac{1}{2}$ and $\frac{1}{3}$ 27. $-\frac{1}{4}$ and $\frac{4}{5}$

29. 256 and 5   31. 2 and 7   33. $\frac{1}{3}$ and $-\frac{1}{2}$ 35. the 7th

37. $\frac{7,609}{81}$ feet   39. $1,395   41. $6,516   43. 14th

Exercise 10.5

1. 120   3. 5,040   5. 3,628,800

7. 70   9. 3,003   11. 132

13. $x^7 + 7x^6y + 21x^5y^2 + 35x^4y^3 + 35x^3y^4 + 21x^2y^5 + 7xy^6 + y^7$

15. $81x^4 + 216x^3y + 216x^2y^2 + 96xy^3 + 16y^4$

17. $32x^5 - 240x^4y + 720x^3y^2 - 1,080x^2y^3 + 810xy^4 - 243y^5$

19. $x^9 + 9x^8y + 36x^7y^2 + 84x^6y^3 + 126x^5y^4 + 126x^4y^5 + 84x^3y^6 + 36x^2y^7 + 9xy^8 + y^9$

21. $243x^5 - 810x^4y + 1,080x^3y^2 - 720x^2y^3 + 240xy^4 - 32y^5$

23. $x^{12} + 12x^{11}y + 66x^{10}y^2 + 220x^9y^3$

25. $x^{13} + 39x^{12}y + 702x^{11}y^2 + 7,722x^{10}y^3$

27. $a^{12} + 48a^{11}b + 1,056a^{10}b^2 + 14,080a^9b^3$

29. $1,024a^{10} + 15,360a^9xy + 103,680a^8x^2y^2 + 414,720a^7x^3y^3$

31. $2,048a^{22}b^{11} - 33,792a^{20}b^{10}z^2w + 253,440a^{18}b^9z^4w^2 - 1,140,480a^{16}b^8z^6w^3$

33. $2{,}048a^{11} - 11{,}264a^{10}b^2 + 28{,}160a^9b^4 - 42{,}240a^8b^6$

35. $792x^5y^7$ **37.** $6{,}435x^7y^8$ **39.** $-12{,}285x^{12}y^3$ **41.** $2{,}099{,}520a^7b^3$

Exercise 11.3

1. $-2 + 0 \cdot i$ **3.** $-1 + 0 \cdot i$ **5.** $0 - 4i$ **7.** $2 + 0 \cdot i$

9. $0 + 2i$ **11.** $0 - 6i$ **13.** $7 - 4i$ **15.** $0 + 0 \cdot i$

17. $1 + 0 \cdot i$ **19.** $0 - i$ **21.** $-1 + 0 \cdot i$ **23.** $0 - i$

25. $1 + 0 \cdot i$ **27.** $\frac{1}{2} + (\sqrt{2}/2)i$ **29.** $(a + c) + (b + d)i$

31. $4 + 3i$ **33.** $10 + 0 \cdot i$ **35.** $4 - 7i$

37. $-3 + 4i$ **39.** $-3 + 4i$ **41.** $-46 - 9i$

43. $\frac{3}{2} + (\frac{1}{12})i$ **45.** $(3\sqrt{2})/2 - (\frac{3}{2})i$ **47.** $0 + i$

49. $0 - i$ **51.** $0 - i$ **53.** $-\frac{1}{2} + (\frac{5}{2})i$

55. $\frac{12}{13} - (\frac{5}{13})i$ **57.** $0 - i$

Exercise 11.4

1. $2i$ **3.** $2\sqrt{6}\,i$ **5.** $\sqrt{2}\,i$ **7.** -3

9. -35 **11.** $-24\sqrt{3}\,i$ **13.** $1 - 2i$ **15.** $10 + 2\sqrt{3}\,i$

17. $-14 + 2\sqrt{15}\,i$ **19.** $-24\sqrt{3}\,i$ **21.** $-53 - 45\sqrt{2}\,i$

Exercise 11.5

1. 0; one real   **3.** 33; two real

5. -3; two complex, not real   **7.** 20; two real

9. -4; two complex, not real   **11.** -68, two complex, not real

13. -36; two complex, not real   **15.** 0; one real

17. -84; two complex, not real   **19.** -248; two complex, not real

21. 1 **23.** $\dfrac{-5 \pm \sqrt{33}}{2}$ **25.** $\dfrac{1 \pm \sqrt{3}\,i}{2}$ **27.** $\pm\sqrt{5}$

29. $\frac{1}{2} \pm (\frac{1}{2})i$ **31.** $-3 \leqq x \leqq 3$ **33.** all x **35.** $-1 \leqq x \leqq 1$

37. all x **39.** all x

Exercise 11.6

1. $(x + 2 + i)(x + 2 - i)$ **3.** $\left(x - \dfrac{1 + \sqrt{33}}{4}\right)\left(x - \dfrac{1 - \sqrt{33}}{4}\right)$

5. $(x + 1 + \sqrt{3})(x + 1 - \sqrt{3})$ **7.** $\left(x - \dfrac{3 + \sqrt{57}}{6}\right)\left(x - \dfrac{3 - \sqrt{57}}{6}\right)$

9. $\left(x - \dfrac{-1 + \sqrt{19}\,i}{2}\right)\left(x - \dfrac{-1 - \sqrt{19}\,i}{2}\right)$

11. $\left(x - \dfrac{k + \sqrt{k^2 - 8}}{2}\right)\left(x - \dfrac{k - \sqrt{k^2 - 8}}{2}\right)$

13. $\left(x - \dfrac{7a + \sqrt{49a^2 + 72a}}{12a}\right)\left(x - \dfrac{7a - \sqrt{49a^2 + 72a}}{12a}\right)$

15. $1, 5, -2$ 17. $-2, \dfrac{7 \pm \sqrt{69}}{2}$ 19. $-\frac{1}{2}, -3, 2$

21. $\dfrac{1}{2}, \dfrac{1 \pm \sqrt{3}\, i}{2}$ 23. $-\dfrac{2}{3}, \dfrac{7 \pm \sqrt{11}\, i}{10}$

Exercise 11.7

1. $\begin{pmatrix} 2 & 0 \\ 0 & 1 \end{pmatrix}$ 3. $\begin{pmatrix} 3 & -1 \\ -1 & 1 \end{pmatrix}$ 5. $\begin{pmatrix} 0 & 0 \\ 0 & 0 \end{pmatrix}$ 7. $\begin{pmatrix} a+1 & a \\ b+2 & 0 \end{pmatrix}$

9. $\begin{pmatrix} 9 & 5 \\ -1 & -1 \end{pmatrix}$ 11. $\begin{pmatrix} 1 & 0 \\ 0 & 1 \end{pmatrix}$ 13. $\begin{pmatrix} 2 & -3 \\ 6 & -3 \end{pmatrix}$ 15. $\begin{pmatrix} 3a & 4-a \\ ab & 2b \end{pmatrix}$

Exercise 11.9

1. $\begin{pmatrix} 1 & -1 \\ 1 & 1 \end{pmatrix}$ 3. $\begin{pmatrix} \sqrt{2} & 3 \\ -3 & \sqrt{2} \end{pmatrix}$ 5. $\begin{pmatrix} 0 & 1 \\ -1 & 0 \end{pmatrix}$

7. $\begin{pmatrix} \sqrt{2} & \sqrt{3} \\ -\sqrt{3} & \sqrt{2} \end{pmatrix}$ 9. $\begin{pmatrix} \sqrt{3} & 0 \\ 0 & \sqrt{3} \end{pmatrix}$ 11. $1 + 2i$

13. $1 + i$ 15. $\sqrt{2} - i$ 17. $-3i$

19. $-\sqrt{3}$ 23. $(\sqrt{2} - 3) + 3i$ 25. $3 - i$

27. $(2 + 3\sqrt{2}) + (\sqrt{2} - 6)i$ 29. $4\sqrt{2} + 2i$

31. 3 33. $1 + i$

Index